T0275644

Biological Clocks, Rhythms, and Oscillations

Biological Clocks, Rhythms, and Oscillations

The Theory of Biological Timekeeping

Daniel B. Forger

The MIT Press
Cambridge, Massachusetts
London, England

This book was set in Syntax LT Std and Times New Roman by Toppan Best-set Premedia Limited.

Library of Congress Cataloging-in-Publication Data

Names: Forger, Daniel B., author.
Title: Biological clocks, rhythms, and oscillations : the theory of biological timekeeping / Daniel B. Forger.
Description: Cambridge, MA : The MIT Press, [2017] | Includes bibliographical references and index.
Identifiers: LCCN 2017007620 | ISBN 9780262036771 (hardcover : alk. paper)
ISBN 9780262552813 (paperback)
Subjects: LCSH: Biological rhythms. | Biological rhythms--Data processing.
Classification: LCC QH527 .F66 2017 | DDC 571.7--dc23 LC record available at
    https://lccn.loc.gov/2017007620

To Charlie (Peskin) for teaching me about mathematical biology

To Dick (Kronauer) for teaching me about clocks

And to Willard (Larkin) for help along the way

Contents

Preface

All areas of biology and medicine contain rhythms. These behaviors are best understood through mathematical tools and techniques that are sometimes not easy to find. Some are buried in the specialized literature. Others are passed down through oral tradition, from advisor to student or colleague to colleague. I realized the need for a book on these tools and techniques when learning the basics as an undergraduate, and the explosion of new biological insights and mathematical techniques in this field now speaks to the need for a new text as well.

My approach is different from that of previous authors in this field. They explain and categorize many examples of biological rhythms. While such explanations and categorizations are very important, students and researchers also need a guide to analyzing biological rhythms that is full of practical advice on what equations they should choose and details on what to do with these equations. Many beautiful applications or models could not be described or are only briefly described here owing to limitations of space. However, the tools and techniques I describe should be broadly applicable.

I see five main ways this book might be used by different audiences:

- As a reference for researchers interested in a quantitative and/or mechanistic understanding of biological rhythms. The text is written so that a reader can jump to a particular chapter where a technique can be found without wading through the previous chapters first.

- As a textbook in a course on modeling biological rhythms. I have successfully used the text as the basis of a graduate course at the University of Michigan. Each chapter can be taught in the span of about a week, especially if most of the frontiers sections are omitted or used as the basis of final projects. I have continued to be surprised by the growing demand for courses in biological rhythms from engineering, physics, medicine, chemistry, statistics, and other disciplines.

- As a textbook in an introductory undergraduate course on mathematical biology. Such a course could first cover some mathematical basics from other texts (e.g., Strogatz's *Nonlinear Dynamics*). I have then used the first chapter of each part of this book and selections from the later chapters. A real benefit of this approach is that studying rhythms allows students to see material from all aspects of biology and medicine.

- As a (perhaps secondary) textbook in courses in engineering, physics, systems biology, chemistry, statistics, theoretical biology, or chronobiology. Here, one or two chapters could add interesting techniques and applications to a course. For example, chapter 9 could be used in a control course, chapter 5 in a physics course, chapter 8 in a statistics course, chapter 10 in a nonlinear dynamics course, or chapter 3 in a chemistry course.

- As a reference to provide biological examples for mathematical ideas. Creative applications of mathematical ideas can be found throughout the text.

The book contains material from many disciplines, including mathematical biology, biological rhythms, systems biology, biology, medicine, mathematics, statistics, numerical methods, stochastic analysis, nonlinear dynamics, coupled oscillators, physics, chemistry, engineering, and control theory. This should be very useful for researchers in any field who wish to learn different approaches to tackle problems in biological rhythms. Great care has been taken to reduce the jargon usually used in these fields.

My target student is interdisciplinary. The students on whom I have tried this material in courses at Michigan have come from all over campus. These students are different from previous generations since they are more interdisciplinary and want to be able to use approaches from many fields rather than exhaustively exploring any one.

To help guide the wide and multidisciplinary audience for this book, I have divided it into three parts, organized in terms of increasing mathematical abstraction:

The introductory first chapter contains many ideas and suggestions about modeling meant to be helpful to the reader. These general remarks are not required for future chapters.

Part I, on models, should be accessible to the entire audience for the book. Readers with some (but not much) knowledge of differential equations should still find chapters 1 and 2 approachable, as well as much of chapter 3. Chapter 4 requires more algebra and some concepts, but it presents a method that is more approachable than other methods typically used in the field. The first sections of later chapters (e.g., 8 and 9) should also be approachable.

Part II, on behaviors, focuses more on simpler models and should particularly appeal to readers interested in simple models, for example as championed by physicists.

Part III focuses more on mathematical techniques. This part of the book is most useful for readers who have a specific model in mind, a particular goal, and a stronger (but not necessarily rigorous) mathematical background.

I take readers right up to the edge of modern research in sections highlighted as "frontiers." These sections can be omitted or skimmed by readers interested in more classical or general techniques. I also present some of the most interesting mathematical results in "theory" sections with new and simplified approaches. These can also be omitted or skimmed by some readers.

I also provide the commented MATLAB code that was used to generate many of the figures in the text. Readers are encouraged to run, comment, and modify these codes to get practical experience with the techniques. These codes can be viewed as exercises giving practical experience.

Working knowledge of differential equations, as is typically found in undergraduate engineering, mathematics, physics, systems biology statistics, or theoretical chemistry majors, is a prerequisite of the book. Parts II and III also use some basic linear algebra. Here, I mean knowledge of the meaning of an eigenvector, or the solutions of linear differential equations, rather than existence or uniqueness proofs.

Many clashes exist between different fields in terms of notation; phase response curves have different meanings to mathematicians and to biologists. Some fields represent $\sqrt{-1}$ as i, and others as j. I have sought to be careful but not rigorous—for example, explaining isochrons intuitively rather than rigorously defining them with foliations. The reader is pointed to texts where mathematical rigor can be found. Many useful tools (e.g., details of phase plane analysis in nonlinear dynamics) were not included since they are not specific to biological oscillations and are easily found in other books.

Special thanks to Deb, Olivia, and little Daniel Forger; the students in Math 463, 563, and 564 at the University of Michigan; the students and faculty of the 2014 MBI-CAM-BAM-NIMBIOS Graduate Seminar and the 2016 summer school at the INS of Shanghai Jiao Tong University; my many collaborators; summers in Woods Hole; time in the Blau lab; funding agencies that supported my research, especially the Human Frontier Science Program, the Air Force Office of Scientific Research, and the Biomathematics program at the Army Research Office; my undergraduates, graduate students, and postdocs; the anonymous reviewers; Leah Edelstein-Keshet, Yining Lu, and Jihwan Myung, who carefully read through the manuscript; Divakar Viswanath for many conversations about book publishing, many scholars who offered suggestions on specific chapters, and last but not least, Robert Prior and Katherine Almeida at MIT Press.

Ann Arbor, 2017

Notation

$\bar{X}_n$	steady state of X_n
$[X]$	concentration, # molecules of X/Volume
$\tilde{f}$	fourier transform of f
x	scalar x
$\underline{x}$	vector x
$\lVert \underline{x} \rVert$	L^2 norm or root-mean-squared of $\underline{x}$
$\langle \underline{x} \rangle$	average
$\underline{x}_0$, $\underline{x}_f$	initial and final states of $\underline{x}$
$\underline{\underline{x}}$	matrix x
α_j, β_j	forward and backward rates as used in the Hodgkin–Huxley or light model
δ	small parameter
ε	small parameter
ε_t	noise at time t
ε_1^0	small tolerances used in coordinate search
ζ_t	log-likelihood of observation
θ	changes to angle or phase on a slow timescale
ϑ	cumulative distribution function
λ_j	eigenvalues
$\underline{\lambda}(t)$	influence functions
μ	factor determining the period of an oscillator
$\xi(t)$	wiener process
ξ_t	noise at time t
π	3.14 …
Π	product
σ	standard deviation

Σ	sum
$\underline{\underline{\Sigma}}$	variance matrix
τ_i	random time to next reaction
τ	period
τ_x	period of human circadian clock
ϕ	phase
φ	cost at final time
χ	angle off plane
ω	frequency
A_i	activation energy
$\underline{B}(x)$	nonlinear terms
B	drive of light
$\underline{c}$	parameters of a model
C^l	continuous derivative
C_E	exterior concentration
C_I	interior concentration
d_t	variance of error in Kalman Filter
e	$2.718\ldots$
$\underline{\underline{E}}$	eigenbasis
E_i	equilibrium potential of i
$f_i(t)$	production rate of ith species
$g_i(t)$	clearance rate of ith species
$g(\underline{x}|\underline{c})$	a model of $\underline{x}$ given parameters $\underline{c}$
$G(\underline{x}(t, c_1, c_2, \ldots, c_n))$	goodness of fit of a model with parameters $c_1, c_2, \ldots, c_n$
$H(t)$	Hamiltonian
i	$\sqrt{-1}$
I	light level, input current
k_i	reaction rate constant
K_d	ratio of reverse to forward rate constants, dissociation constant
$\underline{l}^{\mathrm{T}}$	observation vector
$\underline{\underline{L}}$	Cholesky factor of a matrix
$L(\underline{x}, u)$	cost function
m	exponent typically used in Hill expressions
$m(t_i)$	model prediction at time t_i
n	size of vector
N	number of molecules, number of oscillators

$O(\varepsilon)$	of order ε
p	probability
$p(\theta_j)$	phase shift to oscillator j
P	protein
P_i	ith protein species
$\underline{q_i}$	eigenvector of a matrix
$\underline{Q}$	library of signals for least squares
r_i	rate of conversion of P_i to P_{i+1}
$r(t)$	radius
R_g	gas constant
R^n	n-dimensional real coordinate space
R	(variable) repressor
$s(t)$	signal
s_r	speed defined at time r
$\underline{\underline{S}}$	variance in Kalman Filter
t	time
T	absolute temperature
T_i	series of interspike intervals
V	volume
v_0	initial value of volume
v	synchrony index
u	control or signal to an oscillator
y_t	measurement at time t
z	(variable) complex state variable
z	(constant) charge of ion

1

Basics

We begin with a brief discussion of the goals of the book and the general principles of modeling, and then provide background knowledge useful in understanding these principles. Several examples of biological rhythms are presented that can be studied using the techniques from this book. We discuss, in particular, parsimony and the level of detail needed in modeling. We then highlight the main problems that we will address throughout the book. Useful statistics for circular data are presented, essential for working with oscillators, since every oscillator is in some sense a circular process.

1.1 Introduction

1.1.1 Goals

Living with and predicting rhythms are as much a part of being human as breathing, sleeping, and paying taxes. Humans have lived by and studied rhythms from their origins. By the fourth verse of the Bible, God separates the light from the darkness dividing day from night. The Mayans created a complex calendar for planting and ritualistic purposes, and, some say, they even predicted the beginning and end of the world (see figure 1.1).

While the ways humans live have changed over time, the importance of biological rhythms has remained. For example, the invention of the light bulb in modern industrial society has made humans less controlled by the daily light cycle and seasonal rhythms than all other living organisms. Yet we are not completely free from these rhythms or the effects of going against them. Shift work has been shown to cause many diseases, decreased productivity, and even major disasters (Dunlap et al. 2004).

Living organisms not only respond to rhythms, but they also create them. Sometimes unintended rhythms are generated with annoying or even dangerous effects (e.g., synchronized neuronal firing during epilepsy). Other times, as in the case of biological clocks, they are beneficial and/or evolved (Dunlap et al. 2004). Numerous generated rhythms are essential to the life of most organisms, from bacteria to man and from biochemistry to mood (Chance et al. 1973; Goodwin et al. 2007; Tiana et al. 2007).

Figure 1.1
A Mayan calendar. Taken from https://en.wikipedia.org/wiki/Maya_calendar.

The scope of this book is limited to the study of rhythms caused by or occurring within biological organisms. Studying biological rather than nonbiological rhythms, however, is not much of a restriction. Properties of nonbiological rhythms are often similar to those of biological rhythms, so the methods we use here for rhythm detection are not exclusive to biological applications. The main difference is the mechanisms of the systems themselves, and the mechanisms of rhythm generation in nonbiological applications are described extensively elsewhere. Some aspects of rhythms caused by nonbiological factors (e.g., the solar day) are relevant to our study, and these will be included.

Rather than studying systems that simply report rhythms, we look closely at systems that generate rhythms (clocks), an exploration that is much more interesting. As you will see, the behavior of biological clocks is complex, and it is unintuitive at first. Clocks not only generate rhythms, but they also respond to signals from their environment. Studying rhythms alone—without thinking about how they could be generated, changed, or controlled—would be boring.

Because of the complex behavior of biological clocks, a theory of timekeeping (i.e., of systems that keep time) is needed. Additionally, theory can also lead to new hypotheses and ideas in a way that is best summed up by Norbert Wiener (1976):

There is one great advantage with the mathematician: he may blunder to his heart's content, waste time in asking questions which he cannot answer, fumble and bungle and muddle. … If he welcomes every ghost of a shadow of an idea that comes his way, and tries it before casting it aside, he suffers no harm but great good; for it is just these waifs of notions that may furnish the new point of view which will found a new discipline or reanimate an old. He who lets his sense of the mathematically decorous inhibit the free flow of his imagination cuts off his own right hand.

A key term here is "mathematically decorous." While ideas need to eventually be rigorous (as in Wiener's work), imagination and free thought are essential. Intuition is also essential, since ideas need to be easily transferred from one discipline to another. Thus, while most of the techniques presented here are mathematical in nature, we will take a middle road. We aim to explain concepts intuitively, as a way of helping to make specialized ideas from a variety of fields accessible to a broad audience. Textbooks and professional papers within these fields can be further consulted for details or for explanations using the terminology of individual fields. The reader should also consult Lander (2010) and Paydarfar and Schwartz (2001) for advice about approaches to interdisciplinary problems.

Oscillations, rhythms, and clocks can all mean the same thing—or many different things. For example, in numerous biological disciplines, a "clock" is defined as temperature-compensated (i.e., keeps the same time at different temperatures), whereas in the literature of physics or mathematics, this requirement is not needed. We refer to a clock as something that keeps time for an organism. Oscillators are the broader class of all rhythmic systems, including those whose rhythms may not be autonomously generated. Rhythms are the outputs of oscillators.

1.1.2 Ten Examples of Biological Rhythms

Here we present ten examples of biological rhythms that can be studied using the techniques from this text. Many other examples are found later in the text.

1. Yeast organisms, in response to changing environmental conditions such as temperature, can show oscillations in their metabolism as they switch between high and low levels of dissolved oxygen (Lloyd et al. 2003). An example of this is shown in figure 1.2. These oscillations, whose function remains a topic of debate, are typically damped and depend on temperature. Section 3.5 studies how temperature affects biological clocks.

2. MinD is an important protein in *E. coli*. When these cells are about to divide, the MinD protein oscillates between the two ends of the cell (Raskin and de Boer 1999). This helps to set up the division and the wall between the new cells (see figure 1.3). One question that arises is how the spatial dynamics affect oscillations. We take up this question in section 2.4.

3. A similar oscillation is found in cell polarity in Yeast (Howell et al. 2012). Before dividing, yeast undergoes a budding process where one front is extended. While this was originally thought to be governed by a positive feedback loop involving the Bem1p protein, Howell et al. (2012) showed that this process is also governed by a negative feedback loop creating oscillations in the Cdc24p protein (see figure 1.4). The role of these oscillations remains to be fully determined.

4. Hes1 is a transcriptional repressor that regulates the fate of cells (Kobayashi et al. 2009). It is found in fibroblasts, embryonic stem cells, and neural progenitor cells.

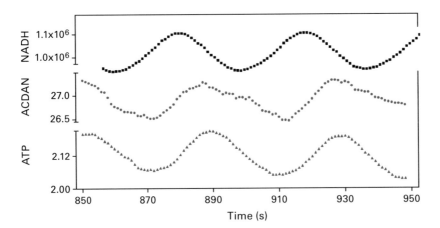

Figure 1.2
An example of glycolytic oscillations in yeast. Shown are three of the oscillating species, NADH (given by flourence intensity), ATP (a source of energy for a cell presented in mM) and 6-acetyl-2-dimethylaminonaphthalene (ACDAN, a flourence marker indicating water transport). Taken from Thoke et al. (2015).

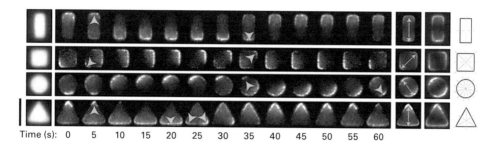

Figure 1.3
Example of rhythms of the MinD protein. This protein oscillates between either pole of an *E. coli* cell just
before division. Cells of different shapes are shown. Taken from Wu et al. (2015).

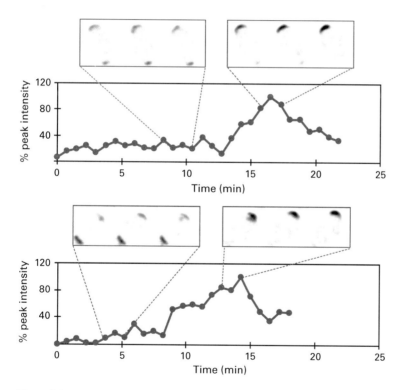

Figure 1.4
Rhythms of the Bem1 protein in yeast. The amount of Bem1 is shown in yeast cells and quantified in the plot
that follows. Taken from Wu et al. (2013).

Hes1 is regulated as part of a transcriptional feedback loop where the protein regulates its own production. It does this by regulating both the transcription of its mRNA and its own degradation. Rhythms of Hes1 are shown in figure 1.5. The Goodwin oscillator described in section 2.6 can model this system.

5. The protein p53 is a key protein in the regulation of cancer and DNA damage (Geva-Zatorsky et al. 2006). It is associated with another protein, Mdm2, in a negative feedback loop where p53 activates Mdm2 while Mdm2 inhibits p53. Oscillations in this system are seen in response to UV light that causes DNA damage. In the absence of DNA damage, p53 oscillations occur as well, although they are of a longer period and tied to the cell cycle. Oscillations of p53 activate a DNA repair mechanism that can correct the deleterious effects of UV light. These rhythms are seen in figure 1.6. The feedback loop structure of p53 can be analyzed through the techniques described in chapter 4.

6. Calcium is one of the key signaling molecules in the cell. Calcium concentrations are kept low in cells since high concentrations can be toxic. Calcium oscillations in the cytoplasm are important for many cellular functions such as insulin release from pancreatic β cells and the fertilization of sea urchin eggs. Oscillations in calcium are typically dependent on the dynamics of calcium entry into the cell, from intracellular

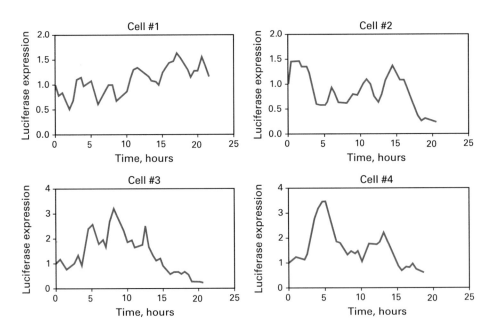

Figure 1.5
Rhythms of the Hes1 protein. Rhythms from four representative cells are shown. Taken from Bonev et al. (2012).

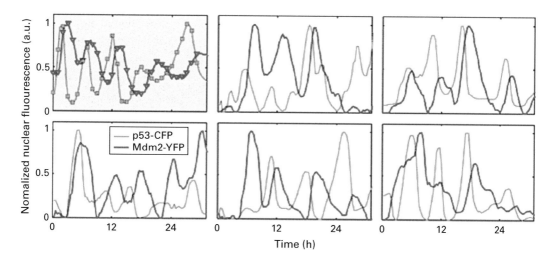

Figure 1.6
Rhythms of the p53 protein and its partner Mdm2. Taken from Geva-Zatorsky et al. (2006).

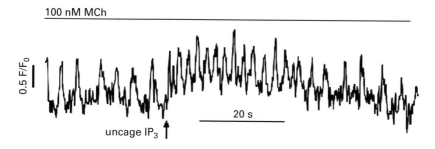

Figure 1.7
Rhythms of the intracellular calcium concentration from pancreatic β cells. Oscillations were stimulated by MCh. Application of IP_3, which releases calcium from intracellular stores, causes a decrease in the period of the rhythms. Taken from Sneyd et al. (2006). Copyright (2006) National Academy of Sciences, USA.

stores (e.g., through IP3; see figure 1.7), cell membrane channels, or other oscillating signaling molecules like cyclic AMP. Calcium levels are an integral component of neuronal oscillations, which are the subject of several models presented in this book, including the Morris-Lecar model studied in section 5.4.

7. One of the most studied problems in oscillating biology is the dynamics between hares and their predators, lynx. Much of the interest in this research was inspired by the great records kept by the Hudson's Bay Company showing ~10-year oscillations in the populations of lynx and hares in Canada (see figure 1.8). This system was modeled

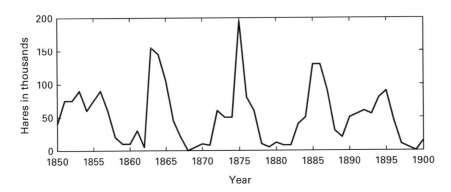

Figure 1.8
Oscillating patterns of the number of hares and lynx as recorded each year by the Hudson's Bay Company.
Digitized from MacLulich (1937).

by one of the best-known examples of an oscillating biological system, the Lotka-Volterra model. Similar models are studied in chapter 5. With x as the population of hares and y as the population of lynx, this model is given by the following equation:

$$dx/dt = \alpha x - \beta xy,$$

$$dy/dt = \delta xy - \gamma y.$$

Here α is the growth rate of the hares, β is the decay rate of the hares due to the lynx, δ is the growth rate of the lynx due to eating hares, and γ is the death rate of the lynx. Unfortunately, this elegant model is not correct (Krebs et al. 2001). The hare population can oscillate without any lynx. This shows the importance of testing models against data, which is the topic of chapter 8.

8. There is a well-known daily variation in heartrate (Vandewalle 2007). The author measured his heartrate over 24 hours using his smartwatch (figure 1.9). The 24-hour rhythm in this data is masked by noise and the author's activity. Chapter 8 studies how to extract signals from noisy data.

9. Sporatic infant apneas: Our breath is rhythmic. Sometimes lapses in our breathing rhythm (apneas) occur. While many factors that cause apneas have been identified, infants sometimes can show unexplained apneas (see figure 1.10). These apneas can be serious or even fatal in the case of sudden infant death syndrome (SIDS). Paydarfar and Buerkel hypothesize that these apneas may be caused by the effects of noise, and that noise, paradoxically, can prevent them. We study how noise can create these behaviors in chapter 5.

10. Bipolar disorder (BD) is one of society's leading mental health challenges. Individuals with BD experience episodes of increased (mania) and decreased (depression) affect,

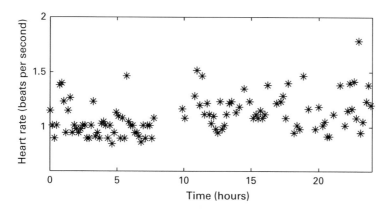

Figure 1.9
Heart rate measurements taken by the author's smartwatch over a 24-hour period.

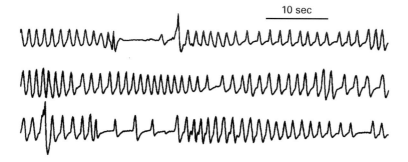

Figure 1.10
Abdominal movement measured from an infant indicating breathing rhythms. In addition to the normal rhythms, periods where the breathing rhythm temporarily stopped (apneas) are shown. Taken from Paydarfar and Burkel (1997).

as well as of more normal affect (euthymia). An example of the timecourse of BD subject is shown in figure 1.11. The cause of BD is unknown, and possible models have been proposed (Cochran et al. 2016).

1.1.3 Overview of Basic Questions

In this book, we address three basic questions:

1. How to build and validate models of biological rhythms;
2. How clocks interact with and adapt to external (environmental) signals;
3. How clocks work at multiple scales.

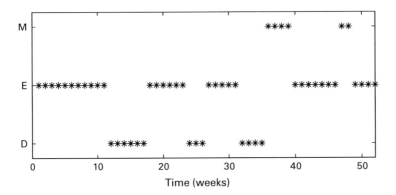

Figure 1.11
Mood of a subject with bipolar disorder enrolled in the Prechter Bipolar Research study. Mood is classified as either manic (M), euthymic (E) or depressed (D) and reported on a weekly basis. See Cochran et al. (2016) for more details on the mathematical study of bipolar disorder and this dataset. Data courtesy of M. McInnis.

To explore these questions, we first consider how to construct mathematical models. Without models, few questions about clocks can be addressed. Unfortunately, we sometimes face a plethora of models with little guidance on which models to choose. Confusion can arise because professionals usually construct models in ways most relevant to their own fields of study. Physiologists, for example, tend to give us a long list of parts that make up systems. A central question for them is how do these genes, ion channels, etc., interact to generate rhythms (chapter 2)? Mathematicians have found general classes of oscillating systems. How can these be used to understand biological systems (chapter 5)? Statisticians have developed methods to extract information from time-series data. How can this approach be used to build and validate models in various professional fields of study (chapter 8)? Our goal is to think critically about how models are built and used.

Adaptability is critical for biological clocks. The human heart needs to adapt to various activities, varying its rate to suit different activities: running, sleeping, etc. Human daily rhythms must be able to shift to new time zones (Dunlap et al. 2004). Populations of organisms change with changing seasons (Koukkari et al. 2006). We consider how clocks adapt to permanent changes in the environment (chapter 3), how they change with respect to transient signals (chapter 6), and how they can be perturbed in optimal ways (chapter 9). Rhythms that cannot adapt to or interact with external signals are useless.

A major challenge in understanding biological rhythms is that they almost always have multiple scales. Even if we understand the individual genes or proteins that form a feedback loop, for example, we are still faced with the question of how these parts interact to produce oscillations (chapter 4). Most biological "clocks" actually consist of many interacting clocks (e.g., one in each cell of tissue). How does the behavior of a multiclock system differ from that exhibited by a single clock (chapter 7)? How can we simulate a multiscale model, or simplify it to study behavior at a single scale (chapter 10)?

This book takes us through three main approaches to these questions:

1. Physiology (chapters 2–4)
2. Behaviors of systems (chapters 5–7)
3. Computational techniques (chapters 8–10)

These approaches are very broad and are meant to indicate where the emphasis will be placed. The chapters on physiology will consider that the many parts that make up systems may or may not be well understood. Yet, models can still be made to study such systems. The chapters focusing on behaviors of systems emphasize the kinds of general principles used by physicists or mathematicians. What matters is the abstraction (or mathematical structure). What unites the final three chapters is the emphasis on computational techniques, whether they be used to extract information from data (chapter 8), calculate optimal stimuli (chapter 9), or simplify or simulate complex models (chapter 10).

This book includes ideas from many different fields as they relate to the three main areas covered in this book:

Physiology: math physiology, temperature compensation, feedback loops;

System behaviors: nonlinear dynamics, phase response curves, coupled oscillators;

Computational techniques: statistics, optimal control, averaging/numerics.

This list is not exhaustive by any means. Additionally, some fields, such as the fields of biological rhythms (e.g., those found in the current Society for Research on Biological Rhythms), computational neuroscience, and systems biology permeate parts of many chapters.

1.2 Models

1.2.1 Fundamentals of Modeling

The most difficult question in mathematical biology is, "Why should I trust the predictions of a model?" Some zealous modelers all but proclaim, "every prediction sacred, every prediction great." Some extremely skeptical biologists also categorically dismiss models offhandedly. The aim of this book is to enable readers to form their own judgments based on the firsthand experience of working with the models themselves.

Let us take a closer look at the three fundamentally different methods by which models can be constructed. Keep in mind that these methods are not mutually exclusive. In fact, most models across the various scientific disciplines use multiple methods.

The first method, based on physiology, takes a bottom-up approach, where the components of the system under study are initially characterized (chapter 2). Physics, engineering, and other quantitative disciplines rely on laws, be they that voltage = current × resistance or the laws of mass action. In such fields, modeling using laws is rarely questioned. Most parts in engineering are built to reliably obey these laws, and even to display key properties like modularity, the ability of a part to show the same behavior in a wide range of systems (Del Vecchio et al. 2008). Likewise, the parts in physics are so fundamental and ubiquitous (e.g., atoms, gravity, mass) that such laws are amenable to almost all problems routinely encountered.

Biological systems consist of many parts of endless diversity. The parts we study could be the ion channels that contribute to the electrical activity of a cell, the proteins in a cellular network, or the muscles in a complex system of locomotion. These parts follow laws, as well. So long as the characterization of the parts we wish to study is complete, the parts can be put together and simulated in a model. This model will enable us to predict the behavior of the overall system.

Years ago, the complexity of so many diverse parts in biological systems called for the abandonment of a "detailed" approach to study biological systems (Levins 1966). Even if we could characterize all the parts, these arguments went, they would be intractable analytically or computationally. Except that a miracle has happened in scientific computing, with each year bringing the solutions to problems that once seemed impossible to solve. More data are collected each day on these parts based on new "high throughput" experimental techniques. New mathematical tools are also being developed to tackle these complex models. So the criticism of these detailed models seems less valid with each passing year.

Now let us look at the second method of model construction: using mathematical principles to match the behavior of a system. Many good mathematical models exist for nonbiological oscillators. Some similar principles may also apply to biological oscillators. This approach gains much more credibility in the context of mathematical theory where classifications of oscillators can be made (chapter 5). Because there are a limited number of ways by which oscillations can be created, certain canonical models can be created and applied to many biological scenarios. In this case, models can be constructed without consideration of the details of the parts.

The third method of constructing models is to fit data or determine models directly from the data. Here, the justification of a model is statistics rather than biological detail or mathematical structure as explained in chapter 8. One chooses the model (or models) that is (are) most likely, given timecourse data. Heavy computation is often required to determine the likelihood of models. Computation can also be used to simplify large detailed models to determine their mathematical structure (chapter 10) or determine which canonical mathematical models best fit some timecourse data.

There could be other justifications for models besides the three presented here. However, the take-home message for the modeler is that there may be many approaches toward modeling. Keep an open mind, while also being prepared to dismiss models that do not have clear support. The take-home message for the biologist is that there is not one specific type of modeler or model. Just as geneticists are different from electrophysiologists or ecologists, physicists are different from applied mathematicians or statisticians.

Be critical, as models are only as good as their justification, but don't be offhandedly dismissive.

1.2.2 Parsimony?

One of the general maxims of modeling is that models should be as simple as possible while being complex enough to capture the essential features. Suppose we want to study the basic mechanisms of a certain biological system by using a time series. Using a time series to collect data involves identifying a single variable we want to study and measuring it repeatedly at fixed intervals of time. How much information can we determine about basic mechanisms from these data alone? This partly depends on the amount of noise in the data. First, we can take a best-case scenario and assume that the time series has no noise. Even in this case, there is a limit to how much we can learn from the data (see chapter 8.2). But, in reality, all experimental time series have noise. One is then left with a dilemma. If we choose a very simple model, for example, one described by just one parameter, many of the important dynamics might be left out. On the other hand, if we allow enough flexibility in the model, one might be able to fit each data point perfectly. However, the data set contains noise that will have been fit as well as the data. This could actually increase the uncertainty of the parts of the model that were fit based on the noise. This risk is what has led to parsimony as one of the general maxims of modeling.

Parsimony is a good idea, but in practice quite difficult. A proposal for how to choose the appropriate level of complexity was made by Akaike (discussed in section 8.9) (Burnham and Anderson 1998; Wasserman 2015). Akaike was able to quantify how much information a model captures. He also found how much additional information one would expect to capture if one added more parameters into a model above the simple goodness of fit, determined by the likelihood of the data given the model. By a comparison of these information estimates, one can determine a parsimonious model, where the model captures a high degree of information. Akaike's analysis has been touted as a mathematical justification for the principle of parsimony (Burnham and Anderson 1998).

These methods work well for high-level models, where basic mechanisms (e.g., biochemical rates) are not known. Most examples presented in key texts on the subject, such as the work of Burnham and Anderson (1998), are from ecology. However, biophysical models like the ones we will study in chapters 2–4 often have large numbers of parameters and seem thereby to violate the principles proposed by Akaike, at least modern interpretations of them. How does one reconcile a 50- or 100-parameter biochemical model with statements like "No model with > 20 parameters can reasonably be trusted," or, perhaps the most abused quote in mathematical modeling, "With four parameters I can fit an elephant, and with five I can make him wiggle his trunk," a remark attributed to von Neumann by Fermi (see Dyson 2004).

Like all things, the devil is in the details. Akaike's theory and the many related theories apply in specific scenarios and require specific assumptions. For example, we will see in chapter 8 that this theory requires many data points, models that are already very good, and parameters that all uniquely affect the fit of the model. It is the last point that is crucially missed by most. When fitting to a time series, many parameters of a biochemical model will be irrelevant (mathematically orthogonal) to the data we study. They may determine transients that are not picked up, or parts of the system that are not measured. Yet we know they exist, so they should be included in the model, even if just to learn that they are unimportant, but they should not count toward the Akaike information criterion (AIC).

We also can model based on mathematical structure, as proposed in chapter 5. We know that models typically have certain mathematical structures, and we can use these to make predictions. However, the likelihood of a mathematical structure is something that is hard, if not impossible, to quantify.

Yet, in general, one cannot determine the flexibility of a model (i.e., the number of model behaviors) simply by counting parameters. One can easily add parameters into a model without changing the behaviors. For example, consider

$$dx/dt = -ax$$

and

$$dx/dt = -(b-c)x.$$

Regardless of the choice of b or c, no new behaviors will be seen in the second model that cannot be reproduced by the first.

However, this example may be viewed by some as too simple. Yet, it may be that b and c have special biological meaning requiring they be kept in the model. For example, b could represent a forward reaction and c a reverse reaction, each tied to a specific protein. An actual experiment in the lab may be able to change only one at a time, so keeping this complexity may help us match data. This gets much less trivial when one considers the multiscale nature of many biological problems.

Consider the act of transcription. This requires hundreds, if not thousands, of steps as an mRNA is produced. The parsimony police quickly arrive on the scene, declare this too complex, and censor the model in favor of a simple approximation: that the amount of mRNA produced is simply the rate of the activation of a gene with a delay. This delay can be reliably measured (e.g., from data on the difference between the binding of transcription factors to a gene and the mRNA concentration), whereas each step surely cannot be measured reliably from the data on a specific gene, at least not in the way accounted for by the criterion proposed by Akaike. Yet this simplistic approach could give erroneous results. We will see in chapter 4 that delay equations such as described before overpredict the occurrence of oscillations in genetic networks. Any detailed model, no matter the parameters (except in the case of infinite parameters) would show this. Moreover, we could construct a ridiculously complex model (at least in the eyes of some) with many unknown parameters, and still be able to make some predictions, such as what would happen if an amino acid were in low supply, or how temperature variations might affect transcription by increasing or decreasing all basic biochemical reactions.

I therefore see little harm in using a detailed model for transcription, so long as its overall rate matches the overall production rate of mRNA. Perhaps one could argue that a detailed model might be too difficult to simulate, but the cogency of such arguments is disappearing quickly with increases in computing power. One also must be careful not to overinterpret multiscale models. Just because the model has rates for adding 215 amino acids does not mean that these rates are accurately known. Such a model, however, can answer useful questions, such as what the possible behaviors that can be seen by such a model are.

1.2.3 Scaling, Nondimensionalization, and Changing Variables to Change Phase

Redefining variables can often simplify problems. Similarly, we often consider redefining a variable, e.g., x as $x' = bx + c$, where b and c are constants. This can remove parameters from the problem, and thus the problem could be solved for x' and then converted to x if the additional parameters are needed. Interesting examples of this are provided in sections 3.5 and 5.8. In general, each variable scaling allows the removal of one parameter, and scaling time can remove one more. Scaling parameters can also help determine the role of parameters. This is seen in the example that follows.

Suppose every time a parameter c appears in a model it appears as $x + c$. We can then remove all instances of $x + c$ with a scaling $x' = x + c$. Next we calculate the period of the system with x'. Since scaling back to x does not change time or the period, the period of the model does not depend on c. Thus, c does not affect the period of the oscillator. Using this technique, we can scale variables to determine if they affect the period.

The units of the variables are an important consideration, and it is often important to make sure that all equations have the right units. For example, if x has units of concentration, and we have

$$dx/dt = ax + bx^2,$$

then a must have units of 1/time and b must have units of 1/(time × concentration). Scaling variables can remove these units when removing parameters, in a process called nondimensionalization.

Finally, we note that a change of variables can sometimes be equivalent to changes in time. For example, consider a wall clock, and now rotate this clock by 90°. This change of variables is equivalent to a change of time of 3 hours, since it shifted all points to other points 3 hours later. This principle is used in section 6.3.

1.3   Period

A key property of many oscillators is that they have an inherent time scale of the period τ. Thus, some oscillations can be used to tell time. If an oscillation has occurred n times, $n\tau$ time units have occurred. From a mathematical perspective, every oscillation could be used to tell time. In the real world, only a few are good enough. Most are too noisy, or show different periods when external conditions are slightly varied. The Dow Jones Industrial Average shows oscillations, but you would not want to time an egg with them.

Clocks have a wide range of biological mechanisms and, correspondingly, a wide range of periods. Here are some examples:

Millisecond	Second	Minute	Hour		Day	Month
Electrical pulses in neurons	Firefly flashing	Calcium waves in cells	Embryonic development	Tidal rhythms	Circadian clocks	Menstrual

Clocks with a period of about a day are particularly important. They comprise a large proportion of biological clock research. These clocks are called circadian, a term developed in the 1950s. The key point is that they have a period of about (circa) a day (dian). We will see that the period of circadian clocks is never exactly 24 hours. The period of the circadian clock in your body is, on average across all humans, 24.2 hours (Czeisler et al. 1999).

While the actual value of the period is crucial for biological applications, it is typically ignored mathematically. Mathematicians, engineers, or especially physicists almost always set the period of a clock to 1 time unit (or sometimes 2π; see the discussion about nondimensionalization in section 1.2.3). Tell this to a biologist, and they will often insist you are missing the point. The key idea is how this is done. If we de facto set all periods to 1, indeed, we have lost something. However, if all we do is change the time unit to a value close to the period (i.e., study circadian clocks in days, calcium waves in minutes, etc.), then there is still room for models to make predictions. So while the period is crucial, the time unit can be dispensable.

1.4   Phase

A good clock should be able to time events quicker than its period. Thus, we are interested in what fraction of a period has elapsed between two events. Within a cycle, time is called the clock's phase and is measured modulo the clock's period. Typically, the phase is expressed as a fraction of the period (or a fraction of 2π). However, biological clocks are not like the digital clocks that read off what fraction of a period has elapsed. *The phase of biological clocks must be interpreted from their state.* Mathematically, this can be represented as $\phi(\underline{x})$ where the state of the system is described by $\underline{x}$. Sometimes this is simple. For example, on a wall clock, we can measure the angle the minute hand has moved around the origin. However, this interpretation works only if the minute hand moves at constant angular velocity. Some simple clock models have this feature (e.g., see the discussion of the radial isochron clock in chapter 6).

However, no biological clocks are this simple. They resemble a roller coaster (which is actually a very accurate clock). One might want to interpret the phase (or time) from the distance the cars of the roller coaster have traveled, but the cars proceed at a very uneven speed. The time it takes for the cars to proceed the first 100 feet (up the hill) is much longer than the next 100 feet. Indeed, the "mapping" between state and phase requires thought.

Likewise, the tracks of a roller coaster follow a path in three dimensions, whereas a wall clock has just two and phase involves just one. The state of the biological system that acts as a clock often involves many state variables, possibly hundreds or thousands. Phase just involves one. So how can we extract this one important variable out of many? The problem presents itself in two ways: (1) Given data, how can I extract phase? and (2) Given a complex biological mechanism, how can I determine the most essential reactions to determine phase? The answers to these questions are largely unknown but some tools to answer them are presented in the next section, which can be skipped by less mathematically inclined readers. Properties of phase will be explored in depth in chapter 6 as part of an elegant theory by Art Winfree (Winfree 1980).

1.5   Frontiers: The Difficulty of Estimating the Phase and Amplitude of a Clock

Determining the phase of a biological clock can be surprisingly difficult. The best way to determine phase experimentally is to measure the time from some marker of the rhythm, for example the peak of the rhythm. Yet, there are other mathematical ways to determine phase in models. For example, let us consider that one has a cycle in that the same position is reached after a period τ. Consider a parametric plot of two of the system variables. If only one is measured, one can plot it against its derivative. One should then have a cycle of some form. The phase can be read as one reads a wall clock, i.e., the angle from the vertical (12:00) position and the center of the cycle. However, this cycle may not be a circle, and the clock may not move through all angles at the same speed. This complicates things, since the phase would then be determined by the position of the origin and by which variables are plotted. Choosing some reference point (e.g., the peak of one of the rhythms) and defining phase as the time since one was last at that reference point solves these complications.

Now consider the case where the system is not on a cycle, but approaching one (a limit cycle, see figure 1.15 in section 1.7 for an example). Phase can easily be defined by the preceding arguments when the system in on a cycle. We would like to define it for all points that eventually approach the limit cycle. This can be done in the following way: Assume we have point $\underline{x}_0$ and point $\underline{x}'_0$ on the limit cycle. Let their trajectories be $\underline{x}(t)$ and $\underline{x}'(t)$ respectively with $\underline{x}(0) = \underline{x}_0$ and $\underline{x}'(0) = \underline{x}'_0$. The points $\underline{x}_0$ and $\underline{x}'_0$ have the same phase if these two trajectories approach each other, or more formally, for every small ε, we can find a t' such that for all $t > t'$,

$$|\underline{x}(t) - \underline{x}'(t)| < \varepsilon.$$

In this way, transients do not matter when defining phase. Figure 1.12 shows an example of how this definition of phase can be used and how it can differ from what we would normally expect. It plots lines that all have the same phase by this definition (isochrons).

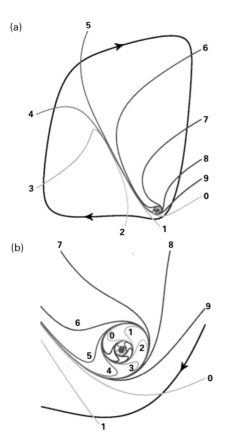

Figure 1.12
Isochrons of the Fitzhugh-Nagumo model as calculated by Langfield et al. (2014). The bottom plot is a magnification of the region around the fixed point of the upper plot. While some systems show remarkably simple isochrons, this is an example of a simple system that has remarkably complex isochrons.

Similar difficulties occur for defining the amplitude of an oscillator. We could define the amplitude as the maximum vs. the minimum of the rhythm. This could present difficulty if the rhythm has multiple variables. Which variables should one then use for this definition? Should the definition of the amplitude be based on just two points? A better definition of the amplitude is the L^2 norm (Euclidean amplitude) discussed in chapter 4.

1.6 Plotting Circular Data

There are also many ways to plot circular data, for example, data that are phases. First, all phases are plotted along the circle by some marker. The data can then be grouped, coarse-graining the data in order to potentially see the overall trends better. Sometimes data are collected in such a way that the individual phases are not recorded, but instead the number of events in a range of phases is measured. These can be plotted as shown in figure 1.13, a figure with some interesting historical context. In general, the radius of each section should be proportional to the square root of the number of elements in each section (Fisher 1993). A simple example helps illustrate this. Imagine a uniform distribution of phases, which would be a circle in this diagram. As we increase the number of data points, the radius would increase by the square root of the number of data points, and the area of the circle would increase linearly with increasing number of data points.

Another way to plot cyclic data is through an actogram. Here the occurrence of an activity is plotted as a linear function of phase. Each separate cycle is also plotted on a new line. Examples of this include that seen in figure 1.14, where the time asleep is plotted in black. In these plots, it is often useful to double-plot or triple-plot the data by showing either two or three days for each line. This allows the patterns to be better seen. For example, in figure 1.14, we see the case of an elderly man who is removed from time cues. He normally sleeps twice a day, one time for a nap and another for the night. In this case, the two rhythms persist when removed from time cues. However, they slowly switch roles, so that once external time is reintroduced, the nap period becomes the nighttime sleep bout.

1.7 Mathematical Preliminaries, Notations, and Basics

Here we introduce many of the important concepts needed for the rest of this book. Any variable in a model is scalar unless it is underlined, indicating it is a vector, or double underlined, indicating a matrix. The variable t will refer to time and $\underline{x}$ the state variables of the system.

Almost all models will be governed by differential equations of the following form:

$$d\underline{x}/dt = \underline{f}(\underline{x},\underline{c},u(t)),$$

which states that the rate of change, $\underline{f}$, of the state of the clock, represented by vector $\underline{x}$ of length n, depends on a set of parameters, $\underline{c}$, which do not depend on time, and on an

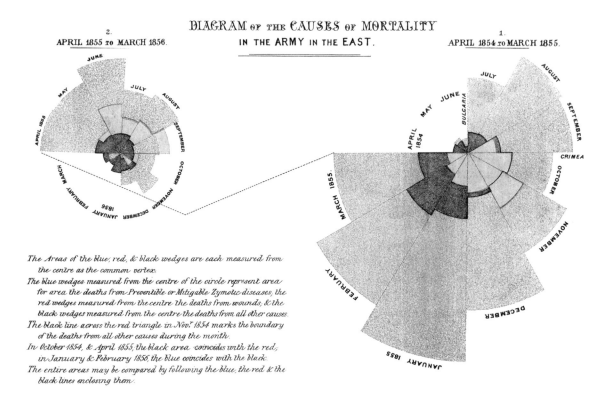

Figure 1.13
A circular histogram drawn by Florence Nightingale showing the causes of death in the Crimean War. Taken from https://simple.wikipedia.org/wiki/Florence_Nightingale/. This illustrates the importance of seasons to the survival of humans and other organisms.

external signal, $u(t)$, which does vary with time (occasionally more than one external signal is considered). We almost always assume f is continuous, and has continuous derivatives. To simulate the model, we need a starting value of $\underline{x}$ (which we call $\underline{x}_0$), and we need $\underline{c}$ and $u(t)$ to be specified. Such equations are called ordinary differential equations, and partial differential equations (where derivatives are taken with respect to variables other than time) are mentioned only in passing in this book, with the exception of chapter 9.

A formula for the solution of these equations, $\underline{x}(t)$, can sometimes be determined without the use of computers. Since the formulas often contain $\underline{c}$ and $u(t)$, they can give a complete description of the behavior of the system. While exact formulas can only rarely be computed, approximate formulas can also be found and can be quite useful. However, with the ever-growing power of computers, most problems we describe will need some form of simulation, and sample code is provided.

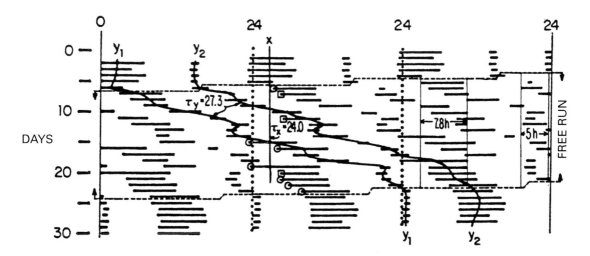

Figure 1.14
Sample actogram. Sleep patterns of an 81-year-old man. The individual was placed in constant lighting conditions in the middle of the experiments ("free run"). Time is shown in hours horizontally and days vertically. Taken from Kronauer (1987).

Much research on clocks (and most of our analysis) focuses on their autonomous behavior where no zeitgebers (time-giving signals, such as light, from outside the organism) are present. Mathematically, this can be represented as

$$\frac{d\underline{x}}{dt} = \underline{f}(\underline{x}, \underline{c}).$$

The parameters, $\underline{c}$, are quantities (e.g., rate constants) that could vary depending on the environment, system, or implementation of the model. If no parameters can be varied, $\underline{c}$ can be removed:

$$\frac{d\underline{x}}{dt} = \underline{f}(\underline{x}).$$

The behavior of any system can be divided into two parts: transient and sustained. Most systems approach a steady state that is sustained until some external signal perturbs the system. Mathematically, we say that a steady state is where

$$d\underline{x}/dt = 0, \quad \underline{f}(\underline{x},\underline{c},u(t)) = 0.$$

Much of science studies steady states, where oscillations are not present. Transients are ignored. However, a steady state is not the only possible sustained behavior that a system can exhibit. Many systems show sustained oscillations. Mathematically we mean:

$\underline{x}(t + \tau) = \underline{x}(t)$ and $\underline{x}(t') \neq \underline{x}(t)$ for some $t < t' < t + \tau$.

Thus, the system leaves and returns to its original state in a certain amount of time, τ. Since $\underline{x}(t + \tau) = \underline{x}(t)$, after the return $\underline{f}(\underline{x}, \underline{c})$ is the same, and the system will then change at the same rate. Thus, if $\underline{x}(t + \tau) = \underline{x}(t)$ for one value of t, it is true for all values of t, assuming that there is no external forcing $u(t)$.

Rhythms may be gained or lost as the parameters, $\underline{c}$, change. If so, we say that the system undergoes a bifurcation at the values of $\underline{c}$ where the change happens (see chapter 5). Likewise, a system that would not oscillate by itself can show rhythms if $u(t)$ is rhythmic. If $\underline{x}(t)$ and $u(t)$ have the same period, regardless of whether $\underline{x}(t)$ could have autonomously generated rhythms, we say that the system described by $\underline{x}$ is entrained.

When a system oscillates, many states are visited, $\underline{x}(t_1), \underline{x}(t_2), \underline{x}(t_3), \ldots.$ Thus, sustained behaviors need to be described by a set of points rather than a single point. If the same set of points are revisited after a certain time lag τ (called the period), i.e.,

$$\underline{x}(t_1 + \tau) = \underline{x}(t_1), \ \underline{x}(t_2 + \tau) = \underline{x}(t_2), \ \underline{x}(t_3 + \tau) = \underline{x}(t_3), \ldots,$$

the collection of points that the system visits is called a cycle, as described previously.

Not all states of a system are part of a cycle, but it is possible that all states of a system tend toward a cycle. If a cycle is approached by at least some state of the system that is not part of the cycle, it is called a limit cycle (see figure 1.15).

The capacity of a system to oscillate does not guarantee that it will oscillate in all cases. For example, a system can have some initial conditions attract to one state, and others attract to another. This is one example of a bistable system. Thus, oscillations depend on the starting state of the system. Examples of this will be seen later.

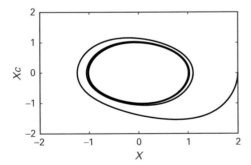

Figure 1.15
An example of a limit cycle. Here two state variables (x and x_c, the variables of a model of the human circadian clock described in section 5.11) are plotted as a parametric plot. Starting from an initial condition off the cycle, trajectories in the phase space approach the limit cycle.

1.8 Key Problems in the Autonomous Case

Here we state *some* of the key problems we will discuss, as well as which chapters they will appear in. To get the reader in the habit of understanding mathematical notation, we state them mathematically. We first consider the case where $u(t) = 0$, i.e., $dx/dt = f(x,c)$.

1. With c unknown and perhaps f only partially known (e.g., whether certain variables inhibit or excite others), can we determine if oscillations are possible (chapters 4, 5)? What structural designs of systems often lead to oscillations?

2. If oscillations are possible, can we determine which values of c yield oscillations (chapters 4, 5)?

3. Assuming $u(t) = 0$, determine the mapping

   $$c \to \tau$$

   or how parameters determine the period (chapters 4, 5).

4. A variation of this is the problem of temperature compensation. As temperature increases, all rate constants in biochemical reactions increase. Thus, every element in c should increase. However, we would like τ to be relatively unchanged in many applications, e.g., since the length of a day does not depend on temperature (chapter 3).

5. Often we observe the behavior of a clock and would like to model it. This is the following problem:

   Given $x(t)$ with $t' \leq t \leq t' + \tau$, find c or f (chapter 8).

6. Determine $\phi(x)$, the phase at each system state as defined in section 1.4.

7. If x has many components, can we determine a simplified system that captures the essential dynamics of x (chapter 10)?

1.9 Perturbations, Phase Response Curves, and Synchrony

Again, we consider a system

$$dx/dt = f(x,c,u(t)),$$

where $u(t)$ is an external signal that confers timekeeping information, perhaps with some noise. For example, when traveling to Europe or Asia, our circadian clocks sense light, which indicates the phase of the external world and tells those internal clocks what the correct phase should be. Each day these signals must correct the period of our circadian clocks to be exactly 24 hours.

Another example with a much shorter period is shown in figure 1.16. Here we simulate the Hodgkin–Huxley model predicting the electrical activity of a neuron. At time 30, we give two stimuli, which change the phase of the neuronal firing. The black curve shows an advance of the phase, which causes the next firing to occur quicker. The red curve shows a delay of the phase, which causes the next firing to occur later. This figure also shows an experiment where a pulse of current starts rhythms. This is another possible effect of a stimulus.

Once a clock's behavior in time isolation is understood, we can begin to understand how stimuli can change the behavior of a clock. A key principle of biological rhythms is that *the effect of a stimulus depends on what phase of a cycle it was applied to.* Think of pushing a child on a swing. Pushing the child forward when he or she is ascending yields an opposite result from that when pushing him or her forward when the child is descending.

Often we are interested in how stimuli change the phase of the clock. In these cases, a simple experimental protocol is often used:

Assess the clock phase > Apply the stimulus at the desired phase > Reassess the phase

The stimulus may cause transients, so the reassessment of the phase is typically done after enough time has passed for these transients to disappear.

To fully characterize the effects of a stimulus, one should apply the stimulus at all possible phases. One can then plot the phase before the stimulus has been applied, along with how much the stimulus changed the phase of the oscillator. Such plots are called phase response curves (PRCs). An example of a phase response curve is shown in figure 1.17.

There is much more to be said about these plots (see chapter 6).

Many biological applications use more than one clock to time events. Within your brain, the suprachiasmatic nucleus (SCN) contains about 20,000 neurons, each with a clock inside it, and these clocks acting together comprise your central circadian clock. Females living in close proximity can synchronize their menstrual cycles (although this has been questioned; see Harris and Vitzthum 2013). Thus, issues of how these clocks synchronize can be particularly important, and these are discussed in chapter 7. In particular, different types of synchronization exist. Synchronization of clocks with different periods occurs when they agree on a common period. They can also agree on the same phase (in-phase synchronization), show opposite phases (antiphase synchronization), or any possibility in between. When many clocks interact, there is a wide variety of possible behaviors.

1.10   Key Problems when the External Signal $u(t) \neq 0$

We continue the listing of key problems begun in section 1.8, extending our consideration now to cases where there is a nonzero external signal, $u(t) \neq 0$.

(a)

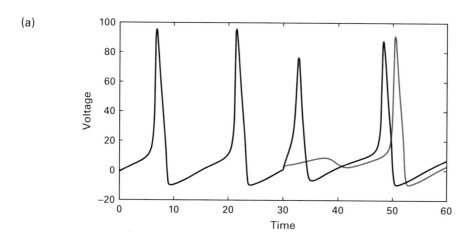

(b)

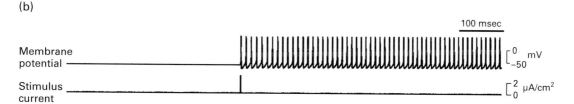

Figure 1.16
(Top) Simulations of the Hodgkin–Huxley model (see chapter 2) representing the voltage of a squid giant axon. At time 30, two separate perturbations were applied to the model. One causes the rhythm to be phase-advanced (black), another to be phase-delayed (red). (Bottom) Experimental data taken from Paydarfar et al. (2006) showing how a pulse of current can start rhythms in a squid giant axon.

8. How can we determine the PRC to a stimulus (chapter 6)?

9. If we are entrained to one signal (e.g., living in a particular time zone) and wish to be entrained to another (go across time zones), how can this be done in the most efficient way (i.e., to minimize jet lag)? This problem is often called reentrainment (chapter 9).

10. Given two (or more) oscillators that communicate, can we predict when they will agree on a period (synchrony)? What other types of behaviors will be seen by coupled oscillators (chapter 7)?

11. How can oscillators respond to zeitgebers, while ignoring other, irrelevant environmental signals (chapter 3)?

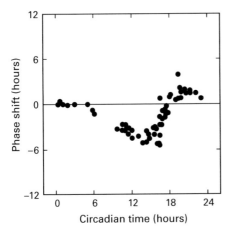

Figure 1.17
An example of a phase response curve. This shows the change in phase of the circadian clock in mice exposed to a 6-hour pulse of bright light. Taken from Vitaterna et al. (2006). Copyright (2006) National Academy of Sciences, USA.

1.11   Frontiers: Probability Distributions with a Focus on Circular Data

Here we summarize some basic principles of probability and statistics. While some of these principles may be familiar to many readers, how to apply them to data on phases, which are defined modulo the period, is less likely to be familiar. We typically discuss phase, but the techniques we mention here can be applied to any circular variable as well as to phase. In fact, these techniques were actually developed mainly to study the directions animals move (Batschelet 1981).

We will provide some background on these techniques to be used in later chapters. Probability theory is used heavily in chapter 8. Gaussian and Poisson random numbers are used in chapters 2, 3, and 5, as well as the idea of independent random variables. Readers who are not working with real data on phases, and those who are already familiar with these basic concepts, can move on with little lost.

A random variable is mathematically characterized by a probability distribution function (pdf). A pdf is a function $f(x)$ that tells us the probability that a random variable, x, is within a certain range:

$$\Pr(a < x < b) = \int_a^b f(x')dx'.$$

Normally, we let a and b take any values between $-\infty$ and ∞; however, if we are considering a circular variable, such as phase, we let a take values between 0 and τ, where τ is the period. We can also determine the cumulative distribution function as

$$\mathrm{cdf}(b) = \int_0^b f(x')dx'.$$

Two probability distributions that occur often are the Gaussian distribution and the Poisson distribution.

Gaussian random variables are random variables, x, with standard deviation σ, whose pdf is given by

$$\frac{1}{\sigma\sqrt{2\pi}}e^{-\frac{x^2}{2\sigma^2}}.$$

In chapter 8, we consider a vector of n Gaussian random variables, which requires not only the n standard deviations of its variables, but also a covariance matrix, $\underline{\underline{\Sigma}}$, which tells how each variable is correlated with each other variable. Given this, the pdf, used in chapter 8, and with $\|$ referring to the determinant, is given by

$$\frac{1}{\sqrt{(2\pi)^n|\underline{\underline{\Sigma}}|}}e^{-\frac{1}{2}\underline{x}^T\underline{\underline{\Sigma}}^{-1}\underline{x}}.$$

Gaussian random variables are popular since the sum of any n independent identically distributed random variables (of virtually any distribution) converges to a Gaussian random variable, by the central limit theorem, as $n \to \infty$ (Chung 2001).

Poisson random variables are, on the other hand, good for measuring processes that occur at fixed rates: for example, for elements to be produced. A Poisson random variable is an integer random variable, x, with mean h (in the typical case the mean can be thought of as a rate), whose pdf is given by

$$\frac{h^x e^{-h}}{x!}.$$

A key property of a Poisson random variable is that its mean and variance are the same.

Sometimes we wish to indicate that choosing one random variable, x, would not affect the choice of another random variable, y. In this case the random variables x and y are called independent. We can then multiply the probabilities in that the probability of $x = a$ and $y = b$ is (probability of $x = a$) times (probability of $y = b$).

Before jumping to circular statistics, it is helpful to consider the data set. If we truly collect circular data (e.g., the direction of flight of a bird, or the phase of a minimum of a waveform), then circular statistics should be used. However, sometimes we infer phase from noncircular state variables (e.g., finding the phase from the voltage and current of a neuron), so regular statistics can be used on the state variables (e.g., error estimates, correlations) to answer key questions. For example, in figure 1.18, we calculate the standard

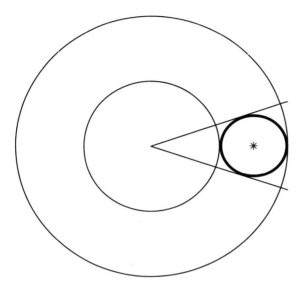

Figure 1.18
Calculation of the error in phase and amplitude from Cartesian error. We start with a point (*) for which we calculate its Cartesian error (dark circle). From this, the error in phase (radial lines) and error in amplitude (thin circles) are calculated.

errors for a two-state system. From these, errors in phase and amplitude can be found. In what follows, however, we assume that circular statistics are needed.

The Poisson distribution is easy to apply to circular variables. We can assume that there is a uniform rate, c, at which events can occur over the cycle. We then have that the probability of an event occurring between phases a and b is $\lambda = c(b - a)$. The Poisson distribution then tells us what the probability is of having 1, 2, 3, etc., events occur.

A more difficult problem is to apply the Gaussian distribution to circular data. One possibility would be to "wrap" the Gaussian distribution in the following sense (Fisher 1993):

$$\Pr(x = a) = \ldots + g(a - 2\tau) + g(a - \tau) + g(a) + g(a + \tau) + g(a + 2\tau) + \ldots,$$

where g is a Gaussian distribution and τ is the period of the circular variable. However, it turns out this is often not the most natural distribution. Instead, most times the von Mises distribution is used. It can be motivated in the following way (Batschelet 1981).

Consider a Gaussian distribution of two variables in a plane where the probability of being at a particular point is a Gaussian distribution of the distance from the origin. In the phase space of these two variables, we choose a circle. The probability of being at any point on this circle is the von Mises distribution. See figure 1.19.

To derive the von Mises distribution, let us assume, without loss of generality, that the center point of the circle is at the point $(a, 0)$, that the circle has radius 1, and the center

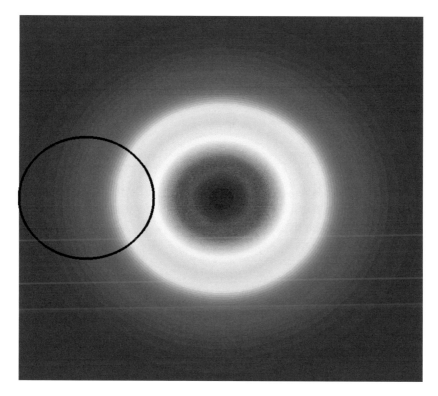

Figure 1.19
Illustration of the derivation of a von Mises distribution. A two-dimensional Gaussian is plotted. We consider a circle in the plane, and the probability of being at each of the phases on this circle is the von Mises distribution.

of the Gaussian distribution is at (0, 0). We also require that $a>1$. The probability of the center point is

$$\frac{1}{\sigma\sqrt{2\pi}}e^{-\frac{a^2}{2\sigma^2}}$$

Any point on the circle has coordinates $(a-\cos(\phi), \sin(\phi))$. We see that the distance from this point to the origin is

$$\sqrt{(a-\cos(\phi))^2 + \sin(\phi)^2} = \sqrt{a^2+1-2a\cos(\phi)}.$$

The probability of this point occurring is

$$\frac{1}{\sigma\sqrt{2\pi}}e^{-\frac{a^2+1-2a\cos(\phi)}{2\sigma^2}} = be^{k\cos(\phi)},$$

where b is a parameter that is typically chosen to scale the probability so that the total probability of being on the circle is 1. This is the von Mises distribution.

Unless a large number of samples are present, it can be difficult to distinguish between the wrapped Gaussian distribution and the von Mises distribution (Fisher 1993). However, the von Mises distribution does have a key property not shared with other distributions: The most likely phase of the von Mises distribution is also the mean phase, a fact that we will find is most useful in section 1.12.

1.12   Frontiers: Useful Statistics for Circular Data

The mean of circular variables is more difficult to calculate than that of noncircular variables. Consider the following example. Assume that we wish to take the mean of two phases measured with respect to a day. One is 11:00 p.m. or 23:00, and the other is 1:00 a.m. By standard methods, the mean of these two numbers is $(23 + 1)/2 = 12:00$, which is nonsensical. We instead represent each of the phases as points on a circle, as in the preceding circular histograms. We can then sum these points as

$$\sum_{j=1}^{n} e^{i\phi_j} \equiv r e^{i\phi},$$

where ϕ_j are the individual phases measured in radians. Then ϕ is the mean phase and the angular variance (Batschelet 1981) of this phase is $2(1-r)$. The angular variance has properties similar to the variance used in noncircular variables (Batschelet 1981), and its square root has units of radians. In the case of grouped data, we can use

$$\sum_{j=1}^{n} n_j e^{i\phi_j} \equiv r e^{i\phi},$$

where n_j is the number of events in each group with mean phase ϕ_j and the angular variance is $2\left(1 - r\dfrac{\lambda/2}{\sin(\lambda/2)}\right)$ where λ the length in radians of each of the groups (Batschelet 1981).

The median phase in circular variables is best described by a picture, as in figure 1.20 (Batschelet 1981; Fisher 1993). First, the phases are plotted on the circle. We then draw a line individually from each of these points through the center point, and count the number of points that lie on each side of the line. If there are an equal number of points on either side of such a line, then the point used in the line is called a median. This can also be calculated as follows. Redefine the phase to be in radians, with phase 0 defined to be one of the points. If half of the remaining points have phases between 0 and π and the other half of the remaining points have phases between π and 2π, that point is a median. Note that there can be more than one median.

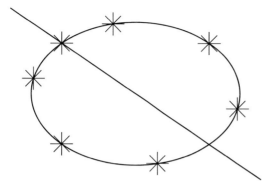

Figure 1.20
Demonstration of the median in circular statistics. First, all phases are drawn on the circle. The median is the
phase that lies on a radius that divides the remaining phases into two equal groups.

The Rayleigh test gives values of the angular variance that would indicate whether the
phases are statistically different from a uniform distribution (Batschelet 1981). Another
similar test is Rao's spacing test (Batschelet 1981) Here we find the differences between
successive phases, $\eta_i = \phi_i - \phi_{i-1}$. Next we consider the differences between these values and
their values if the phases were equally spaced:

$$\frac{1}{2} \sum_{i=1}^{n} \left| \eta_i - \frac{2\pi}{n} \right|$$

The levels of significance of this quantity have been worked out (Batschelet 1981).

Sometimes we wish to determine if a group of phases clusters around a particular phase.
Here, one can use the V test (Batschelet 1981). We first determine the mean phase and
angular variance. Together, they form a vector $re^{i\phi}$. We then take the projection of this
vector in the direction of the phase of interest, a, yielding $v = r\cos(\phi - a)$. The test then uses
$(2n)^{1/2}v$, where n is the number of phases, and checks this against levels of significance.

We can also test whether the data are different from any possible distribution, using the
chi-squared test (Batschelet 1981). We first group the data into k equally spaced regions, as
done for the circular histograms. With n_i as the number of elements in group i, and e_i as the
expected number of elements in each group, we then find

$$\sum_{i=1}^{k} (n_i - e_i)^2 \Big/ e_i$$

and compare this value against levels of significance. Another similar test is Kuiper's test
(Batschelet 1981). Here we calculate the cumulative distribution of the data, and find the
maximum and minimum deviation from a test distribution. The sum of these deviations is
then checked against levels of significance.

Another problem we may encounter is to determine if two circular data sets are different. Again, there are several possible tests. The simplest may be the run test (Batschelet 1981). Plot both data sets on the circle. Then proceed along the circle and count how many times a data point from one data set is followed by a data point from the other data set. Another way to measure this is to determine the cumulative distribution function of each data set and find the maximum upward and downward deviations between these cumulative distribution functions. This sum of these deviations can be tested against significance values; this is a variant of Kuiper's test described earlier (Batschelet 1981).

We mention a final statistical test that can be quite useful. Suppose we are monitoring a biological rhythm with a marker of a particular phase (e.g., the minimum of the rhythm, or the time an animal wakes up). The time between these markers is then recorded as $\tau_1, \tau_2, \ldots, \tau_n$, where each of these measurements is the period plus some noise. We wish to determine if this noise is inherent to the rhythm (e.g., a sloppy oscillator) or inherent to the measurement (e.g., a perfect oscillator with poor measurement that contains noise). Pittendrigh and Daan (1976) propose the following serial correlation test, which continues to be used (Ko et al. 2010). Find the correlation between $\tau_1, \ldots, \tau_{n-1}$ and $\tau_2, \ldots, \tau_n$. If the measurement is the source of the noise, then one would expect a negative correlation, i.e., that if the marker comes earlier (respectively, later), on subsequent cycles it will come later (respectively, earlier). This would not be the case for noise inherent to the rhythm, where one would expect a correlation close to zero.

Exercises

General Problems
For these problems pick a biological system of interest, preferably one you study.

1. Provide a brief description of the underlying biology behind the system.

2. Of the 11 problems listed in sections 1.8 and 1.10, which ones are most applicable to your system?

3. How is the system measured in the lab or in the wild?

4. Give a definition of phase that can be accurately measured in the lab.

Specific Problems
1. Think of five biological clocks. For each, write down some of their key variables ($\underline{x}$), parameters ($\underline{c}$), and zeitgebers ($u(t)$).

2. Prove that a linear system can never show a limit cycle.

3. Given the linear system

$$\frac{dx}{dt} = \underline{\underline{A}}\,\underline{x}, \; \underline{\underline{A}} = \begin{pmatrix} a_{11} & a_{12} \\ a_{21} & a_{22} \end{pmatrix},$$

find the answer to the seven questions posed in section 1.8 based on a_{ij}.

4. Consider one of the most classic models of an oscillator, the van der Pol equation: $\frac{dx}{dt} = ax_c + \varepsilon(x - cx^3), \; \frac{dx_c}{dt} = -ax$. By computer simulations, determine the role of a, c, and ε. Choose appropriate values such that the limit cycle has a radius of about 1 and the period has a value of about 24.

5. Give three ways one can define phase.

6. When we linearize an oscillating system around a fixed point and find a pair of imaginary eigenvalues, what can be said about the oscillations' amplitude and period?

7. Show that the most likely phase of the von Mises distribution is also the mean phase.

I
MODELS

2

Biophysical Mechanistic Modeling: Choosing the Right Model Equations

Here we describe how to write down equations for a model. We cover several important classes of models, including biochemical models and neuronal models. Our goal is not to make a comprehensive presentation of all known types of models, but to study in depth the building blocks of these models so that the reader can learn to develop their own models. Additionally, we present several potential pitfalls to modeling.

2.1 Introduction

This chapter focuses on how to develop biochemical and electrophysiological models of biological clocks. We present all the details necessary to develop such models and show the process by which model equations are chosen. By doing so, we can carefully determine the key assumptions that the models make, and that often go unreported, and potential pitfalls in modeling. Many researchers fall into these pitfalls. Key assumptions can be subtle but important. For these reasons, readers who are already familiar with these types of models may benefit from this chapter, even if some of these details are not needed for the remaining chapters.

Let us begin by looking at the types of models we will build. Biophysical models use the underlying physiology to try to reproduce the known data on the system as a whole. In building this kind of model, we take the parts, put them together, and see if they are sufficient to reproduce the known system behaviors. If the model falls short of this goal, the way to improve the model, in the biophysical modeling approach, is to collect more data on the parts, if additional data from experiments are available. Another name for this approach is "parametric" modeling, and it is contrasted with the "black box" modeling approach in which only the input and output are taken into account and the inner workings of the system are not investigated.

Unfortunately, complete knowledge of the parts is rarely possible, and moreover, particularly in cellular biology, we seldom know which parts (e.g., which proteins) are important and which are dispensable for the behaviors we are modeling. Yet, with every passing year, more data on the parts have become available, and the models we construct now ought to take advantage of these new data and adhere more closely to the biophysical modeling approach.

Having said this, we must acknowledge that we are taking a potentially purist view of modeling. Often modelers mix black-box models and biophysical models, using parametric models for those parts that are well understood and black boxes for the parts of the modeled system that are not well understood. As we will see later, this can be very risky. In such mixed models, it is hard to determine where error should be attributed (in black-box models it is lack of understanding of the system's behavior, and in parametric modeling it is lack of knowledge of the system's parts). So while many modelers (including myself) will mix and match approaches, we must heed the warning: Modelers Beware!

Nevertheless, taking this potentially purist point of view, we seek to determine the *best* set of equations possible based on the available data. To do this, we will look critically at the ways in which models are constructed and the assumptions on which they are based. One will find that most models in the literature have inconsistencies between what they claim occurs and the actual equations they use. I have collected, in this chapter, some of the key misunderstandings to watch for in creating models, and I hope the reader, by keeping these in mind, will be able to critically evaluate models, including my own, and to create new and better models.

Both the biochemical models and the electrophysiological models we study are based on laws: the law of mass action and Ohm's law. We begin by giving some biological preliminaries, and then we explain the laws and consider them critically to determine when they apply in the cellular environment.

Modelers often start with higher-level expressions, such as Hill expressions for transcription, Michaelis-Menten expressions for phosphorylation, delay equations for unknown parts, or high-level neuronal models such as the Fitzhugh-Nagumo model. These could also be used as nonparametric models whose parameters are fit (as discussed in chapter 8), but only if rigorous statistical consideration of the data is presented. However, our hope in biophysical modeling is that the mathematical expressions we choose are derived from underlying biophysical mechanisms and therefore impart predictive ability to the model.

2.2   Biochemical Modeling

Before we begin, it is helpful to list some of the key cellular rhythms that are modeled using biochemical feedback loops. These include models of the cell cycle (Tyson and Novak 2001), circadian rhythms (Kim and Forger 2012), glycolytic oscillations (Goldbeter 1996, 2002), and rhythmic firing of neurons. While a large percentage of the genes in a cell show rhythmic patterns of expression (Panda et al. 2002), the following proteins and cellular signaling components can play a key part in generating rhythms that are separate from those just mentioned:

- calcium (Goldbeter 2002),

- cyclic AMP (Goldbeter 2002),

- Hes1 (Hirata et al. 2002; Monk 2003),

- MinD (Kruse et al. 2007),

- Msn2 (Garmendia-Torres et al. 2007),

- Nrf2 (Xue et al. 2014),

- peroxidase (Goldbeter 2002),

- p53 (Kim and Jackson 2013; Lev Bar-Or et al. 2000).

The types of neurons that show rhythmic firing are too numerous to list here. Some of these rhythms were described in section 1.1.2.

Cellular biology relies on the following "central dogma": DNA makes mRNA makes protein.

DNA contains the instructions for making a protein. DNA must be transcribed, a process whereby an RNA copy is made. How much RNA is made is a tightly controlled process. Proteins called transcription factors bind to DNA at promoters around the instructions (coding sites) and regulate how much RNA is transcribed. These transcription factors can be activators or repressors. Once produced, mRNA molecules and proteins must also degrade or be diluted out as the cell grows and divides.

From the RNA instructions, proteins are made by ribosomes, a process known as translation. Proteins can be modified posttranslationally, particularly by kinases that add phosphate groups to the protein or phosphatases that remove phosphate groups. Most events in the cell (e.g., transcription, degradation, nuclear transport, etc.) can be controlled by phosphorylation or other protein modifications. In addition, proteins may form complexes by dimerization, including homodimerization where one protein binds with another of the same chemical species.

In eukaryotes (e.g., mammalian cells), DNA is localized in the nucleus of the cell and not in the cytoplasm. In eukaryotes, proteins and mRNAs can be transported into the nucleus or the cytoplasm. Such a division does not exist in prokaryotes.

2.3 Law of Mass Action: When, Why, and How

The basic principle in forming equations to model biochemistry is the law of mass action. This states that the rate of any reaction is proportional to the concentrations of the reactants (we denote concentrations by []). Thus, if A and B bind together to form a complex that we denote AB, we write this reaction as

$$A + B \rightarrow AB,$$

and the rate of this reaction is given by

$k[A][B]$,

where k is a rate constant. This term is added to the appropriate differential equations based on the stoichiometry, or how many molecules of each species are gained or lost in each equation. For example, we would model the preceding reaction as

$d[A]/dt = -k[A][B]$,

$d[B]/dt = -k[A][B]$,

$d[AB]/dt = k[A][B]$,

where $[AB]$ is the concentration of the complex. Different terms from different reactions can be added to form the differential equations for a model. One should be careful to get the stoichiometry correct. For example, in the reaction of C homodimerizing,

$C + C \rightarrow CC$,

we have

$d[C]/dt = -2k[C]^2$

and

$d[CC]/dt = k[C]^2$,

since two molecules of C are lost to create one CC homodimer. Many models in the literature miss this subtle but important point.

While the law of mass action is a good place to start, it does make some key assumptions. It assumes that the number of interacting molecules is large enough that individual molecular interactions do not significantly affect the behavior of the system. Mass action can be formally derived from basic molecular interactions. For example, Kurtz in 1972 showed how well-mixed molecular interactions can lead to a mass-action description. Several authors, such as Smoluchowski (1917), show how diffusion-limited reactions in three dimensions yield mass-action kinetics.

Sometimes the number of molecules in a system is particularly low. If so, another approach is needed to account for the randomness of molecular interactions. This is described in chapter 3. Diffusion-limited reactions in a crowded cellular environment, or diffusion-limited reactions restricted to a two-dimensional membrane, or a one-dimensional strand like DNA may show time-varying rates (Berry 2002; Grima 2010; Schnell and Turner 2004). While we will explore this topic in more detail in the next section, we should mention here that this fact has led some authors to modify mass action. They incorporated non-integer exponents in their equations, particularly since it is easier to fit simulations of complex systems or experimental data with this freedom (Savageau 1995). The difficulty with this approach is that it is often difficult to know how many molecules are actually being used in the reactions.

Since our goal is not simply to fit data, but to predict it from basic biochemical principles, we advocate using mass action unless there is a good reason to question it (e.g., low molecule numbers or diffusion-limited reactions on low-dimensional surfaces). If such reasons exist, simulations of individual molecular reactions can be done to account for complex spatial geometries. An example of this is shown in the next section.

2.4   Frontiers: The Crowded Cellular Environment and Mass Action

Many cellular reactions are isolated to surfaces, for example, the cell membrane, or the membranes surrounding organelles. Recent work has highlighted how proteins in the cell can be sequestered to specific areas of the cytosol, nucleus, or cell membrane. Proteins can even be restricted to one-dimensional surfaces. For example, transcription factors such as the lac repressor often are localized around DNA, where they diffuse (Hammar et al. 2012). Recent work in the 4D nucleome has suggested that the nucleus of the cell contains pockets where genes and transcription factors colocalize, and that this spatial segregation is important for gene function (Rajapakse et al. 2011). Thus, rather than freely diffuse in three dimensions, proteins tend to be restricted to small areas where they can freely diffuse and react. Along longer spatial scales, proteins can jump between these regions.

To illustrate the limits of mass action, we consider a simple reaction, which we assume happens on a 2-D surface. We model this reaction by considering a model similar to beads on a necklace or candy dots on paper. Within small regions along a 1-D surface (DNA), or a 2-D surface (a cell membrane), or 3-D (cell cytosol or nucleus), proteins can freely react. Consider a protein A that reacts with itself to degrade. An analogous situation would be that when A reacts with itself, the resulting homodimer leaves the surface and is no longer considered. This gives

$$A + A \rightarrow \varnothing.$$

We also assume that proteins can switch among different regions where they interact. In 1-D they can move left or right, in 2-D left, right, up, and down, and in 6 directions for 3-D simulations. We also choose periodic boundary conditions simulating the circular DNA in bacteria, or a closed membrane surface. We set the rate of moving any protein in any direction to occur with Poisson rate 1, and the rate of binding between proteins in a bin to be $100*A_i*(A_i -1)$, where A_i is the number of proteins in one region (i.e., the product of a constant rate of 100 and the number of possible interactions between molecules in a region).

Interestingly, Isaacson and Peskin (2006) show that this method can also work when proteins are freely diffusing and are not confined to specific regions. Thus, the bins can be thought of as discretizations of the 1-D, 2-D, or 3-D reaction space and the transit times between bins describe the time it would take a protein to diffuse from one bin to the next.

The possible events are as follows: (1) the bin could lose a molecule to any of its neighbors, and (2) a dimerization could take place in the bin, which leads to the removal of two molecules of A. If a reaction occurs with a Poisson rate h, we can generate a random time to the next event by

$(1/h)\log(1/\text{rand})$,

where rand is a random number drawn uniformly between 0 and 1 (see chapter 3). We simulate the model reaction by calculating the random time to each of the next events. We then execute the reaction that takes place first, and generate new times to the next reaction. This process is repeated until all reactions occur. This method is called the Gillespie method. It will be described in further detail in chapter 3.

The key question to ask is whether mass action still holds. If the reaction proceeds at a rate proportional to A^2, where A is the total number of proteins in the simulation (in all bins), mass action holds. To calculate the reaction rate, we calculate the time it takes to degrade 50 molecules, then an additional 50 molecules and so on until all are degraded. From this, the rate can be calculated as 50/(time to degrade 50 molecules), and we divide this rate by A^2. Thus, we plot:

$50/((\text{time to degrade 50 molecules})\, A^2)$.

If this number remains constant, mass action holds. If it does not, then mass action does not hold.

As one can see from figure 2.1, initially the law of mass action holds, and the rate remains constant. Then, a transition occurs where the reaction drastically slows. This occurs when there is approximately 1 protein per bin, and the reaction starts to be diffusion-limited. In the 2-D simulation, the law of mass action basically holds, even in the diffusion-limited case, though the rate does decrease over time. However, a much more dramatic change in the rate is seen for the 1-D case. Code for these examples is shown in code 2.1, found at the end of the chapter.

In figure 2.2, we lower the rate of the reaction to 1 and the number of molecules in each bin at the beginning of the simulation to 1 per bin. As one can see, this dramatic decrease in the rate is not due to the lower reaction rate and mass action continues to hold.

2.5   Three Mathematical Models of Transcription Regulation

In this section, we consider three mathematical expressions representing transcription regulation. Modelers should determine if their systems match the assumptions of any of these three expressions, or if a specialized model needs to be constructed. We present the details

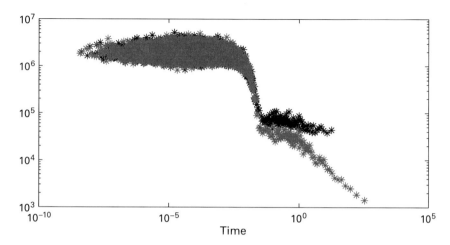

Figure 2.1
Testing the law of mass action. Proteins were placed on a periodic 2-D lattice (black) or 1-D closed curve (red) representing proteins that react on membrane surfaces, or DNA. Each surface was discretized with 10,000 bins. The simulation was started with, on average, 100 proteins per bin randomly distributed among all bins. Proteins could dimerize, which we assumed led to degradation, or move to the adjacent 2 (red) or 4 (black) bins. The dimerization reaction rate was initially set at 100, and the probability of moving to another bin was given a Poisson rate of 1. Shown is the rate of reaction measured in the simulation divided by A^2, where A is the total number of proteins in the bin. This is an estimate of the rate constant, which would be fixed if mass action holds.

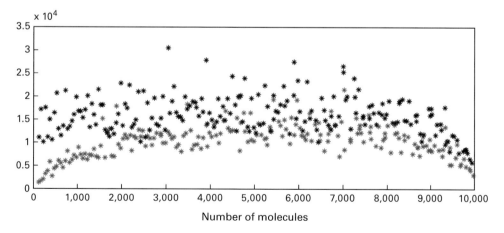

Figure 2.2
Same as figure 2.1 but with a 100-fold lower reaction rate and a 100-fold reduction in protein level. Here we plot the rate constant with respect to the number of molecules rather than time.

for constructing these three; other expressions can be constructed following similar techniques. Some examples of models of circadian clocks using these techniques can be found in work of Forger and Peskin (2003), Goriki et al. (2014), and Kim and Forger (2012).

Model "a": **A Model for Transcription Regulation with Independent Binding Sites**

Let us start with a model of repressor molecules (R) binding to a promoter. We assume m independent binding sites (G) for a repressor. In this case we write the forward (binding) and reverse (unbinding) reactions as

$$G + R \xleftrightarrow[k_r]{k_f} GR.$$

Let G be the probability that a particular site is free. Mathematically denoting concentration as [], we have

$$\frac{dG}{dt} = k_r(1-G) - k_f G[R].$$

If we assume that the binding and unbinding of R to G are fast compared with other cellular processes, we can look at the steady state and assume $dG/dt \approx 0$. We then have

$$k_r = G(k_r + k_f[R])$$

or

$$G = \frac{1}{\left(1 + \frac{k_f}{k_r}[R]\right)}.$$

This is the probability that a particular site is free. Now, suppose that all m sites must be free for transcription to occur. The fraction of time this will occur is G^m (which assumes the bindings to different sites are independent) or $1/(1+(k_f/k_r)[R])^m$. When all m sites are free, transcription will occur with a maximal rate k_{max}. Thus, the overall rate at which transcription will occur is

$$\frac{k_{max}}{\left(1 + \frac{k_f}{k_r}[R]\right)^m} \tag{2.5.1}$$

The preceding expression can be used if we can make the following assumptions:

* There are m binding sites on a promoter and binding to any of these sites stops transcription.
* The repressor protein binds rapidly with the promoter.

We plot this function in figure 2.3.

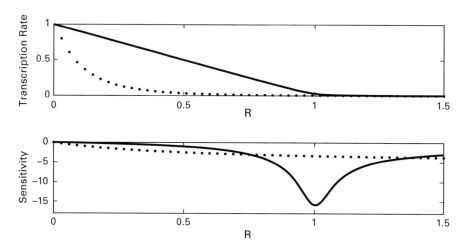

Figure 2.3
Transcription regulation functions as a function of the concentration of repressor (R) derived previously. The upper plot shows the rate of transcription as a function of a repressor. The dotted curve is $f(R) = (1/30)(1/(.05 + R)^5)$. The solid curve is $f(R) = 1 - (1 + R + 0.001 - ((1 + R + .001)^2 - 4R)^{1/2})$, which considers a revised model where a repressor binds to an activator to inactivate it. The lower plot shows the sensitivity of these functions, $d\log f/d\log R = (df/dR)(R/f(R))$, which tells us the relative change in transcription for a small change in R. The importance of this sensitivity will be seen in chapter 4.

Model "b": The Hill Expression

In this model, it is assumed that m molecules of a protein act together. This can occur in several ways. First, a complex of m proteins forms rapidly and stably. This m-mer (combination of m proteins) then rapidly binds and unbinds to the DNA. A second possibility is that the only two stable states of the DNA are the state with m molecules bound to the DNA and the state with no molecules bound to the DNA. This second case could arise because the binding of repressor molecules may cause conformational changes to the DNA (that is, changes to its structure). In either case, we have

$$G + mR \xleftrightarrow[k_r]{k_f} GR^m,$$

so the differential equation for the probability G is

$$\frac{dG}{dt} = k_r(1 - G) - k_f G[R]^m.$$

With rapid binding, we are interested only in the steady states, so

$$k_r = G(k_r + k_f[R]^m)$$

or

$$G = \frac{1}{1 + \frac{k_f}{k_r}[R]^m},$$

and so transcription occurs at the rate

$$\frac{k_{max}}{1 + \frac{k_f}{k_r}[R]^m} \tag{2.5.2}$$

This is called a Hill expression. The Hill expression can be used if we can make the following assumptions:

• On the promoter, either no proteins are bound and transcription occurs, or m proteins are bound and no transcription occurs.

• The repressor protein binds rapidly with the promoter.

• Binding of the protein occurs only on the promoter (i.e., unbound proteins go to bound proteins on the promoter)

As these assumptions are very restrictive, it is very surprising how often the Hill expression is used.

Model "c": Activator and Repressor

The third model considers the effects of an activator and a repressor. Assume that an activator can bind to a site to activate transcription. Thus, we have

$$G + A \underset{k_r}{\overset{k_f}{\longleftrightarrow}} GA,$$

which yields the following equation

$$\frac{dG}{dt} = k_r(1 - G) - k_f G[A].$$

Solving for G as before, we find

$$G = \frac{1}{1 + \frac{k_f}{k_r}[A]}.$$

In this model, transcription is proportional to GA or $1-G$, so the expression for the transcription rate becomes

$$1 - G = \frac{[A]}{[A] + \frac{k_r}{k_f}}.$$

Now let us assume that the activator can be in one of two states: either free A that can bind with DNA, or bound to a repressor, AR. The binding to the repressor is given by the following equations:

$$A + R \xleftrightarrow{\substack{k'_f \\ k'_r}} AR.$$

From this, we have

$$[A][R] = K_d[AR],$$

where $k'_r / k'_f \equiv K_d$. Assume we know the total concentration of A, $[A_T]$, and the total amount of R, $[R_T]$. We then have

$$([A_T] - [AR])([R_T] - [AR]) = K_d[AR].$$

Rearranging this in terms of [AR] gives the quadratic equation

$$0 = [AR]^2 - ([R_T] + [A_T] + K_d)[AR] + [R_T][A_T],$$

which admits the following solution:

$$[AR] = \frac{([R_T] + [A_T] + K_d) \pm \sqrt{([R_T] + [A_T] + K_d)^2 - 4[A_T][R_T]}}{2}. \tag{2.5.3}$$

Since [AR] is less than $[A_T]$, we must choose the − sign before the square root to maintain appropriate ordering. The overall transcription rate is then proportional to the amount of free A, which is

$$[A_T] - [AR].$$

This expression is plotted in figure 2.3, and it can be used if we can make the following assumptions:

- Transcription is proportional to the amount of free activator.
- The repressor protein binds rapidly with activator molecules, and stops them from initiating transcription (e.g., by preventing them from binding to DNA, or stopping their ability to recruit transcriptional machinery).

It is interesting to note that if $K_d = 0$ (i.e., A and R bind very tightly), then

$$[AR] = \text{either } [A_T] \text{ or } [R_T], \text{ whichever is smaller,}$$

since [AR] cannot be greater than $[A_T]$ or $[R_T]$. With this assumption, solving for $[A] = [A_T] - [AR]$, we find it is equal to either $[A_T] - [R_T]$ (if $[A_T] > [R_T]$) or 0 (if $[R_T] > [A_T]$).

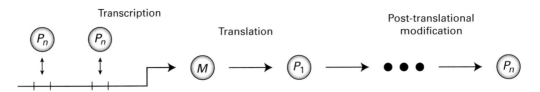

Figure 2.4
General schematic of the Goodwin and related models. Adapted from DeWoskin et al. (2014).

2.6 The Goodwin Model

We can now form a generalized "Goodwin" model (Goodwin 1966), which is shown in figure 2.4. All variables in this model will be concentrations, so we will drop the [] notation. Let M be the concentration of mRNA. The protein translated from this mRNA can be in several states (e.g., phosphorylation levels or proteins in particular compartments). The last state of the protein is that which acts as a repressor. Note that all reaction rates are positive. Goodwin uses the model "b" of transcription, and thus the expression of the concentration of mRNA is

$$\frac{dM}{dt} = \frac{k_{\max}}{1 + \frac{k_f}{k_r} P_n^m} - d_0 M,$$

where the first term describes transcription using the original Hill function with Hill coefficient m, and where P_n is the concentration of the repressor. The second term represents degradation that occurs with rate d_0. Translation occurs with rate k_l, and we assume a rate of r_i for the conversion of one state of the protein to the next state, such as, for example, modification due to phosphorylation. Thus, the rest of the model is

$$\frac{dP_1}{dt} = k_1 M - d_1 P_1 - r_1 P_1,$$

$$\frac{dP_2}{dt} = r_1 P_1 - d_2 P_2 - r_2 P_2,$$

$$\dots,$$

$$\frac{dP_n}{dt} = r_{n-1} P_{n-1} - d_n P_n.$$

This is the basic Goodwin oscillator, except that Goodwin lets $q_i = d_i + r_i$ and considers only two states of the protein (Goodwin 1966). See figure 2.5 and code 2.2 for model simulations.

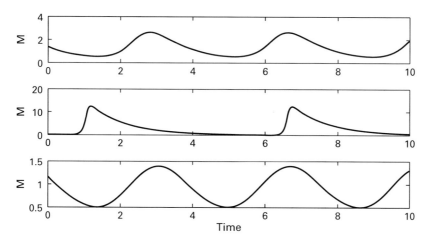

Figure 2.5
Simulations of feedback oscillations in biochemical feedback loops. Top, simulations of the mRNA "M" from the Goodwin model with a Hill coefficient (m) of 9. Middle, simulations of the Goodwin model with a Hill coefficient of 25. Bottom, simulations of the Kim-Forger model. See code 2.2 for more details.

As we will see in chapter 4, Goodwin's original model contained a crucial error. However, replacing the transcription regulation term by that derived in model "c" in section 2.5 gives an accurate model of the mammalian intracellular circadian clock (Kim and Forger 2012), which also has nice mathematical properties. We refer to this model as the Kim-Forger model:

$$dM \,/\, dt = \alpha\left(A_T - \left((R + A_T + K_d) - \sqrt{(R + A_T + K_d)^2 - 4A_T R}\right)\right) - d_0 M,$$

$$dP/dt = k_l M - q_1 P,$$

$$dR/dt = \mathrm{r}P - d_2 R.$$

This model oscillates when $A_T \approx R$ and K_d is small. See figure 2.5 and code 2.2 for model simulations.

2.7 Other Models of Intracellular Processes (e.g., Michaelis-Menten)

2.7.1 Nuclear Transport
As formulated, the Goodwin oscillator can represent a transcription-translation feedback loop in a prokaryote (e.g., bacteria). In such a case, the cell is not organized into compartments, and molecules can, in theory (although actually not in practice), diffuse freely throughout the cell. However, many of the applications we will consider are in eukaryotes, where there is a separate nuclear compartment. In this case, we need to keep track of the

transport of mRNA and proteins into and out of the nucleus. This presents some problems, since the variables in our model are concentrations that are numbers of molecules per unit volume, and the nucleus usually does not have the same volume as the cytoplasm. Thus, as one molecule moves from the nucleus to the cytoplasm (or vice versa), the change in concentration of the chemical species in the nucleus does not equal the change in concentration of the chemical species in the cytoplasm. A common solution to this problem, which to my knowledge was first presented by Goldbeter, is to *define all chemical concentrations with respect to the total cell volume* (Goldbeter 1995). When this is done, the equations for the model look very much like the original Goodwin system:

$$\frac{dM_n}{dt} = \frac{k_{\max}}{1 + \frac{k_f}{k_r} P_{n,\text{nuc}}^m} - d_0 M_n - k_m M_n,$$

$$\frac{dM_c}{dt} = k_m M_n - d_1 M_c,$$

$$\frac{dP_1}{dt} = k_l M_c - d_2 P_1 - r_1 P_1,$$

$$\frac{dP_2}{dt} = r_1 P_1 - d_3 P_2 - r_2 P_2,$$

$$\cdots,$$

$$\frac{dP_n}{dt} = r_{n-1} P_{n-1} - d_{n+1} P_n - k_P P_n,$$

$$\frac{dP_{n,\text{nuc}}}{dt} = k_P P_n - d_{n+2} P_{n,\text{nuc}},$$

where M_n is the mRNA in the nucleus, M_c is the mRNA in the cytosol, P_i are states of the protein in the cytosol and $P_{n,\text{nuc}}$ is protein in the nucleus. However, this approach requires that we scale the rates of all reactions where two or more molecules bind together (since this depends on the volume of the cell).

2.7.2 Michaelis-Menten Dynamics

Catalysts act to speed up chemical reactions. Enzymes are proteins that act as catalysts. For example, kinases are important catalysts that help add phosphate groups to proteins. The most commonly used model for this process was developed by Michaelis and Menten. Maud Menten, who jointly developed the Michaelis-Menten equation we will study, was one of the first women to earn a PhD in Canada. Millions know the name, but very few know her remarkable story.

Suppose chemical species B helps convert A to C. The chemical reactions that describe this are the following:

$$A + B \xleftrightarrow[k_r]{k_f} AB \xrightarrow{k_a} B + C$$

The differential equations for this system are

$$d[A]/dt = -k_f[A][B] + k_r[AB],$$

$$d[B]/dt = -k_f[A][B] + k_r[AB] + k_a[AB],$$

$$d[AB]/dt = k_f[A][B] - k_r[AB] - k_a[AB].$$

There have been many studies about how to simplify these equations (Briggs and Haldane 1925). Here we choose the most widely accepted (and perhaps least rigorous) way of analyzing these equations. Let us assume that $d[AB]/dt \approx 0$, which we call the rapid equilibrium assumption. We then find that

$$\frac{[A][B]}{[AB]} = \frac{k_r + k_a}{k_f}.$$

Let $K = (k_r + k_a)/k_f$. Also note that $[B] + [AB] = [B_{tot}]$, where $[B_{tot}]$ is constant. Thus, we have

$$\frac{[A]([B_{tot}] - [AB])}{[AB]} = K.$$

Solving for [AB], we have

$$[AB] = \frac{[B_{tot}][A]}{[A] + K}.$$

and the rate of conversion of A to C is $k_a[B_{tot}][A]/([A]+K)$, the celebrated Michaelis-Menten expression (Briggs and Haldane 1925; Michaelis and Menten 1913). Thus, so long as the rapid equilibrium assumption is valid, it seems that we do not need to solve the differential equations for the binding of A to B. However, $k_a[B_{tot}][A]/([A]+K)$ depends on [A] and not on [A_{tot}]. We could keep track of [A] and [A_{tot}] as well, but then we would have to solve additional differential equations, negating the benefits of this analysis. Instead, it is commonly assumed (sometimes without valid justification) that [B_{tot}], and thus [AB], is low. Then [A_{tot}] $\approx$ [A], and one does not need differential equations for the binding of A to B.

Let us assume that there is some catalyst for degradation. Assuming that the last element, P_n, degrades according to Michaelis-Menten kinetics, the equations for the Goodwin system then become

$$\frac{dM}{dt} = \frac{k_{\max}}{1 + \frac{k_f}{k_r}P_n^m} - d_0 M,$$

$$\frac{dP_1}{dt} = k_l M - d_1 P_1 - r_1 P_1,$$

$$\frac{dP_2}{dt} = r_1 P_1 - d_2 P_2 - r_2 P_2,$$

$$\ldots,$$

$$\frac{dP_n}{dt} = r_{n-1}P_{n-1} - d_n \frac{P_n}{K + P_n}.$$

In summary, the Michaelis-Menten expression assumes the following:

- Enzymes rapidly bind and unbind to their targets.
- The total enzyme concentration is much less than the total substrate concentration.
- The enzyme does not bind to the product.

2.7.3   Reversible Reactions

Many biochemical reactions are reversible. Good candidates for reversible reactions in the Goodwin system are the following: (1) phosphorylation, (2) nuclear transport, and (3) binding to DNA. The third example is already reversible in our model. Other reactions, such as transcription, translation, or degradation, are not reversible. We can add in the reverse of a reaction in the Goodwin system in a straightforward way. In the following example, the reverse of reaction r_1 is incorporated with rate constant $r_{1\mathrm{rev}}$:

$$\frac{dM}{dt} = \frac{k_{\max}}{1 + \frac{k_f}{k_r}P_n^m} - d_0 M,$$

$$\frac{dP_1}{dt} = k_l M - d_1 P_1 - r_1 P_1 + r_{1\mathrm{rev}}P_2,$$

$$\frac{dP_2}{dt} = r_1 P_1 - r_{1\mathrm{rev}}P_2 - d_2 P_2 - r_2 P_2,$$

$$\ldots,$$

$$\frac{dP_n}{dt} = r_{n-1}P_{n-1} - d_n P_n.$$

2.7.4 Delay Equations

Many models use delay equations where some variables depend on the state of other variables at previous times. However, very few biochemical processes are accurately modeled by a pure delay (an example is given in chapter 3), and, unless one studies a system with a time machine, I find they often should be avoided unless rigorously derived, at least when used in biophysical models. Delay equations are more difficult than ordinary differential equations to numerically solve or to mathematically analyze, and their predictions can be problematic: for example, they tend to overpredict the occurrence of oscillations in biological systems (see chapter 4). Instead, biological delays can be modeled by considering the underlying biological processes and using ordinary differential equations. Using delay equations in the present context would also put us in a paradoxical situation: Should one assume a perfect timekeeping mechanism (i.e., a delay equation) in determining how timekeeping is generated?

2.8 Frontiers: Bounding Solutions of Biochemical Models

Variables in biochemical models are typically concentrations, which, by definition, are ≥ 0. Therefore, any clearance term in these models (any term with a $-$ sign) will go to zero when the species goes to zero. Thus, if the concentration of this species is ever 0, its rate of change will be ≥ 0. This is true for many nonbiochemical models as well, and is sometimes termed the axiom of parenthood, based on terminology from models of ecology.

A reasonable question is whether solutions of the Goodwin model can grow without bound. Biologically, this does not make sense because at some point physical limitations (e.g., cell size, availability of amino acids, etc.) will come into play. A refined model should then be used. Nevertheless, solutions could grow very large. An example of this is the following:

$$dM/dt = k_{max} - M/(M + K).$$

Here the production rate is fixed at k_{max}, but the degradation rate has a maximum value of 1. If k_{max} is greater than 1, solutions will grow without bound.

One way of testing this is to construct a bounding box in the phase space of the model from which no solutions can leave. Again, consider the standard Goodwin system:

$$\frac{dM}{dt} = \frac{k_{max}}{1 + \frac{k_f}{k_r} P_n^m} - d_0 M,$$

$$\frac{dP_1}{dt} = k_l M - d_1 P_1 - r_1 P_1,$$

$$\frac{dP_2}{dt} = r_1 P_1 - d_2 P_2 - r_2 P_2,$$

…,

$$\frac{dP_n}{dt} = r_{n-1}P_{n-1} - d_nP_n.$$

Lower bounds are relatively easy to construct. Consider the planes $M = 0$, $P_1 = 0$, …, $P_n = 0$. The flow of the system normal to these planes is given by

$$\frac{dM}{dt} = \frac{k_{max}}{1 + \frac{k_f}{k_r}P_n^m} > 0, \frac{dP_1}{dt} = k_l M > 0, \frac{dP_2}{dt} = r_1 P_1 > 0, \ldots, \frac{dP_n}{dt} = r_{n-1}P_{n-1} > 0,$$

which always flows into the region where $M > 0$, $P_1 > 0$, …, $P_n > 0$. This, again, is the axiom of parenthood. Next, let us find an upper bound on the solutions. We first note that if $M > k_{max}/d_0$, then

$$\frac{k_{max}}{1 + \frac{k_f}{k_r}P_n^m} < d_0 M.$$

It follows that $dM/dt < 0$, so the amount of mRNA must decrease. Thus, solutions starting with $M < k_{max}/d_0$ can never grow to $M = k_{max}/d_0$, and an upper bound for M is k_{max}/d_0. Choose one plane of the box at M^* where $M^* > k_{max}/d_0$. Since M must be less than M^* when solutions start in the box, we can choose $P_1^* > k_l M^*/(d_1 + r_1)$, and flow starting in the box will never grow past P_1^*. This process can be continued until all variables are bounded. Thus, solutions are bounded within $M^* > M > 0$, $P_1^* > P_1 > 0$, …, $P_n^* > P_n > 0$.

2.9 On Complex Formation

The formation of complexes can have a great effect on the dynamics of a biochemical feedback loop. Here, we illustrate the differences with some simple examples. First, consider a system that contains just one degradation reaction. If the degradation occurs linearly, we have

$$dA/dt = -aA.$$

The solution to this equation is

$$A(t) = A_0 e^{-at},$$

where A_0 is the initial condition ($A_0 = A(0)$). Now consider the case where A forms a dimer that degrades. We can represent this as

$$dA/dt = -aA^2$$

or

$dA/A^2 = -a \, dt.$

Integrating, the solution of this equation is

$(1/A(t)) - (1/A_0) = at,$

which is quite different from the linear degradation case.

Next, note that within many biochemical feedback loops, complexes of many proteins form. It has been shown that the order in which these proteins form a complex has a great influence on whether oscillations are seen in biochemical feedback loops. See the study by DeWoskin et al. (2014) for more details.

Finally, note that complex formation can allow a system to be more accurate in the presence of noise. While noise will be discussed more in chapter 3, we consider a simple case. Imagine the number of molecules of A and B are governed by a Poisson distribution with mean h. The variance of AB (the rate of a complex forming between A and B) is

$2h^3 - h^2,$

whereas the rate of complex formation of A with itself (i.e., $A(A - 1)$) can be derived to be

$4h^3 - 2h^2,$

which is exactly double the variance of the rate of A forming a complex with B. So by having A and B bind together, twice the number of molecules are available, and the rate has half the variance.

2.10   Hodgkin–Huxley and Models of Neuronal Dynamics

Neurons are the basic computational units within the body. They communicate by electrical signals. When a neuron receives electrical signals from other neurons, it must decide whether or not to send a signal (called an action potential) to the other neurons to which it is connected. The basic work describing the electrical behavior of neurons mathematically was conducted by Hodgkin and Huxley in the late 1940s and early 1950s (Hodgkin and Huxley 1952). They took a long time to publish their results, but the model they developed is one of the most widely studied models in all of mathematical biology. Hodgkin and Huxley won the Nobel Prize for this work in 1963, making it arguably the first Nobel Prize for mathematical biology.

Neurons maintain different concentrations of particular ions within the cell than what is found outside of the cell. This creates a voltage difference between the outside and inside of the cell, or membrane potential. We often refer to the voltage difference between the inside and outside of the cell as the voltage of the cell. When a neuron fires an action potential, channels are opened that selectively let certain ions into or out of the cell. This causes a change in the electrical potential around the membrane and serves as the electrical signals neurons use to communicate.

Neurons can signal to other connecting neurons repetitively (repetitive firing of action potentials). When oscillating, neurons also synchronize to signals from neighboring cells. In addition to synchronization, other interesting dynamics can be seen. Before we get to oscillations, it is important to describe the physiology of neurons, as seen in figure 2.6, and how it informed the Hodgkin–Huxley model.

The effect of an action potential in the presynaptic cell is to transiently increase or decrease the voltage of the postsynaptic cell. Stores of neurotransmitters are held within presynaptic cells in synaptic vesicles that fuse with the cell membrane and release neurotransmitter (see figure 2.7). This neurotransmitter diffuses across the synapse and binds to receptors on the postsynaptic cell opening ion channels that increase or decrease the cell's membrane potential. An excitatory postsynaptic potential (EPSP) increases the postsynaptic cell's membrane potential. An inhibitory postsynaptic potential (IPSP) decreases the postsynaptic cell's membrane potential. We model how the voltage changes in detail later. It is also important to note that some receptors trigger a signaling cascade within the cell rather than directly opening ion channels.

Neurons are typically thought of as binary in that either they send an action potential or they do not (although see Belle et al. 2009 for other types of behaviors). When the neuron is off, it remains at a hyperpolarized (lower voltage with respect to the outside) state. Typically, the voltage difference in this case is about –60 to –70 mV. This voltage difference is maintained because of the different ion concentrations inside and outside of the cell.

When the neuron fires an action potential, the action potential starts at the axon hillock and then proceeds down the axon. At any point on the axon, the change in voltage looks like the curve shown in figure 2.8.

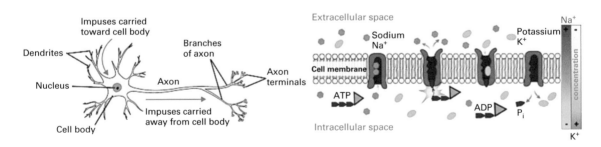

Figure 2.6
(Left) The basic structure of a neuron. Taken from https://www.wpclipart.com/medical/anatomy/cells/neuron/neuron.png.html. (Right) Sodium (Na$^+$) is kept in higher concentrations outside of the cell than inside of the cell. Potassium is kept in higher concentrations inside of the cell than outside of the cell. This is done via a sodium-potassium exchanger. Taken from https://en.wikipedia.org/wiki/Na%2B/K%2B-ATPase.

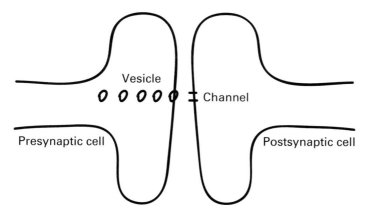

Figure 2.7
The structure of a synapse.

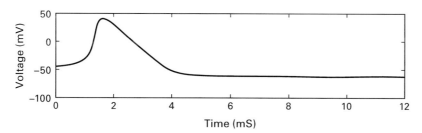

Figure 2.8
An action potential.

First, the sodium channels open, and the membrane voltage potential rises as sodium ions flow into the cell. The higher voltage causes the potassium channels to open, which then causes the membrane potential to lower as potassium leaves the cell. Finally, after a refractory period (when the neuron will not fire an action potential even if signals are present from other neurons), the neuron goes back to equilibrium.

Hodgkin and Huxley developed a special experimental preparation for studying the behavior of axons. In this technique, an axon is dissected from a squid (because squid have a huge axon that can be seen with the naked eye). A wire is run through the inside of the axon, which allows current to be sent through the axon membrane. This wire is connected with electronic devices (or a computer) that can regulate the current or voltage traveling through the axon. In particular, the voltage can be set to a particular level and currents applied from the electronic device fix the voltage (voltage clamp). Alternatively, the device could be set to apply a fixed current (current clamp). Chemicals (e.g., tetrodotoxin) can then be added to block certain channels from carrying current. In this way, the voltage can be set to a particular level, and the current flowing through particular channels (e.g., those that allow sodium ions to pass) can be measured.

The Hodgkin–Huxley model begins with basic physics. First, we have the standard Ohm's-law relationship between voltage, V, current, I, and the conductance $g = 1/resistance$, is

$$gV = I.$$

The membrane of the neurons is a lipid bilayer, and thus it acts as a capacitor. The formula for the current change due to a capacitor is

$$C\,dV/dt = I,$$

where the capacitance of the membrane, C, is 1.0 μF/cm^2 in squid. There will be currents flowing through the membrane from sodium and potassium ions as well as general nonselective leak (L) ions. The balance from all these currents must be equal to zero:

$$C\,dV/dt + I_{Na} + I_{K} + I_{L} = 0.$$

Each of the ionic currents will obey

$$gV = I.$$

Actually, this is a linear approximation, and certain ionic currents (e.g., calcium) are best described by a more complicated relationship between the voltage difference and the amount of current that flows. That relationship is described by the Goldman–Hodgkin–Katz equation found in section 2.11 (Goldman 1943; Hodgkin and Katz 1949; Peskin 2000).

Because of the differences in concentration between the ions inside and outside of the membrane, current will still flow at zero voltage through channels that selectively allow only one ion to pass. The equilibrium potential is the level of the voltage difference, which will counteract the effects of the different concentrations. It is given by the following formula by Nernst:

$$E = (R_g T/zq)\log(C_E/C_I),$$

where

R_g = Boltzmann's constant,

T = absolute temperature,

z = charge of ion,

q = charge of a proton,

C_E = exterior concentration, and

C_I = interior concentration.

The equilibrium potentials for the ions we are considering are

$E_{Na} = 45$ mV,

$E_K = -82$ mV,

$E_L = -59$ mV.

Plotting these equilibrium potentials on the graph of the action potential shows the influence of different ions during different parts of the action potential, as shown in figure 2.9.

The voltage-gated currents are

$I_{Na} = G_{Na}(V-E_{Na})$,

$I_K = G_K(V-E_K)$,

$I_L = g_L(V-E_L)$.

The key assumption to the Hodgkin–Huxley model is that the conductances for the sodium and potassium ions are dynamic and voltage-sensitive. Hodgkin and Huxley found that $g_L = 0.3$ (μA/mV)/cm^2.

We first begin with a description of the potassium channels. They are assumed to have four subunits. Each subunit must be in the "open" state for current to flow through the ion. We let n be the probability that a particular subunit is in the "open" configuration, and we assume that the rate at which subunits are opened or closed depends on the voltage. This is represented schematically in figure 2.10.

Or, using differential equations,

$dn/dt = \alpha_n(V)(1-n)-\beta_n(V)n$

Assuming four subunits, we have that the conductance of potassium is

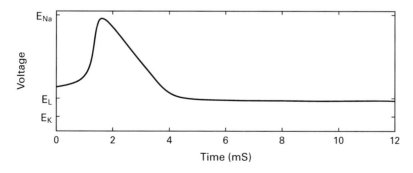

Figure 2.9
An action potential noting the equilibrium potential of ions.

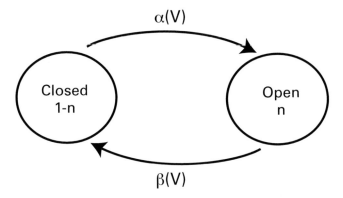

Figure 2.10
Hodgkin–Huxley model of channel subunits.

$$G_K = g_K n^4,$$

since the probability of all subunits being open equals the product of each one of the four subunits being open (n) assuming they are identically distributed. This yields

$$I_K = g_K n^4 (V - E_K),$$

where $g_k = 36$ (μA/mV)/cm^2. Now the rates at which the subunits are opened or closed are governed by

$$\alpha_n(V) = \frac{-\left(\dfrac{60+V}{100}\right)}{\exp\left(-\dfrac{60+V}{10}\right) - 1}$$

$$\beta_n(V) = (0.125)\exp\left(-\frac{V+70}{80}\right),$$

At the steady state, n will become

$$n_\infty = \alpha/(\alpha + \beta).$$

Thus, looking at the deviation from n_∞, we have $n = n' + n_\infty$. Substituting this in the dn/dt equation, we find

$$dn'/dt = -(\alpha + \beta)n'.$$

If the voltage is held constant, n changes exponentially until it reaches n_∞. The steady state n_∞ grows to 1 as the voltage increases, and the time constant $1/(\alpha + \beta)$ is highest for moderate voltages. Thus, if we let $1/(\alpha + \beta) = \tau_n$, we can rewrite dn/dt as

$$dn/dt = (n_\infty - n)/\tau_n \equiv \alpha_n(V)(1 - n) - \beta_n(V)n.$$

Next, we consider the sodium channels. It is again assumed that the sodium channel has four parts. Of these, three parts are the same (the m subunits) and have fast dynamics, and one part is slow (the h subunit). The equations are similar to those for the n subunits of the potassium channel:

$$dm/dt = \alpha_m(V)(1 - m) - \beta_m(V)m,$$

$$dh/dt = \alpha_h(V)(1 - h) - \beta_h(V)h.$$

Assuming 4 subunits, we have

$$I_{Na} = g_{Na}m^3h(V - E_{Na}) \tag{2.9.1}$$

where $g_{Na} = 120$ (μA/mV)/cm^2. The rates at which the subunits are opened or closed were fit by Hodgkin and Huxley:

$$\alpha_m(V) = \frac{\frac{V+45}{10}}{1 - \exp\left(-\frac{V+45}{10}\right)},$$

$$\beta_m(V) = 4\exp\left(-\frac{V+70}{18}\right),$$

$$\alpha_h(V) = 0.07\exp\left(-\frac{V+70}{20}\right),$$

$$\beta_n(V) = \frac{1}{1 + \exp\left(-\frac{(V+40)}{10}\right)}.$$

The steady value of m, $m_\infty \equiv \alpha_m/(\alpha_m+\beta_m)$, increases with increasing voltage, whereas h_∞ decreases with increasing voltage. This is illustrated in figure 2.11.

The steady state m_∞ is much larger than the steady state h_∞ if $V > 25$. During an action potential, the m units open, depolarizing the membrane, followed by the h units that later close. The final equations for the Hodgkin–Huxley model are as follows:

$$C\frac{dV}{dt} + g_{Na}m^3h(V - E_{Na}) + g_Kn^4(V - E_K) + g_L(V - E_L) + I_{Applied} = 0.$$

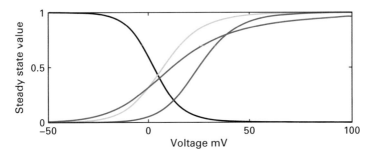

Figure 2.11
Steady-state values for the gating variables for the Hodgkin–Huxley model. The fast sodium subunit, m_∞, is shown in blue. The potassium channel, n_∞, is shown in red for the original model, and green is used for the revised model (see the following). The slow sodium subunit, h_∞, is shown in black.

The *m* subunits are fast with respect to the other subunits. So when an action potential is initiated, we can consider *h* and *n* to be constant. This leaves a two-variable system where we can look at the two-dimensional phase space. Likewise, when the neuron returns to the rest potential, its behavior can be described by a phase space in *V* and *n*. This will be discussed further in chapter 5.

In figure 2.12 we simulate the model with an applied current of 8 μA/cm^2. Note that the model is bistable for the applied current: one initial condition leads to repetitive firing of action potentials, whereas the other leads to a steady state.

Finally, we note that this basic prediction of the Hodgkin–Huxley model is incorrect. The model predicts that when sufficient current is applied, action potentials will be repetitively fired, when, in actuality, the squid giant axon will only fire one action potential. Based on new measurements from squid giant axon, Clay et al. (2008) determined the source of this error. Replacing the rate of closing of the K channel, β_n, by the expression

$$\beta_n(V) = (0.125)\exp(-(V + 70)/19.7)$$

actually gives a very accurate representation of the electrical behavior of the squid giant axon. This is illustrated in figure 2.12. Some historical context is helpful. Hodgkin and Huxley were mainly interested in the dynamics of the sodium channel. Without computers, simulating their model was tedious. For these reasons, it is hard to be too critical of their work.

Recent modeling efforts combine models of many neurons coupled together. Some examples are (DeWoskin et al. 2015; Hill and Tononi 2005; Myung et al. 2015). Many models have been created based on the Hodgkin–Huxley formalism; see Sim and Forger (2007), for example.

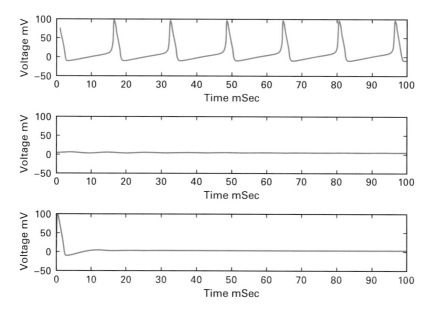

Figure 2.12
Simulations of the Hodgkin–Huxley model with an applied current of 8 μA/cm². Top, simulation starting at an initial condition that leads to repetitive firing of action potentials. Middle, simulation starting at an initial condition that leads to quiescence. Bottom, the Hodgkin–Huxley model, when corrected, does not show repetitive firing for any initial conditions, consistent with data from squid giant axon. See code 2.3.

2.11 Frontiers: Rethinking the Ohm's Law Linear Relationship between Voltage and Current

In the preceding analysis, we assumed that there was a linear relationship between the voltage across a channel and the current that flows. Reality is much more complex. Goldman, Hodgkin, and Katz (GHK) worked out a more precise expression for how much current should flow (Goldman 1943; Hodgkin and Katz 1949) based on drift and diffusion of an ion in a channel, as well as on Maxwell's equations. In this derivation, one sees that the current is

$$I = aV \frac{C_E \exp(-bV) - C_I}{\exp(-bV) - 1}.$$

See Ermentrout et al. (2010) and Peskin (2000). Here, a is a proportionality constant that depends on the permeability of the channel, C_E and C_I are the concentrations of the ion outside and inside the cell, respectively, and $b = zq/R_gT$, where z is the charge of the ion, q is Faraday's constant, R_g is the gas constant, and T is the absolute temperature. The ratio q/R_gT can be approximated as 1/25 (Clay 2009). In practice, the constant a is fit from electrophysiological data rather than derived from the channel biophysics.

We first note that the current is zero when $C_E \exp(-bV) - C_I = 0$. Solving for V, one recovers the Nernst equation, which gives the equilibrium potential for the channel. When bV is very positive, the current is approximately

$I = aVC_I$.

When it is very negative, the current is approximately

$I = aVC_E$.

Thus, in the actual relationship between the voltage and the current, the slope varies by a factor of C_I/C_E as the voltage increases. In practice, neuronal models treat a as both voltage- and time-dependent (e.g., $a \sim m^3 h$ as in (2.9.1)). This can account for any scalings caused by the GHK equation. Sometimes fitting the GHK current can help (Clay 2009), although this will not affect the dynamics of a model. However, the notion of a reversal potential itself may cause confusion, as some ions (e.g., calcium) naturally flow only in one direction (rectification).

2.12   Ten Common Mistakes to Watch for When Constructing Biochemical and Electrophysiological Models

Biochemical models

- Missing stoichiometric coefficients (e.g., factor of 1/2 or 2 in dimerization)
- Using mass action for low concentrations, not well-mixed environments or one-dimensional diffusion-limited environments
- Using a Michaelis-Menten formalism automatically without verifying if enzyme concentrations are low, if rapid binding occurs or if the enzyme is not bound to the product
- Use of delay equations without derivation or to determine mechanisms of oscillations generation
- Using Hill expressions automatically for transcription (rarely does this approximate reality)

Electrophysiological models

- Overreliance on simple models without consideration of what simplifications of the Hodgkin–Huxley formalism were used
- Forgetting that the valence of the ion affects the GHK equation
- Using data from ion channels from different cell types blindly
- Forgetting about the effects of temperature
- Not checking if the numerical method to solve the equations is accurate

2.13   Interesting Future Work: Are All Cellular Oscillations Intertwined?

Rhythms are almost always studied in isolation, yet many rhythms coexist in a cell. It is tempting to think of them as separate for theoretical simplicity, and they may be physically separated in different parts of the cell. Nevertheless, a growing volume of literature shows how these rhythms affect each other. Here are some examples:

- Cell divisions are carefully timed with respect to the circadian clock (Bieler et al. 2014; Feillet et al. 2014)
- Neuronal firing regulates circadian rhythms (Belle et al. 2009; Diekman et al. 2013)
- Metabolic rhythms have been proposed to be part of the mechanism of circadian time-keeping (Morre et al. 2002)
- Min oscillations are part of the cell cycle (Kruse et al. 2007)
- p53 oscillations involve circadian proteins (Kim and Jackson 2013)

Many interesting future modeling projects could explore the relationship between these rhythms. Principles of synchrony (see chapter 7) could be helpful.

Code 2.1 Spatial Effects

(see figures 2.1, 2.2)

```
% Some sparse comments are provided in this code. The reader is
%encouraged to figure out this code and to provide additional comments
clear
numbox = 10000;
rowlen = sqrt(numbox);
numperbox = 100;
ratebin = 100;
x = zeros(numbox, 1);
for ij = 1:(numbox*numperbox)
box = ceil(rand*numbox);
x(box) = x(box) + 1;
end
reac(1:numbox) = (1./x).*log(1./rand(numbox, 1));
reac((numbox+1):(2*numbox)) = (1./x).*log(1./rand(numbox, 1));
reac((2*numbox+1):(3*numbox)) = …
(1./(ratebin.*x.*max(x-1,0))).*log(1./rand(numbox, 1));
reac((3*numbox+1):(4*numbox)) = (1./x).*log(1./rand(numbox, 1));
reac((4*numbox+1):(5*numbox)) = (1./x).*log(1./rand(numbox, 1));
% set the above two rates to Inf for a 1-D simulation
tim(1) = 0;
Xtot(1)= sum(x);
for ijj = 2:200000000%I choose this to be a large number
```

```
[T, I] = min(reac);
if T == Inf%no more molecules left the to next reaction is Inf
        break;
end
tim(ijj) = tim(ijj-1) + T; %updates the time
reac = reac-T;
Xtot(ijj) = sum(x);
if I < numbox + 1%move to left
movefrom = I;
moveto = I-1 +(1-min(mod(I-1, rowlen), 1))*rowlen;
%moveto = mod(I-2, numbox)+1;
%use this line instead for the 1-d problem
elseif I < 2*numbox + 1%move to right note in lattice,
movefrom = I-numbox;
moveto = movefrom + 1 - (1-min(mod(movefrom, rowlen), 1))*(rowlen);
%moveto = mod(I, numbox)+1;%use this line instead for the 1-d problem
elseif I < 3*numbox + 1 %dimerization
movefrom = I - 2*numbox;
x(movefrom) = x(movefrom) -2;moveto = movefrom;
elseif I < 4*numbox + 1 %move up
movefrom = I- 3*numbox;
moveto = mod(I+rowlen-1, numbox)+1;
else %move down
movefrom = I- 4*numbox;
moveto = mod(I-rowlen-1, numbox)+1;
end
x(movefrom) = x(movefrom) -1; %move the molecules if this occurred.
x(moveto) =x(moveto)+1;
for ij = [movefrom moveto]% update reactions that changed
reac(ij) = (1./x(ij)).*log(1./rand);
reac(ij+numbox) = (1./x(ij)).*log(1./rand);
reac(ij+2*numbox) = (1./(ratebin.*x(ij).*max(x(ij)-1,0))).*log(1./rand);
reac(ij+3*numbox) = (1./x(ij)).*log(1./rand);
reac(ij+4*numbox) = (1./x(ij)).*log(1./rand);
end
end
%The following code finds the rates, times and average numbers of
%molecules after every 50 degradations of protein.
[aa bb] = size(Xtot);
XrecXX = 50;
xxind = 1;
for ij = 1:(bb - 1)
if Xtot(bb-ij) > (50+XrecXX)
recind(xxind) = bb-ij;
xxind = xxind+1;
XrecXX = XrecXX + 50;
end
```

```
end
for tindt = 1:(xxind-2)
xamt = …
sum(Xtot((recind(tindt+1)+1):recind(tindt)).*(tim((recind(tindt+1)+1 …
):recind(tindt))-tim((recind(tindt+1)):(recind(tindt)- …
1))))/(tim(recind(tindt))-tim(recind(tindt+1)+1));
rateX(tindt) = (50/(xamt^2))/(tim(recind(tindt))-tim(recind(tindt+1)+1));
rateXtim(tindt) = mean(tim((recind(tindt+1)+1):recind(tindt)));
xamtrec(tindt) = xamt;
end
figure(1)
loglog(rateXtim,(numbox*numbox)*rateX, '*k')
xlabel('Time')
ylabel('Rate')
figure(2)
semilogy(xamtrec, (numbox*numbox)*rateX', '*k')
xlabel('Number of Molecules')
ylabel('Rate')
```

Code 2.2 Biochemical Feedback Loops

(see figure 2.5)

```
%The following code is provided for readers who may want to see how
%biochemical models are simulated in MATLAB
function Y = good(t, X) %code for the goodwin model, these lines should
%be entered into a m-file good.m
M = X(1); P = X(2); R = X(3);
Y(1) = 10/(.001^25 + R^25) - M;
Y(2) = M-P;
Y(3) = P-R;
Y = Y';
end
%the following two lines should be entered into the command line
%[T, X] = ode45(@good, 0:0.01:100, [1 1 1]);
%plot(T, X(:,1))
function Y = kimforger(t, X) %code for the Kim-Forger model, these
%lines should be entered into a m-file kimforger.m
M = X(1); P = X(2); R = X(3);
kd = .001;
A = 1;
Y(1) = 10*(A-((A+R+kd)-((A+R+kd)^2-4*R*A)^(1/2))/2) - M;
Y(2) = M-P;
Y(3) = P-R;
Y = Y';
end
%the following two lines should be entered into the command line
```

```
%[T, X] = ode45(@kimforger, 0:0.01:100, [1 1 1]);      .
%plot(T, X(:,1))
```

Code 2.3 The Hodgkin–Huxley Model

(see figures 2.11, 2.12)

```
function Y = HH(t, X) %code for the revised Hodgkin-Huxley model, these
%lines should be entered into a m-file HH.m
% Note these equations use the HH scaling so that rest is 0 rather than
%-70mV. Thus, the voltage has been increased by 70mV. You can revert
%back to the original equations by using the commented line in the Bn
%function rather than the corrected HH model.
gna = 120; ena = 115; gk  = 36; ek = -12; gl = 0.3; el = 10.613;
appcurr = 8;
V = X(1); M = X(2); N = X(3); H = X(4);
% The follow code corrects for removable singularities at 10, 25 and 30.
if V == 10
      V = 10.000001;
end
if V == 25
      V = 25.000001;
end
if V == 30
      V = 30.000001;
end
Y(1) = appcurr + gna*M*M*M*H*(ena—V) + gk*(N^4)*(ek—V) + gl*(el-V);
Y(2) = Am(V)*(1-M) - Bm(V)*M;
Y(3) = An(V)*(1-N) - Bn(V)*N;
Y(4) = Ah(V)*(1-H) - Bh(V)*H;
Y = Y';
function y = Ah(V)
y = 0.07.*exp(-V./20.0);
end
function y = Am(V)
y = (25.0 - V)./(10.0.*(exp((25.0 - V)./10.0) - 1.0));
end
function y = An(V)
y = (10.0 - V)./(100.0.*(exp((10.0 - V)./10.0) - 1.0));
end
function y = Bh(V)
y = 1.0./(exp((30.0-V)./10.0)+1.0);
end
function y = Bm(V)
y = 4.0.*exp(-V./18.0);
end
function y = Bn(V)
```

```
%y = 0.125.*exp(-V./80.0);
y = 0.125.*exp(-V./19.7);%this is for the Corrected HH model
end
end
%the following two lines should be entered into the command line
%[T, X] = ode45(@HH, 0:0.01:100, [4.21 0.0858 0.3838 0.4455]); This
%goes to a steady state with the original model
%[T, X] = ode45(@HH, 0:0.01:100, [10 0.5 0.5 0.5]); This shows APs with
% the original model
```

Exercises

General Problems

For the first five problems, choose a model representing a biological system of interest, preferably one you study.

1. What parts of the model are consistent with a black-box model, or with a biophysical model? What models in chapter 2 are closest to your model?

2. Are there other similar competing or related models? How do these models differ?

3. List all the assumptions needed in your model. How do these assumptions affect errors?

4. How could noise enter the system?

5. Are there some states that the system cannot be in (e.g., negative values for concentrations)? If so, is this enforced in the model?

Specific Problems

1. Assume the number of molecules of A and B are all governed by a Poisson distribution with mean μ. Show that variance of AB (the rate of a complex forming between A and B) is $2h^3 + h^2$ and the rate of complex formation of A with itself (i.e., $A(A-1)$) is $4h^3 + 2h^2$.

2. Derive a version of the standard Goodwin oscillator where an enzyme, whose total concentration $[S_{tot}]$ is constant, acts as a catalyst for degradation of P_2. Do not assume rapid equilibrium, and assume that, in addition to the s mediated degradation, there is also a low level of degradation as originally formulated in the model. Be sure to justify your answer.

3. Describe the difference between the following terms:

    $K^m/(K^m + R^m)$ and $K^m/(K + R)^m$

4. Consider the system

$dP_1/dt = f_0 - g_1(P_1)$

...

$dP_i/dt = f_{i-1}(P_{i-1}) - g_i(P_i)$

...

Prove that solutions of this system are bounded if $f_{i-1}(P_{i-1})$ and $g_i(P_i)$ are monotonically increasing and unbounded.

5. Consider the following model and parameter list. Identify at least five errors:

$dM/dt = a_1/(K^m + P_2{}^m) - a_2M - k_lM$

$dP_1/dt = k_lM + a_3P_1 - a_4P_1{}^2$

$dP_2/dt = a_4P_1{}^2 - a_5P_2$

a_1 = average rate of transcription

K = (binding rate of repressor to gene)/(unbinding rate of repressor to gene)

a_2 = clearance rate of mRNA

k_l = translation rate of M

a_3 = clearance rate of P_1

a_4 = rate of dimerization of P_1

a_5 = clearance rate of P_2

6. Add in, individually, the following three processes in the standard Goodwin oscillator. In each case determine, either through simulations or analysis, whether the change encourages or discourages oscillations.

    a. Michaelis-Menten degradation of P_2.

    b. Michaelis-Menten translation.

    c. A step where P_2 dimerizes, and it is this dimer that acts as a transcription factor.

    Problems 7–9 are based on the following background: Consider the following simplified model of the circadian clock of the fruit fly *Drosophila*. The mRNA of two proteins, PER and TIM, are transcribed in the same way. Each protein is translated and phosphorylated (assume two successive phosphorylations for each protein). After these phosphorylations, PER and TIM dimerize. The PER and TIM dimer then enters the nucleus of the cell where it acts as a repressor of the transcription of per and tim mRNA.

7. Develop a mathematical model for the *Drosophila* circadian clock based on the preceding description. Assume all processes are linear (as in the original Goodwin

oscillator) except for dimerization, and transcription (transcription can be thought of as a Hill-type function as in the Goodwin oscillator). For simplicity, assume all reactions are not reversible. Find parameters where the system produces 24-hour oscillations.

8. In the tim^{01} mutant, where no functional tim mRNA or protein is made (which might be modeled by setting transcription of tim equal to zero), high per mRNA levels, but lower PER protein levels, are seen compared to the levels in the wild-type (unmutated) *Drosophila*. Give a biological explanation for this fact and demonstrate it with simulations (you may need to choose new parameters for the system).

9. Find an additional set of parameters that gives 24-hour oscillations, and for which each new parameter is greater than the corresponding value in problem 7.

   Problems 10–12 consider the Hodgkin–Huxley model with an applied current of $8\ \mu A/cm^2$.

10. Show (through simulations) that this model is bistable (the model can show repetitive firing of action potentials or quiescence depending on the initial state).

11. In addition to the applied current, assume that the current can be raised (or lowered) for 1 mS only, and then returns to $8\ \mu A/cm^2$. You can vary when this additional "pulse" is given, its sign (whether current is raised or lowered), and its amplitude. Determine, through simulations, which amplitudes and signs cause action potentials when the model is started in the quiescent state.

12. Starting with the neuron repetitively firing, determine the phase, sign, and amplitude of stimuli that will cause repetitive firing to stop.

3

Functioning in the Changing Cellular Environment

Most biological systems must cope with a fluctuating environment. In this chapter, we discuss three cases of how modeling can help to determine how cells adapt to fluctuating environments. In particular, we consider molecular noise, cell division, and temperature compensation.

3.1 Introduction

Most mathematical models do not explicitly consider the cellular environment, yet the effects of that environment can be enormous. Timekeeping is affected by the stochasticity of chemical reactions, by the dependence of cellular reactions on temperature, which can vary widely, and by ongoing activities, such as cell division. How can accurate timekeeping be maintained amid these fluctuating demands? This chapter will consider that question.

3.2 Frontiers: Volume Changes

The biochemical models we have developed are formulated as a rate of change of concentration. For convenience, we normally consider the volume fixed, and thus regard changes in concentration as depending solely on the gain or loss of molecules of a chemical species. However, when the cell volume changes, there are changes in concentration that occur as a result, and these may be significant. When taking these changes into account, it becomes important to distinguish between the number of molecules of a chemical species and the concentration of that chemical species. To make this distinction, let us use the notation

$A = $ # of molecules of the chemical species A, and

$[A] = A/V$ is the concentration of A,

where V is the volume. We consider several problems. First, assume we are given a mathematical model in terms of rate of change of concentration, and thus an expression for $d[A]/dt$ is given. This model is assumed to have the effects of volume change incorporated already. We now wish to calculate dA/dt. For simplicity, let us assume that we have three terms:

$$d[A]/dt = k_0 + k_1[A] + k_2[A]^2. \tag{3.2.1}$$

The first term could represent transcription, the second a degradation process, and the third dimerization. If we have more complex terms than these, for example a Hill coefficient, we assume that they could be decomposed into basic zero-, first-, and second-order reactions, yielding terms as before. We now have

$$d[A]/dt = d/dt(A/V) = -(A/V^2)(dV/dt) + (1/V)(dA/dt)$$

and

$$k_0 + k_1[A] + k_2[A]^2 = k_0 + k_1A/V + k_2(A/V)^2.$$

Equating the two sides yields

$$-(A/V)dV/dt + dA/dt = k_0V + k_1A + k_2A^2/V,$$

or

$$dA/dt = k_0V + (k_1 + d\ln V/dt)A + k_2A^2/V, \tag{3.2.2}$$

where $d\ln V/dt = (1/V)(dV/dt)$. Now if $dV/dt = \rho V$, then $d\ln V/dt = \rho$ and $V = v_0e^{\rho t}$, so

$$dA/dt = k_0v_0e^{\rho t} + (k_1 + \rho)A + (k_2/(v_0e^{\rho t}))A^2$$

Thus, zero-order terms in the concentration equation (3.2.1) mean that, in terms of numbers of molecules, reaction rates increase with cell volume. In contrast, as volume increases, second-order reaction rates decrease.

A less complex picture appears when we are given dA/dt and wish to find $d[A]/dt$. Again, we have

$$d[A]/dt = -(A/V)(dV/dt)(1/V) + (1/V)(dA/dt).$$

Let us separate out the first-order terms in dA/dt and write the equation as:

$$dA/dt = -d_0A + f(A).$$

Also assume that $dV/dt = qV$ or $(dV/dt)(1/V) = \rho$. This yields

$$d[A]/dt = -[A](\rho + d_0) + f(A)/(v_0e^{\rho t}).$$

Note that the growth of volume has two effects. Its primary action is as a dilution of A, exerting the same effect as a degradation of A. However, there are also secondary effects on the other reactions coming from the $f(A)/(v_0e^{\rho t})$ term.

What happens if the cell divides? We assume that the cell momentarily stops growing during division, or that the division event is fast with respect to the normal growth of the cell. During the act of division, we can consider the total cell volume (either that of the original cell, or that of the combined mother and daughter cells after division) constant. For example, in the dA/dt equation (3.2.2), this yields

$$dA/dt = k_0 V + k_1 A + k_2 A^2 / V.$$

When cells divide, we replace V with $V/2$ and A with $A/2$, assuming a perfectly symmetrical division. This yields

$$dA/dt = k_0 V + k_1 A + k_2 A^2 / V,$$

which is the same as the original equation. Similar results are seen for the $d[A]/dt$ equation. This might suggest division itself has little effect on the modeled reactions. However, this notion is indeed oversimplified. When cells divide, the nuclear envelopes disappear, DNA is replicated, proteins are actively transported, ... Thus, our analysis certainly leaves out many biological events that could affect the dynamics.

These can be very important. For example, assume that a cell is dividing every 5 hours, and that for a period of 30 minutes during each cycle, transcription cannot occur, e.g., due to DNA replication. Imposing this limitation on the Goodwin oscillator, in circumstances when oscillations normally take place, causes very disrupted (see figure 3.1), and even chaotic (see figure 3.2), rhythms.

3.3   Probabilistic Formulation of Deterministic Equations and Delay Equations

So far, all equations we have considered have been deterministic, meaning that each time they are simulated, the same answer is found. Methods for analyzing stochastic systems can be very powerful (Gardiner 2004; Horsthemke and Lefever 1984). However, these stochastic methods may not have the advantages over deterministic methods that one might expect. For instance, although one commonly hears that deterministic equations do not account for any effects of molecular noise, this statement is not true: deterministic models do account for randomness, but do so by taking a limit. They are akin to using a probability distribution to understand a random variable.

For example, consider a reaction where the probability of a degradation of a molecule of A between time t and $t + dt$ is $d_0 A dt$. Thus, $dA = -d_0 A dt$, or $dA/dt = -d_0 A$. The solution of this equation is a decaying exponential. Likewise, let us assume that the probability another reaction occurs over dt is $\rho(t)dt$. We then have

$$dA/dt = \rho(t) - d_0 A.$$

This can be expressed as an integral equation,

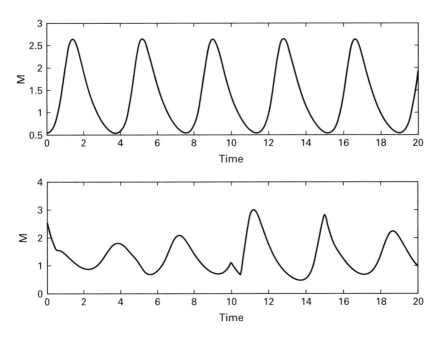

Figure 3.1
Simulations of the Goodwin model without (top) and with (bottom) transcription disrupted each cell cycle.
We assume that cells divide every five hours, and that transcription is not available for 30 minutes each cycle.
Time is in hours. This uses code 2.2 with $n = 9$. We simulate the disruption of transcription by changing the line
simulating mRNA to $Y(1) = (\mathrm{mod}(t, 5) > 0.5) * (10/(.001^n + R^n)) - M$.

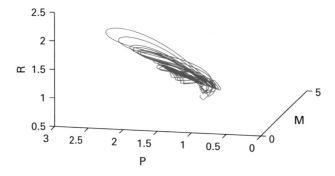

Figure 3.2
Parametric plot of the simulations in the bottom frame of figure 3.1.

$$A(t) = \int_{t_b}^{t} \rho(t') e^{-d_0(t-t')} dt' + B_1,$$

and in this way, the total amount of A is seen to be a weighted average of $\rho(t)$. Again these equations are, like all our others, based on a probabilistic formulation. However, we take the limit as the number of reactions approaches infinity.

Assuming that all reactions were not stochastic would yield a different type of equation. For example, assume that degradation occurred at a fixed time t_0 after production. If we started with A molecules produced at time zero, we would have A molecules until time t_0 and then none afterward. This is quite different from the solution of our deterministic equation (see figure 3.3).

However, there are times when a fixed time reaction rate is indeed justified. For example, proteins and mRNA can have thousands of individual components, which are assembled step by step, so another way to model transcription and translation accounts for each of these individual steps. Assume that all steps proceed with the same dynamics.

The time to transcription or translation is the sum of the times, τ_i, required for each individual step, which, in the preceding model, we can choose from an exponential distribution with mean 1. When we sum all these times we get

$$\sum_{i=1}^{w} \tau_i = w \sum_{i=1}^{w} \frac{\tau_i}{w}.$$

For the former expression, as $w \to \infty$, we can invoke the law of large numbers and see this converges to w. However, if we allow each τ_i to be scaled by $1/w$, meaning we decrease the time of each step as we add more steps, we then find that there is a fixed time that it takes to make an mRNA or protein, or a delay, t_d, for transcription or translation:

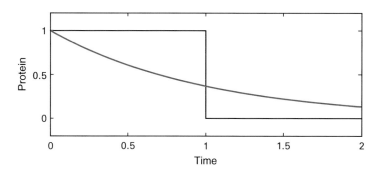

Figure 3.3
Standard degradation of a protein with rate 1 (red = e^{-t}) as well as noiseless degradation where all proteins degrade after 1 time unit (black). Time is in hours.

Pr(transcription or translation occurring between t and $t + dt$) = d_0 $M(t-t_d)dt$

where M is the number of mRNA or factors that promote transcription.

This yields what is called a delay equation. These equations can be dangerous. In a field where telling time is the goal, they assume a perfect timekeeping mechanism. If properly justified, as, for example, in the preceding model of transcription or translation elongation, a delay equation can be useful. Otherwise, assuming a perfect timekeeper in a model to understand timekeeping is a risky business. For example, because delay equations are very likely to oscillate, they can make predictions of sustained oscillations that the actual system does not generate (see chapter 4).

3.4   The Discreteness of Chemical Reactions, Gillespie, and All That

Although our deterministic equations do account for the stochasticity of chemical reactions, as explained in section 3.3, they do not account for their discreteness. When chemical reactions occur, the resulting numbers of molecules change by discrete amounts (e.g., from A to $A+1$, or from A to $A-1$). However, the equations for the number of molecules described earlier are ordinary differential equations, and so the quantities in them change continuously. We could, instead, evolve our number of molecules discretely. To do this we would need to know, at each timepoint, which reactions occur next, and how much time elapses between reactions. Gillespie (1977) is often credited for developing a methodology for accurately simulating this process, although the essential idea of the method appears in many forms in other fields. Gillespie's method could best be described as a kinetic Monte Carlo (KMC) method. However, here we will use the convention in the field and refer to it as the Gillespie method.

A rigorous comparison between these the Gillespie method and the deterministic equations is given by Kurtz (1972). Many simulational studies, e.g., that of Forger and Peskin (2005), also compare deterministic and stochastic approaches. For our purposes, three points are of particular interest:

1. Deterministic and Gillespie approaches do not have to agree. In fact, there are many cases where oscillations are seen in a Gillespie simulation of a system, but not in a deterministic treatment of the system (see figure 3.4 and Forger and Peskin 2005).

2. It is often said that as the cell volume increases, while keeping the concentrations fixed, the Gillespie simulations approach their deterministic limit. This is true to an extent, but increasing cell volume would not necessarily increase the number of genes. Thus, the real comparison should be made as the number of individual reactions increases. This could occur either because the cell volume increases (leading to larger numbers of molecules) or because some reactions, whose reactants have a low number of molecules, occur reversibly on a fast timescale.

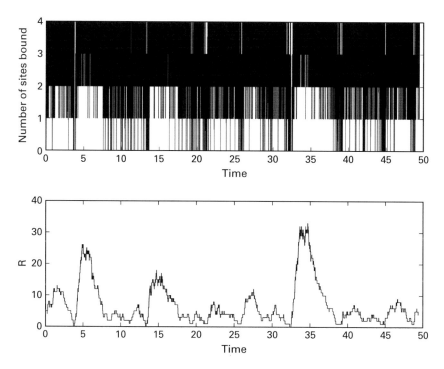

Figure 3.4
Simulations of a stochastic version of the Goodwin model. The number of sites bound is shown (top), as well as the timecourse for repressor levels (bottom). Stochastic oscillations are seen even when the deterministic model does not show oscillations. Time is in hours. See code 3.1 for more details.

3. As the number of reactions increases, we expect the standard deviation of our results to decrease like $1/N^{1/2}$ where N is the number of reactions. This relationship is predicted from basic probability theory. An example is shown later in the section in figure 3.6.

Details on the derivation of the Gillespie method are best left to his original paper (1977). However, it is important to keep in mind his assumption that reactants are well mixed. By making this assumption, we avoid any need to track individual molecules, and can regard each molecule as just as likely to react as the next. This allows us to track only the total number of molecules.

The first step in the method is to list all reactions, with their rates and their stoichiometries. Here we use the rate for number of molecules as described in section 3.2. Following are some examples, where G indicates the number of binding sites on a gene that are occupied by a transcription factor, M is the number of mRNA molecules, P is the number of protein molecules, and P^* is the number of protein dimer molecules:

Reaction	Rate	Stoichiometry
Transcription	$\rho_1 = a_1 G V$	$M \rightarrow M+1$
Translation	$\rho_2 = a_2 M$	$P \rightarrow P+1$
Dimerization	$\rho_3 = (a_3/2)P(P-1)/V$	$P \rightarrow P-2, P^* \rightarrow P^*+1$
Clearance	$\rho_4 = a_4 P^*$	$P^* \rightarrow P^*-1$
Binding	$\rho_5 = a_5 P^* G/V$	$G \rightarrow G-1, P^* \rightarrow P^*-1, GP^* \rightarrow GP^*+1$

If each reaction occurred independently from all others, we could find the time until the next reaction by the formula

$$\tau_i = (1/\rho_i)\log(1/\text{rand}),$$

where rand is a random number with a uniform distribution between 0 and 1. Solving for rand, we find

$$\text{rand} = \exp(-\rho_i \tau_i),$$

or that τ_i comes from an exponential distribution matching the Poisson probabilistic form described earlier. The probability of each reaction happening will stay the same until the next reaction occurs. Thus, we can find all τ_i. The smallest τ_i is the one that will occur next. We can then update the reactants and products, the ρ_i values, and the time ($t \rightarrow t + \tau_i$). This process, which can be repeated, is called Gillespie's first-reaction method. It is actually one of the methods used in code 3.1, although, as we will see, other variants of the method may be more efficient.

Another way to proceed is to choose randomly among all the reactions but weight the reactions' chances of being chosen by their reaction rates. To do this, we choose a random number between 0 and ρ_T, the sum of all the ρ_i. If this random number is between 0 and ρ_1, we say that the first-listed reaction occurs next. If between ρ_1 and $\rho_1 + \rho_2$, we say the second reaction occurs next, and so forth. Once we find the reaction that occurs next, we find the time until this reaction by

$$\tau_T = (1/\rho_T)\log(1/\text{rand}).$$

Time can then be updated ($t \rightarrow t + \tau_T$), as well as the reactants, the products, and the ρ_i values. This is called the direct method. It is equivalent to Gillespie's first-reaction method, and this equivalence is described in Gillespie's original papers (1977). We will refer to both of these methods as the Gillespie methods.

Most numerical methods use a fixed time step, and their error is determined by this time step. A remarkable property of the Gillespie methods is that they are exact, in that no time step is used. These methods can be used even with reaction rates that depend explicitly on time (Anderson 2007). When no fixed time step is used, the only errors will be due to rounding and drawing random numbers. One drawback, however, is that these methods are computationally costly, since many random numbers need to be generated. This computational cost can be reduced by fixing a time step (see, e.g., Petzold and Gillespie 2003), but this will add errors into the simulations.

Useful speed-ups have been proposed by Gibson and Bruck (2000) through the use of an adapted method known as the next-reaction method, which allows some of the random numbers to be reused (a practice generally taboo in numerical simulations). We first generate random times for when each reaction would occur if in isolation, and then choose a next reaction. Time is updated to the time of this next reaction, meaning that all the other reaction times are decreased by this amount. A new time is generated for this chosen reaction, and any other reactions whose rates were changed by the chosen reaction have their times scaled by the ratio (new rate)/(old rate).

As mentioned earlier, deterministic and Gillespie approaches do not necessarily agree, and generally the dynamics of a stochastic system can be quite different than that of the corresponding deterministic system. In particular, stochastic simulations can show oscillations even when deterministic models do not. This phenomenon has been shown to be experimentally important for the mammalian circadian clock (see figure 7.8 and (Ko et al. 2010)). We now turn to an example that exhibits this divergence between stochastic and deterministic treatments.

We simulate the Goodwin model for four binding sites using the Gillespie method (see code 3.1). Code for both the direct and first-reaction methods is provided. Results are plotted in figure 3.4. One will note that stochastic oscillations are seen even though there are only four sites. The deterministic model, in contrast, quickly approaches a steady state, and indeed, as discussed in chapter 4, the situation with only four sites is one where oscillations are not seen in the deterministic model.

We also note that the stochastic oscillations seen in our model decrease as the number of reactions increases (figure 3.5). The standard deviations of the timecourses of mRNA, protein, and repressor all decrease as well (figure 3.6) as the number of reactions increases.

This method can easily be extended to neuronal systems. In neurons, individual ion channels open and close stochastically. This phenomenon can be treated in a way similar to the one proposed by Gillespie (Clay and DeFelice 1983). Here we consider a finite number of channels, and therefore a finite number of gating variables. The voltage V can be simulated where, for m, n, and h, we use the respective ratios of the number of m, n, or h gating subunits that are open to the total number of m, n, or h gating subunits. We then consider as Gillespie reactions the opening and closing of the m, n, or h gating variables. The opening rate is $\alpha_j(V)(1-j)$ and the closing rate is $\beta_j(V)j$, where $j = m$, n, or h. We find the next reaction, implement it, and move time forward by Δt, the time to the next reaction. We also integrate the dV/dt equation to find the voltage Δt later, for example, by taking one time step using the Euler method.

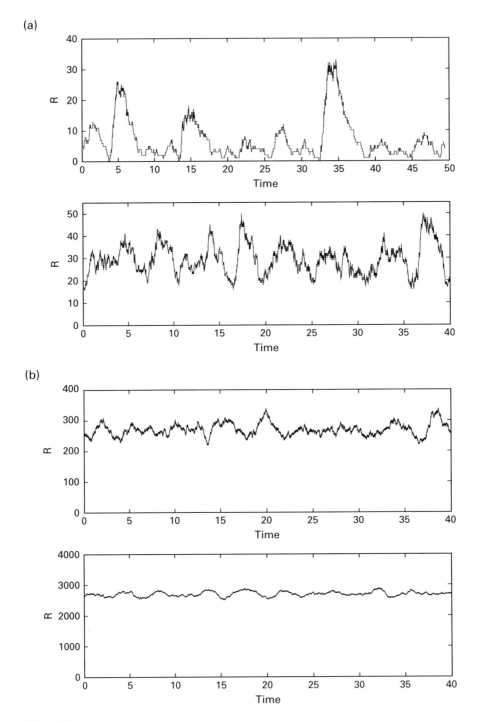

Figure 3.5
Simulations of the Goodwin model with increasing numbers of molecules and increasing rates of binding to and unbinding from the gene. The top simulation is as in figure 3.4, followed in successive frames by a scaling of the cell volume to 10, 100, 1,000, and 10,000 (see code 3.1). As more molecular events occur, one sees that R reverts to its mean, which is the behavior seen in the deterministic system. Time is in hours.

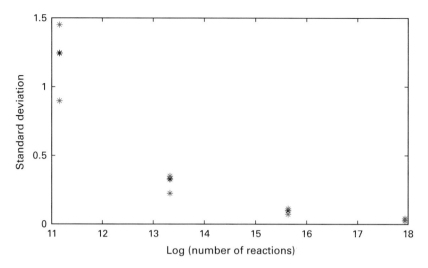

Figure 3.6
Scaling of standard deviation divided by the mean of the mRNA (blue), protein (black), and repressor (red) timecourses with increasing numbers of reactions. This shows the predicted inverse square root scaling: $1/(\text{\# reactions})^{0.5}$.

One difficulty with this method, as noted earlier, is that it is computationally expensive. The Langevin approach offers another simplification: simulating the differential equation with an added noise term. For example, for the stochastic Hodgkin–Huxley equation, the gating variables could be simulated by

$$dj/dt = \alpha_j(V)(1-j) - \beta_j(V)j + \xi(t).$$

The noise term $\xi(t)$ is a Wiener process, a way of introducing randomness that we interpret here by describing how it is simulated. Without this term, the standard Euler method would be to march forward in time by the following rule:

$$j(t + \Delta t) = j(t) + (\alpha_j(V)(1-j) - \beta_j(V)j)\Delta t.$$

In the Langevin formulation, we would add a random number ξ, drawn from a Gaussian distribution, at each time step. The size of this random number scales, surprisingly, with $\Delta t^{0.5}$, and is otherwise independent of the number of channels. We call this simulation process additive noise. However, the noise may also be proportional to the state of the system variables, in which case it is called multiplicative noise. For the stochastic Hodgkin–Huxley equation, the scaling works out so that

$$j(t+\Delta t) = j(t) + (\alpha_j(V)(1-j) - \beta_j(V)j)\Delta t + \alpha_j(V)\beta_j(V)/(\alpha_j(V)+\beta_j(V))\Delta t^{0.5}\xi,$$

where ξ is a random Gaussian number (Fox and Lu 1994). The variance of this number can be scaled according to the size of the noise. Further details can be found in Bodova et al. (2015).

A final note about the Gillespie method is that each simulation is different, and it may be difficult to determine statistics or the distribution of states, because one would need to run many simulations. However, sometimes one can derive a partial differential equation for the distribution of states of the system. The analysis of this partial differential equation, which is called a Fokker-Planck equation, is beyond the scope of this book.

3.5   Frontiers: Temperature Compensation

You have probably observed that chemical reactions proceed at very different rates depending on what the external temperature is. We freeze food to slow down any chemical reactions that might cause it to spoil. Many organisms are cold-blooded (poikilotherms) and thus do not regulate their internal body temperature. During the day, their body temperature can vary by 20° C or more, and reaction rates within their bodies should therefore vary greatly. This could cause reactions to not function properly unless some sort of compensation, called temperature compensation, comes into play.

Here, we are concerned with biological clocks: How can they be reliable in the face of temperature fluctuations? Temperature compensation, or the ability of biological clocks to keep a fixed period regardless of temperature, was first discovered by Colin Pittendrigh in 1952, while he was counting flies in an old outhouse in the Rocky Mountains (Pittendrigh 1992). Since these organisms were already known to exit their pupal case (eclosion) at a specific time of day, avoiding the hot midday sun that could cause dehydration and death, Pittendrigh was able to study the circadian clock in *Drosophila* by determining how many flies emerged over time. He postulated that the flies must have an internal clock mechanism governing the timing of eclosion, and if such a clock were to be of any use, it had to generate rhythms independent of temperature. Using the outhouse as a darkroom with ambient temperature, and using a pressure cooker as a second darkroom immersed in a cold trout stream, he observed the timing of eclosion in both darkrooms. Two days later, eclosion occurred close to the normal time even in the cold darkroom.

When he subsequently tried to replicate his findings in the lab, raising flies in 26° C and then switching the flies to 16° C, he initially failed. As shown in figure 3.7, on the first day in a new temperature, eclosion did not occur 24 hours after the previous eclosion but had a different period, which in the case of the cold temperature was about 12 hours later than normal. Luckily, however, he continued the experiment and found that future cycles returned to the original period: the second period of eclosion did occur 24 hours after the first period of eclosion in the new temperature. Thus, Pittendrigh concluded that circadian rhythms were temperature-compensated. Although affected by temperature, they always returned to a 24-hour rhythm. His experiments were a major influence in establish-

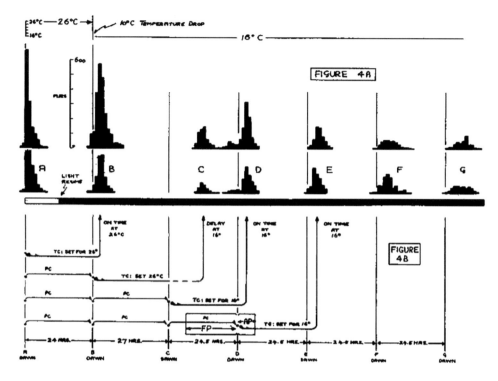

Figure 3.7
The eclosion rhythm of two populations of flies kept in darkness (to avoid signals to the circadian clock) when transferred to conditions 10° C lower. The numbers of eclosing flies in two samples are plotted with respect to time. Taken from Pittendrigh (1954).

ing circadian rhythms as a fruitful field of study, prompting further substantial research in the study of temperature compensation, such as work done by Hastings and Sweeney on luminescent algae (1957). The mechanisms of temperature compensation are still debated, and are one of the great unsolved mysteries of circadian timekeeping. As for biological rhythms other than circadian rhythms, some, but not all, are temperature-compensated as well (Morre et al. 2002).

The temperature dependence of chemical reactions is typically measured by the Q_{10} quotient, which is defined as

Q_{10} = (reaction rate at temperature $(T+10)°C$)/(reaction rate at temperature $T°C$).

Reactions typically have a Q_{10} of 2 to 4, which means their rates double to quadruple for every 10° C temperature increase.

A physical model for this temperature dependence proceeds as follows. Molecular reactions occur when reactants collide (1) in the proper orientation and (2) with enough energy for the reaction to take place. Although the final state of a reaction will have a lower potential energy than the initial state, the intermediate states during the rearrangement of the atoms may have higher potential energies. Thus, some extra energy is needed to achieve the intermediate stages of the reaction. This extra energy is called the activation energy. From statistical mechanics, one finds that the probability that a given molecular collision is above this activation energy is proportional to $e^{-(E_i/R_gT)}$ where E_i is the activation energy of the ith reaction, R_g is the gas constant and T is the absolute temperature. Thus, we find that the rate at which a particular reaction proceeds is

$$k_i = A_i e^{-\frac{E_i}{R_gT}},$$

which is called the Arrhenius relations. What if all reactions were to increase by some factor b? Each term in our biochemical models is of the form $k_i[R_1]\cdots[R_n]$, except for terms where the mass action assumption was used. These can be converted into individual reaction terms as before without any loss of rigor (in fact, since this assumption was not included, more rigor is seen). Thus, at a new temperature all terms are of the form: $bk_i[R_1]\cdots[R_n]$. Each differential equation can then be written as

$$\left(\frac{1}{b}\right)\frac{dX_i}{dt} = f_i(\underline{X}).$$

By scaling time $t' = bt$, we now have the same differential equations as before:

$$\frac{dX_i}{dt'} = f_i(\underline{X}).$$

As one would intuitively imagine, increasing temperature is the same as causing time to proceed faster.

There is a biological problem with this. Certain reactions cannot proceed faster at higher temperature. In particular, the period of the external day is 24 hours regardless of temperature, so biological circadian clocks must also keep a 24-hour period regardless of temperature as well. We saw that this is not possible if all reactions speed up at the same rate. For this to be possible, some reactions must be more sensitive to temperature than other reactions.

We will use the Hodgkin–Huxley equations for neuron activation as an example. Hodgkin and Huxley noted that the ion channel openings and closings are similar to other chemical reactions and have a Q_{10} of approximately 3 (Hodgkin and Huxley 1952). Thus, all α and β in the model should be increased by a factor of 3 for each 10°C temperature increase. Interestingly, the voltage equation does not have an appreciable temperature dependence, since it is determined by the diffusion of ions across the membrane. Diffusion-limited reactions are typically proportional to absolute temperature, and absolute temperature does not change much for a 10°C temperature change. Therefore, although all channels are assumed to increase by the same rate, the effect of temperature in this model is more than simply a scaling of the period.

Hastings and Sweeney (1957), realized another problem with temperature compensation. The basic assumption is that all reaction rates increase with increasing temperature. If we supposed that increasing any reaction rate always causes a clock to speed up, then temperature compensation could never be achieved. Thus, there must be some reaction rates within the clock that, when increased, due to increasing temperature, in some way slow down the clock. We will refer to these reactions as temperature compensation elements.

It is possible that nonlinear interactions come into play with regard to the period of the clock, which could complicate this model. In this way, increasing any particular reaction rate would increase the period of the clock, however, if several reaction rates were increased together, the period might decrease. For instance, consider a clock whose period is determined by

$$\tau = \frac{k_1^2 + k_2^2}{k_1^2 + k_1 k_2 + k_2^2}.$$

Around the point $k_1 = k_2 = 1$, increasing k_1 or k_2 individually would cause τ to increase, but increasing them both by the same factor would not change τ. Thus, in mathematical terms, talking about temperature compensation elements might not make sense as it depends on thinking about reaction rates in a linear way.

The linear theory predominates, and it has been championed in many papers by Peter Ruoff (Rensing et al. 1997; Ruoff et al. 1996; Ruoff et al. 1997). He finds that the period τ is given by

$$\tau = \tau(k_1, \ldots, k_n),$$

and letting T denote temperature,

$$\frac{\partial \tau}{\partial T} = \sum_i \frac{\partial \tau}{\partial k_i} \frac{\partial k_i}{\partial T}.$$

We can now take advantage of the Arrhenius temperature dependence and find that

$$\frac{\partial \tau}{\partial T} = \sum_i \frac{\partial \tau}{\partial k_i} e^{-\frac{E_i}{R_g T}} \frac{E_i}{R_g T^2} = \sum_i \frac{\partial \tau}{\partial \ln k_i} \frac{E_i}{R_g T^2}.$$

We now multiply by $1/\tau$ and find

$$\left(\frac{1}{\tau}\right)\frac{\partial \tau}{\partial T} = \frac{1}{R_g T^2} \sum_i \left(\frac{\partial \tau}{\partial \ln k_i}\right)\left(\frac{1}{\tau}\right) E_i,$$

$$\frac{\partial \ln \tau}{\partial T} = \frac{1}{R_g T^2} \sum_i \left(\frac{\partial \ln \tau}{\partial \ln k_i} \right) E_i,$$

and

$$\eta_i \equiv \frac{\partial \ln \tau}{\partial \ln k_i}.$$

Now let us define new variables $\tau' = \ln \tau$ and $k_i' = \ln k_i$. Thus, τ' is a function of k_1', ..., k_n'. We know the magnitude of the derivative of τ' in the direction $(\partial \tau'/\partial k_1', \dots, \partial \tau'/\partial k_n')$ is -1, since if we scale k_1, ..., k_n by a factor b, we scale the period by a factor $1/b$. This gives the following equation:

$$\sum_i \eta_i = -1.$$

And thus, we find that if $E_i = 1$, we have

$$\frac{\partial ln \tau}{\partial T} = -\frac{1}{R_g T^2}.$$

However, for temperature compensation to be achieved, we want

$$\frac{\partial ln \tau}{\partial T} = 0,$$

and this is satisfied only by choosing $\sum_i (\partial ln \tau / \partial ln k_i) E_i = 0$. The only way this can be achieved is if some $\eta_i < 0$ while others are > 0. This is equivalent to Hastings and Sweeney's statement that some reaction rates within the clock, when increased, slow down the clock, which was the basis for the idea of temperature compensation elements.

Ruoff assumes that, because of the structure of a feedback loop, there must exist some $\eta_i < 0$. We now show that this assumption may not be true. For example, consider the harmonic oscillator, arguably the most studied model of an oscillator: $dx/dt = k_1 y$, $dy/dt = -k_2 x$, or $d^2 y/dt^2 = -k_1 k_2 y$. Solutions of this system are $y = \sin((k_1 k_2)^{0.5} t)$. The period of this oscillator can be calculated as $\tau = 2\pi / \sqrt{k_1 k_2}$. Thus, there are no η_i that are negative. Let us come back to this fact later.

In terms of our equation for the period, written in terms of the η_i,

$$\frac{\partial ln \tau}{\partial T} = \frac{1}{R_g T^2} \sum_i \eta_i E_i,$$

Ruoff now assumes that η_i is constant regardless of the period, essentially a linearity assumption. Adopting this assumption, we integrate both sides of the equation and find

$$\int_{\tau_0}^{\tau_1} d\ln\tau = \sum_i \eta_i E_i \int_{T_0}^{T_1} \frac{dT}{R_g T^2},$$

$$\ln\tau_1 - \ln\tau_0 = \ln\frac{\tau_1}{\tau_0} = -\frac{\sum_i \eta_i E_i}{R_g T_1} + \frac{\sum_i \eta_i E_i}{R_g T_0}.$$

Each n_i can be estimated by varying a parameter and measuring the period. The constraint $\sum_i (\partial ln\tau / \partial lnk_i) E_i = 0$ means that we can choose any parameter set in an $(n-1)$-dimensional subspace to achieve temperature compensation. Thus, Ruoff says that by choosing some appropriate E_i, temperature compensation is achieved. He claims that through evolution, the E_is could be tuned to achieve this.

There are two key assumptions in this model that need to be tested. First is the linearity assumption that the η_is are constant with respect to temperature. More fundamental is the assumption that all biochemical oscillators have some $\eta_i < 0$. Even if some $\eta_i < 0$, if the magnitude of η_i is small, we may need unrealistically large activation energies to get temperature compensation. Here, we present two counterexamples first presented by the author at the Kavli Institute for Theoretical Physics (Forger).

Consider the following Goodwin-type model with dimerization:

$$\frac{dm}{dt} = \frac{s}{\left(K + \bar{\bar{p}}_n^*\right)^h} - a_1 m,$$

$$\frac{dp}{dt} = rm - b_1 p - g_1 p^2,$$

$$\frac{d\bar{\bar{p}}}{dt} = \frac{g_1 p^2}{2} - b_2 \bar{\bar{p}} - g_2 \bar{\bar{p}},$$

$$\frac{d\bar{\bar{p}}^*}{dt} = g_2 \bar{\bar{p}} - b_3 \bar{\bar{p}}^* - g_3 \bar{\bar{p}}^*,$$

$$\frac{d\bar{\bar{p}}_n^*}{dt} = g_3 \bar{\bar{p}}^* - b_4 \bar{\bar{p}}_n^*.$$

Figure 3.8 shows that increasing any of the parameters, at least up to 50 percent, only decreases the period, meaning that the clock cannot be temperature-compensated. See code 3.2 for more details.

Another counterexample is the Goodwin model itself:

$$\frac{dM}{dt} = \frac{k_{\max}}{K + P_2^m} - d_0 M,$$

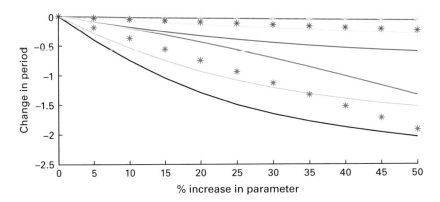

Figure 3.8
The period dependences in the temperature compensation counterexample. Increasing any of the parameters by the amount listed decreases the period, as one would expect. The curves are as follows: g_1 (blue), g_2 (black), g_3 (green), s (red), K (cyan), a_1 (magenta), b_2 (yellow), b_3 (dotted yellow), b_4 (dotted magenta), r (dotted cyan), b_1 (dotted red). The change in period is given in hours.

$$\frac{dP_1}{dt} = k_l M - d_1 P_1 - r_1 P_1,$$

$$\frac{dP_2}{dt} = r_1 P_1 - d_2 P_2.$$

In this model we can redefine $d_0 P'_2 = P_2$, $k_l P'_1 = P_1$, and $(k_{max}/r_1{}^m)M' = M$. In the following analysis, we consider K to be zero, which is an approximation. The reduced system then becomes:

$$\frac{dM'}{dt} = \frac{1}{P'_2{}^m} - d_0 M',$$

$$\frac{dP'_1}{dt} = M' - d_1 P'_1 - r_1 P'_1,$$

$$\frac{dP'_2}{dt} = r_1 P'_1 - d_2 P'_2,$$

and the only variables left are rates of degradation. The period of this system can be estimated, to first order, as $1/(d_0 d_2 + d_0(d_1 + r_1) + d_2(d_1 + r_1))^{0.5}$ (see the linear approximation of the Goodwin model in Chapter 4). Thus, only four parameters are present, and there are no temperature compensation elements.

Lakin-Thomas et al. (1991) propose another interesting theory of temperature compensation. They suggest that as temperature increases, the amplitude of the circadian clock oscillation could increase, and that it would take a longer time to proceed through the cycle. Numerical simulations by Pittendrigh et al. (1991) show that a larger circadian amplitude need not lead to a longer period, and experimental data in *Gonyaulax* shows that with increasing temperature, the amplitude of the output signal decreases (Sweeney and Hastings 1960). However, despite this counterexample, the principle is generally true in biochemical models (see chapter 10, section 10.2.1).

The Lakin-Thomas theory of temperature compensation can be made mathematically rigorous by noting two facts:

1. the period of the clock depends not only on the amplitude of the circadian oscillation, but also on the speed with which the system proceeds through the cycle (which may depend on temperature), and
2. the amplitude of the entire circadian oscillation depends on all variables, not all of which react the same way to temperature increases: the amplitude of the oscillation in some variables may increase while the amplitude of oscillation in other variables may decrease.

Mathematically, this can be represented as, $\tau = \int_{\Omega}\left(1 / \left|\frac{d\underline{x}}{dt}\right|\right)d\underline{x}$, where τ is the period, Ω is the path in phase space, and $\underline{x}$ is the point in phase space. Unfortunately, measuring Ω would require experimental measurements of all variables, and we do not have an estimate of how fast the system proceeds through the circadian cycle. Thus, while this theory can be made mathematically correct, it is difficult to implement experimentally.

We have seen that Ruoff's theory of temperature compensation shows how one can numerically find temperature compensation reactions in mathematical models of circadian clocks. This is done by finding which reaction rates increase the period in these models, and this technique has led to predictions about the temperature compensation mechanism in *Neurospora* (Ruoff et al. 2005).

We note that a similar procedure can be carried out in the lab. Mutation can be introduced into a gene of interest to decrease a particular reaction rate. An extreme, but useful, example of this is null mutants, where the rate of transcription of active protein is set to zero. If the period of this mutant is less than the original mutation, we predict that the reaction is a temperature compensation reaction. The mathematical justification for this follows. Similarly to Ruoff, we can take a linear approximation to the period of the circadian clock. However, as distinct from Ruoff, we represent it as a standard Taylor series evaluated at a particular temperature:

$$\tau = \tau^0 + \frac{\partial \tau}{\partial k_1}\left(k_1 - k_1^0\right) + \ldots + \frac{\partial \tau}{\partial k_n}\left(k_n - k_n^0\right),$$

where τ^0 and $k_1^0, \ldots, k_n^0$ are the period and reaction rates at a particular (e.g., room) temperature. (Note that these reaction rates do not need to be explicitly solved for.) If we create a mutation in the ith reaction, we have $k_1 = k_1^0$, …, $k_n = k_n^0$, except for $k_i < k_i^0$. Thus, if $\partial\tau/\partial k > 0$, $\tau < \tau^0$. Also, if the mutation increases a particular rate, we find that $k_i > k_i^0$ and if $\partial\tau/\partial k_i > 0$, $\tau > \tau^0$. Thus, *if a mutation causes a decreased reaction rate and a shorter period, that reaction rate is a potential temperature compensation reaction.* Likewise, *if a mutation causes an increased reaction rate and a longer period, that reaction is a potential temperature compensation element.*

3.6   Frontiers: Crosstalk between Cellular Systems

Unfortunately for the modeler, cellular systems do not operate in isolation. There is much evidence for crosstalk between systems (e.g., see (Ventura et al. 2010)). Components for one system are often reused for other systems. Calcium, for example, is used as a signaling molecule for many cellular systems. Models, however, often treat systems in isolation. This section will consider how the overall cellular environment can affect individual systems.

Let us take an example like the one simulated in chapter 2, section 2.4. Consider a transcription factor that diffuses along DNA. For simplicity, we assume that there are 1,000 bins along the DNA, two of which are copies of a gene. When the transcription factor finds one of these two sites, transcription of the gene occurs. We consider the "black widow" transcription model, where once a gene is transcribed, the transcription factor is degraded (Kodadek et al. 2006). We also assume that transcription factors are attracted to these genes in that, once they find the site, they have a much lower (1/100) rate of moving away.

As in chapter 2, we calculate the rate at which transcription occurs. We do this both for the case described earlier, and for the case where 10 additional transcription-factor binding sites are included on the DNA. These transcription-factor binding sites do not produce any mRNA, nor lead to degradation of the transcription factor. They instead act as sinks whereby transcription factors are attracted to them in the same way that transcription factors are attracted to the genes but are presumably there for other systems.

Starting the simulation with on average 1 transcription factor per bin on the DNA, we run the simulation with and without the additional sites and plot the rate of transcription as a function of the number of transcription-factor proteins. As seen in figure 3.9, the result of this simulation, like the one in section 2.4, is another example of non-mass-action kinetics. We do see a large (note the log scale) effect of the additional sites in reducing the rate of transcription. This illustrates how additional binding sites, involved in other cellular systems, play an important role in the dynamics of the system we study. As these sites become activated or repressed, as the chromatin of the DNA changes, differing effects on our system of study will be seen.

So should one consider all possible cellular systems? Perhaps, but this will certainly be impractical for many years to come. While our models are not yet up to the job of han-

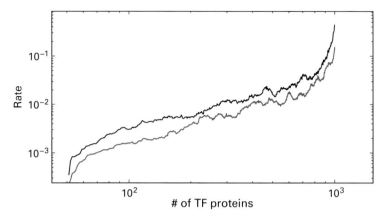

Figure 3.9
The rate of transcription in two models: (black) without or (red) with 10 additional binding sites for the transcription factor.

dling all crosstalk, could it be that cellular systems themselves are designed to withstand crosstalk, just as they are designed to be insensitive to temperature fluctuations? Interesting explorations along these lines, such as those presented by Del Vecchio et al. (2008), are already underway.

3.7   Common Mistakes in Modeling

- Thinking of cellular systems in isolation
- Using additive noise when multiplicative noise is more appropriate
- Assuming all clocks can be temperature-compensated
- Assuming stochasticity has no role in ordinary differential equation modeling
- Using delay equations to determine timekeeping mechanisms
- Ignoring the effects of cell volume
- Assuming that changes in cell volume affect only the dilution of proteins
- Failing to account for the discreteness of chemical reactions
- Assuming mass action for the black-widow, DNA-diffusing transcription factors
- Assuming that clocks slow down with larger amplitude

Code 3.1 Simulations of the Goodwin Model Using the Gillespie Method

(Figures 3.4, 3.5, and 3.6)

```
% This code is provided to give the reader an example of a Gillespie
% code in MATLAB. Some comments are provided and the reader is encouraged
%to figure out this code and to provide additional comments,
clear
%sets the initial conditions and basic rates
G = 0; M = 5; P = 10; Pp = 10; T = 0; sites = 4;
V = 10; %this is increased to 100,1000 etc. in figure 3.5
bin = 100; unbin = 10*V; trans = 100*V;%Normally bin, being a
%bimolecular reaction would scale as 1/V. However, We also scale both
%bin and unbin by V to increasing the number of reactions with
%increasing V.
%Sets the reaction parameters
a(1) = bin*(sites-G)*Pp; a(2) = unbin*G; a(3) = trans*max(0, 1-G);
a(4) = M; a(5) = M; a(6) = P; a(7) = P; a(8) = Pp;
tcnt = 0;% This records how many reactions took place
tic %time the simulation
while T < 100
%The code is with the direct method code
Atot = sum(a);
r = rand*Atot;
cnt = 1;
partsum = a(1);
while r > partsum
  cnt = cnt + 1;
  partsum = partsum + a(cnt);
end
T = T + (1/Atot)*log(1/rand);
%the code below is for the first-reaction method
%for ij = 1:8
% ta(ij) = (1/a(ij))*log(1/rand);
%end
%[telap, cnt] = min(ta);
%T = T + telap;
switch cnt%This implements the chosen reaction and updates reaction
%rates
case 1
        G = G + 1;
        a(1) = bin*(sites-G)*Pp;
        a(2) = unbin*G;
        a(3) = trans*max(0, 1-G);
case 2
        G = G - 1;
        a(1) = bin*(sites-G)*Pp;
        a(2) = unbin*G;
        a(3) = trans*max(0, 1-G);
```

```
case 3
        M = M + 1;
        a(4) = M;
        a(5) = M;
case 4
        M = M - 1;
        a(4) = M;
        a(5) = M;
case 5
        P = P + 1;
        a(6) = P;
        a(7) = P;
case 6
        P = P - 1;
        a(6) = P;
        a(7) = P;
case 7
        Pp = Pp + 1;
        P = P - 1;
        a(8) = Pp;
        a(6) = P;
        a(7) = P;
        a(1) = bin*(sites-G)*Pp;
case 8
        Pp = Pp - 1;
        a(8) = Pp;
        a(1) = bin*(sites-G)*Pp;
end
%The following lines record the results
tcnt = tcnt + 1; %records that a reaction took place
tl(tcnt) = T; Gl(tcnt) = G; Ml(tcnt) = M; Pl(tcnt) = P; Ppl(tcnt) = Pp;
end
toc%ends the timing
figure(1)
plot(tl(floor(tcnt/2):tcnt)-tl(floor(tcnt/2)), Gl(floor(tcnt/2):tcnt))
figure(2)
plot(tl(floor(tcnt/2):tcnt)-tl(floor(tcnt/2)), Ppl(floor(tcnt/2):tcnt))
xlabel('Time')
ylabel('R')
figure(3)
hold on
plot(log(tcnt), …
std(Pl(floor(tcnt/2):tcnt))/mean(Pl(floor(tcnt/2):tcnt)), '*k')
plot(log(tcnt), …
std(Ppl(floor(tcnt/2):tcnt))/mean(Ppl(floor(tcnt/2):tcnt)), '*r')
plot(log(tcnt), …
std(Ml(floor(tcnt/2):tcnt))/mean(Ml(floor(tcnt/2):tcnt)), '*')
```

Code 3.2 Temperature Compensation Counterexample

(see figure 3.8)

```
%This codes provides an example of a system that cannot be temperature
%compensated in that it does not have temperature compensation elements

function Y = tempcompexam(t, X)
%This codes the model
global a
gone = a(1); gtwo = a(2); gthree = a(3); s = a(4); K = a(5);
aone = a(6); btwo = a(7); bthree = a(8); bfour = a(9); r = a(10);
bone = a(11);
A = X(1); B = X(2); CC = X(3); F = X(4); G = X(5);
Y(1) = s/(K + G)^6 - aone*A;
Y(2) = r*A—bone*B—gone*B^2;
Y(3) = 0.5*gone*B^2 - btwo*CC—gtwo*CC;
Y(4) = gtwo*CC—bthree*F—gthree*F;
Y(5) = gthree*F—bfour*G;
Y = Y';
end
%This sets up the basic value of the parameters
%global a
%a(1) = 0.5; a(2) = 0.5; a(3) = 2*0.5; a(4) = 1; a(5) = 0.1;
%a(6) = 0.1; a(7) = 0.1; a(8) = 0.1; a(9) = 0.1; a(10) = 1; a(11) = 1;
%for ik = 1:10
%fac = 1+ik*.05;% How much the parameter will be changed
%for ij = 1:11
%a(ij) = a(ij)*fac;%Changes the parameter
%[T, X] = ode45(@tempcompexam, 0:0.001:1000, [2 2 2 2 2]);
% The next line, which continues to the next line calculates the period
%pers = find((X(500001:1000000, 1) <= mean(X(:,1))).*(X(500002:1000001,…
%1) >= mean(X(:,1))));
%pera(ik, ij) = (mean(pers(2:11)-pers(1:10)))/1000;
%a(ij) = a(ij)/fac;
%end
%end
%figure(1)%Plots the results
%hold on
%plot(0:5:50, [0 (pera(:, 2) - 25.6422)'], 'k')
%plot(0:5:50, [0 (pera(:, 3) - 25.6422)'], 'g')
%plot(0:5:50, [0 (pera(:, 1) - 25.6422)'])
%plot(0:5:50, [0 (pera(:, 4) - 25.6422)'], 'r')
%plot(0:5:50, [0 (pera(:, 5) - 25.6422)'], 'c')
%plot(0:5:50, [0 (pera(:, 6) - 25.6422)'], 'm')
%plot(0:5:50, [0 (pera(:, 7) - 25.6422)'], 'y')
%plot(0:5:50, [0 (pera(:, 8) - 25.6422)'], 'y*')
```

```
%plot(0:5:50, [0 (pera(:, 9) - 25.6422)'], 'm*')
%plot(0:5:50, [0 (pera(:, 10) - 25.6422)'], 'c*')
%plot(0:5:50, [0 (pera(:, 11) - 25.6422)'], 'r*')
```

Code 3.3 A Black-Widow DNA-Diffusing Transcription Factor Model

(Note that this code is similar to code 2.1)

```
% Like in code 2.1 some sparse comments are provided in this code. The
%reader is encouraged to figure out this code, provide comments and
%determine differences from code 2.1.
numbox = 1000;
numperbox = 1;
x = zeros(numbox, 1);
g = zeros(numbox, 1);
g(1) = 100000;
g(numbox/2) = 100000;
for ij = 1:(numbox*numperbox)
box = ceil(rand*numbox);
x(box) = x(box) + 1;
end
transrate = ones(numbox, 1);% the following lines set the rate at which
%protein can jump
transrate(1) = 1/100;
transrate(numbox/2) = 1/100;
%for ij = 1:10% the following lines insert 10 additional binding sites
%box = ceil(rand*numbox);
%transrate(box) = 1/100;
%end
reac(1:numbox) = …
(1./(x.*transrate)).*log(1./rand(numbox, 1));
reac((numbox+1):(2*numbox)) = …
(1./(x.*transrate)).*log(1./rand(numbox, 1));
reac((2*numbox+1):(3*numbox)) = (1./(g.*x)).*log(1./rand(numbox, 1));
tim = 0;
Mtot(1)= 0;
Mtottim(1) = 0;
Mtotcnt = 1;
xtotrec(1) = sum(x);
for ijj = 2:200000000
[T, I] = min(reac);
if T == Inf%no more molecules left so time to next reaction is Inf
break;
end
tim = tim + T;
reac = reac—T;
if I < numbox + 1%move to left
```

```
movefrom = I;
moveto = mod(I-2, numbox)+1;
x(movefrom) = x(movefrom) -1;
x(moveto) =x(moveto)+1;
elseif I < 2*numbox + 1%move to right
movefrom = I-numbox;
moveto = mod(I, numbox)+1;
x(movefrom) = x(movefrom) -1;
x(moveto) =x(moveto)+1;
else
movefrom = I- 2*numbox;
moveto = I- 2*numbox;
x(movefrom) = x(movefrom) -1;
Mtot(Mtotcnt+1) = Mtot(Mtotcnt)+1;
Mtottim(Mtotcnt+1) = tim;
Mtotcnt = Mtotcnt+1
xtotrec(Mtotcnt+1) = sum(x);
end
for ij = [movefrom moveto]%we only update changed rates, Gibson-Bruck
reac(ij) = (1./(x(ij)*transrate(ij))).*log(1./rand);
reac(ij+numbox) = (1./(x(ij)*transrate(ij))).*log(1./rand);
reac(ij+2*numbox) = (1./(g(ij).*x(ij))).*log(1./rand);
end
end
figure(100)
hold off
[aa ab] = size(Mtot)
loglog(xtotrec(1:(ab-50)),-(xtotrec(51:ab)-…
xtotrec(1:(ab-50)))./(Mtottim(51:ab)-Mtottim(1:(ab-50)))),'k')
```

Exercises

General Problems

For these problems pick a biological model of interest, preferably one you study.

1. Calculate or simulate the effects of two of the three environmental changes:
    a. Changes in cell or system volume
    b. Biochemical or system noise via Gillespie
    c. Changing temperature

2. Are there ways your model counteracts these environmental changes?

Specific Problems

1. Which of the following mechanisms of temperature compensation always work?
    a. Having the amplitude of the oscillation increase with increasing temperature

    b.   Balancing the activation energies of reactions whose rates, when increased, increase and decrease the period

    c.   Making all reactions diffusion-limited

2.   Describe the difference between temperature compensation and temperature independence.

3.   Deterministic simulations (circle all that are correct)

    a.   do not account for the discreteness of single molecules.

    b.   do not account for the randomness of molecular reactions.

    c.   always agree with stochastic (Gillespie) simulations when the cell volume is increased sufficiently.

4.   Which is an inherent assumption of the Gillespie method?

    a.   That reactants are well mixed

    b.   That numbers of molecules are low

    c.   That reactions do not occur independently

    d.   That a time step must be chosen

4

When Do Feedback Loops Oscillate?

Biochemical feedback loops consist of many elements. In this chapter, we show how the elements of a biochemical feedback loop determine its behavior. A methodology is presented in section 4.3, using the Fourier transform, to determine whether oscillations appear in a given biochemical model and, if they do, we find a formula for the period of the oscillations. This methodology simplifies much of the linear stability analysis used previously to analyze feedback loops. We use this methodology in several examples in sections 4.4 to 4.7 to show how futile cycles can diminish oscillations, why a sleep disorder has a short circadian period phenotype, and how positive feedback without bistability can still help generate oscillations. A separate methodology is presented to study how oscillations can be generated using biochemical switches. Limitations of the methodology are also presented. The chapter ends with a generalization of our methodology called the global secant condition.

4.1   Introduction

This chapter is organized around the roles of individual components in feedback loops and how they contribute to overall behavior. Learning these methods not only helps us understand naturally occurring rhythms but also helps us design synthetic clocks (Atkinson et al. 2003; Elowitz and Leibler 2000; Stricker et al. 2008). We seek to identify which elements are indispensable for rhythms and which elements carefully regulate the period of rhythms of a feedback loop. Most of our analysis will follow rhythms as they proceed through a feedback loop. Our go-to model will be the Goodwin oscillator described in chapter 2. Later in the chapter we will look at multiple time scales within rhythm-generating systems.

Goodwin's paper on biological rhythms, published in *Nature* (1966), presented the first model showing oscillations from a mathematical model of a biochemical feedback loop. Unfortunately, many subsequent studies analyzing these equations called these oscillations into question (Gedeon 1998; Grodins 1963; Mees and Rapp 1978; Rapp 1979; Sontag 2006; Thomas and D'Ari 1990; Tyson and Othmer 1978). Ultimately, Goodwin resolved this controversy by adding an exponent to one of the terms in this model, indirectly acknowledging that his original simulations were not correct, as described later in the chapter.

But this was just one part of the controversy surrounding oscillations from biochemical feedback loops. In the 1950s, Belousov experimentally observed oscillations from a set of chemical reactions that could be reconstructed in the lab. At the time, many did not believe this to be possible, and he had great difficulty publishing his work (Winfree 1984). An important figure in this debate was Prigogine, who published a model showing oscillations in another biochemical feedback loop shortly after Goodwin's publication. For this and other work, Prigogine received the 1977 Nobel Prize in chemistry. Since then, how oscillations arise in biochemical feedback loops has been an important topic, and one we study here.

Here in chapter 4, we develop a new methodology that allows for the study of biochemical feedback loops. This methodology also easily allows the analytical treatment of more complicated feedback loop networks, as shown in several examples. While the calculations presented here are somewhat involved, they are much less tedious than other methods that involve complex calculations of eigenvalues and eigenvectors.

The key idea in the study of feedback loops is that after proceeding through a feedback loop, one must recover what one started with. So if one assumes that oscillations are present in the first element, after determining the properties of the second element, third element, and so forth, upon returning to the first element, one must recover the same oscillations. If not—if, for example, one recovers oscillations with smaller amplitude—then the original assumption of sustained oscillations must be invalid.

4.2   Introduction to Feedback Loops

Here we present some examples of what happens when we proceed through a feedback loop. Proceeding from one element to the next, we will see that the phase of rhythms is shifted and their amplitude is changed. The total phase shift that occurs from the first element to the last is typically about half the cycle with the other half made up by negative feedback. If the other operations of the feedback loop shift the phase by half the cycle again, we return to the original rhythm having gone through the feedback loop. To be more precise, it is helpful to have a definition of negative feedback.

Negative feedback refers to a signal that is low when the original signal is high, and high when the original signal is low. The definition we use is a process that multiplies the rhythms by -1, thus flipping the rhythms and making the maximum of the rhythms a minimum and vice versa, an operation that is equivalent to a phase shift of half the cycle. However, sometimes a negative value of a rhythm is not physiological, as in the case of a chemical concentration. Thus, we typically assume that the negative feedback only applies to deviations of the rhythm around their mean, perhaps leaving the mean of the rhythm unaffected.

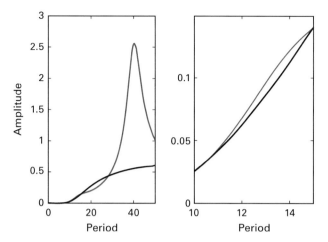

Figure 4.1
Numerical experiments using the Goodwin model. In this feedback loop, 20 steps are used, and the transcription regulation function is assumed to be sinusoidal rather than dependent on the last element of the feedback loop (black curves). As the period of this transcription rhythm is decreased, the amplitude of the last element is attenuated. When we add back a transcription regulation term (red), we see resonance for a period of around 40, canceling of rhythms for a period of around 20, and resonance for rhythms with a period around 13 (right). This shows the effects of the total phase shift of the feedback loop (including negative feedback) being one cycle, one and a half cycles and two cycles, respectively. See code 4.1.

To better see this, we conduct a numerical experiment with the Goodwin oscillator. In this case, we consider the Goodwin oscillator with 20 steps. At first, we remove the transcription regulation function that is typically used and instead drive rhythms from a rhythmic promoter. The period of these rhythms is varied. We record the amplitude (measured as the standard deviation of the last variable) as the period is varied (see code 4.1 for more details). As seen in figure 4.1, this amplitude decreases as the period decreases. This shows that as rhythms become faster, the many steps of the feedback loop further attenuate rhythms. This is called a low-pass filter in engineering.

We now add back the transcription regulation term, along with the driving stimulus, using the mechanism of Kim and Forger (2012). An interesting pattern emerges. When the period of the driving rhythms is around 40, the phase of the rhythm of the last element is about 1/2 a cycle out of phase with the first element, and the transcription regulation function, because of the negative feedback, is in sync with the driving stimulus. The amplitude of rhythms is now very high because of the feedback. In engineering applications this is called resonance.

When the period is around 20, there is a greater change in phase, as one proceeds through the feedback loop, than when the period was 40. Here, rather than proceeding through 1 cycle's worth of phases, we proceed through 1.5 cycles, and the transcription regulation function is out of phase with the driving stimulus. These two signals cancel, and the

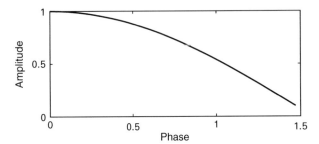

Figure 4.2
The phase-amplitude relationship between two elements in the Goodwin model.

amplitude is diminished. Interestingly, at an even faster period, we proceed through 2 full cycles of the feedback loop, and the transcription regulation function is now in phase with the driving signal. Thus, the rhythms are amplified again, as is shown in figure 4.1 (right).

This numerical experiment shows that, as we proceed through a feedback loop, we shift the phase and change the amplitude of rhythms. Almost always, we are concerned with a phase shift from the first element to the last element of half the cycle. This is because other possibilities would lead to rhythms that are too attenuated for the negative feedback to counteract. Looking at one element, let us assume it obeys the following equation:

$$\frac{dX}{dt} = s(t) - X.$$

We calculate in figure 4.2 the change in phase (in radians) and the change in amplitude between X and s for sinusoidal $s(t)$ of different periods.

The total phase shift is between 0 and $\pi/2$. Achieving any phase shift will mean that the amplitude will be attenuated. Therefore, to get rhythms, some elements of the feedback loop must amplify rhythms and overcome this attenuation. These special elements that amplify rhythms are needed in all feedback loops. We will study this in greater mathematical detail in section 4.3.

4.3   General Linear Methodology and Analysis of the Goodwin Model

This section makes precise the arguments presented in section 4.2. Although it can be applied to any biochemical feedback loop, we use the Goodwin model, developed in chapter 2, as an example. We assume that there are m binding sites to a gene. Following the conventions of chapter 3, k_f and k_r are the forward and reverse rates of binding of P_n to the m sites on a gene. In chapter 2, we sometimes raised these rates to the mth power. For simplicity, we will now just assume that any such power is incorporated in the definition of the rates k_f and k_r. Of the remaining parameters, $k_{\max}$ is the maximum rate of transcription, k_l is the rate of translation, d_0 and q_i are the clearance rates of the chemical species, and r_i is the rate of conversion of P_i to P_{i+1}. We can represent the Goodwin model as

$$\left(\frac{d}{dt}+d_0\right)M = f(P_n),$$

$$\frac{1}{k_l}\left(\frac{d}{dt}+q_1\right)P_1 = M,$$

$$\cdots$$

$$\frac{1}{r_{n-1}}\left(\frac{d}{dt}+q_n\right)P_n = P_{n-1},$$

with $f(P_n) = \dfrac{k_{\max}}{1+\left(k_f/k_r\right)P_n^m}$,

where we have added as many additional steps in the feedback loop as needed. Combining these equations yields

$$\left(\frac{d}{dt}+d_0\right)\cdots\left(\frac{d}{dt}+q_n\right)P_n = \frac{k_{\max}k_l r_1\cdots r_{n-1}}{1+\left(k_f/k_r\right)P_n^m}.$$

In this way, the system has "one degree of freedom," since it can be reduced to one equation. First, let us solve for the fixed points of the system. This is where all the derivatives are set to zero. Here we find

$$d_0 q_1\cdots q_n \bar{P}_n = \frac{k_{\max}k_l r_1\cdots r_{n-1}}{1+(k_f/k_r)\bar{P}_n^m}$$

or

$$\bar{P}_n + \frac{k_f}{k_r}\bar{P}_n^{m+1} = \frac{k_{\max}k_l r_1\cdots r_{n-1}}{d_0 q_1\cdots q_n}.$$

Note that $M, P_1, \ldots, P_n > 0$ since concentrations cannot be negative. The left-hand side of the previous expression is monotonic in $\bar{P}_n$, so there is a unique fixed point.

Let $f(P_n) = 1/\left(1+(k_f/k_r)P_n^m\right)$. Since we are interested in rhythmic solutions, we can now take the Fourier transform of both sides, using the $\sim$ superscript to denote the Fourier transform. Before doing this, we note a few properties of the Fourier transform.

The Fourier transform F of a function $f(x)$ is denoted by $\tilde{f}(w)$ and is given by

$$F(f(x)) = \tilde{f}(w) = \int_{-\infty}^{\infty} f(x')e^{-iwx'}dx'.$$

Likewise, the inverse Fourier transform is defined by

$$\Gamma^{-1}(\tilde{f}(w)) = f(x) = (1/2\pi)\int_{-\infty}^{\infty} \tilde{f}(w')e^{iw'x}dw'$$

Fourier transforms have several important properties. Here are four:

1. The square of the amplitude (L^2 norm or root-mean squared) of a function is the sum of the squares of the magnitudes of its Fourier coefficients.

2. The Fourier transform of the convolution of two functions is the product of the Fourier transforms of the functions.

3. The Fourier transform of $af + cg$ is $a\tilde{f} + c\tilde{g}$.

4. The Fourier transform of the derivative of a function is iw times the Fourier transform of the function. More generally, $\widetilde{d^s f / dt^s} = (iw)^s \tilde{f}$ (Körner 1988).

For these reasons, taking the Fourier transform of a differential equation can often allow us to solve it algebraically. This is exactly what we do here.

This gives

$$(iw + d_0)...(iw + q_n)\tilde{P}_n = t_{max}t_1 r_1...r_{n-1}\tilde{f}(P_n).$$

Calculating $\tilde{f}(P_n)$ based on $\tilde{P}_n$ can be very difficult. So we can approximate f as linear. This could be because we seek solutions near a fixed point, or because f is quasilinear. Anyway, the point is that there exists r_n such that $\tilde{f}(P_n) \approx -r_n\tilde{P}_n$ where $\tilde{P}_n$ is the Fourier coefficient at the frequency at which the system oscillates. Note that the negative sign appears to preserve the structure of a "negative feedback loop," and in particular that more P_n causes less transcription to occur. In this case we have (after canceling $\tilde{P}_n$)

$$(iw + d_0)...(iw + q_n) = -r_n t_{max} t_l r_1...r_{n-1}.$$

Now look carefully at this equation. On the left, we have several complex numbers, for example $(iw + d_0)$, which shift the phase of the rhythms. The effect of these must equal the negative feedback that adds a minus sign on the right.

Suppose the system had just 1 step (here mRNA directly feeds back on itself). We then have

$$iw + d_0 = -r_n k_{max},$$

which cannot be satisfied unless $w = 0$. So no oscillating solutions exist in this case. Again, this does not guarantee that oscillating solutions will not be seen in the original model, since we have approximated f. However, it turns out that this conclusion is valid nonetheless (see section 4.10).

Now suppose there is just one state of the protein ($n = 1$). We then have

$$(iw + d_0)(iw + q_1) = -w^2 + q_1 d_0 + iw(q_1 + d_0) = -r_n k_{\max} k_l.$$

This equation can also never be satisfied unless $w = 0$. The reason for this can intuitively be seen from the simulations in section 4.2. The maximum phase shift of each element is $\pi/2$ and a total phase shift of π is needed for the negative feedback loop. However, a phase shift of π in a two-element loop would mean that then amplitude is zero, meaning that rhythms cannot be seen.

Now consider a three-step loop ($n = 3$) where

$$(iw + d_0)(iw + q_1)(iw + q_2) = -r_n k_{\max} k_l r_1.$$

Setting the imaginary part of this expression equal to zero, we have

$$w = \sqrt{q_1 d_0 + q_1 q_2 + q_2 d_0} \qquad (4.3.1)$$

This is an important formula. It states that the period of oscillation of the linearized system does not depend on the rate of transcription or translation (as neither r_n, $k_{\max}$ nor k_l appears in this equation). It does depend on the transfer rate r_1 since $q_1 = d_1 + r_1$. A similar formula for the period of the Goodwin model exists without any linear assumptions (Forger 2011).

Setting the real part equal to zero and substituting for w^2, we have:

$$(q_2 + q_1 + d_0)(q_1 d_0 + q_1 q_2 + q_2 d_0) = r_n k_{max} k_l r_1 + q_1 q_2 d_0$$

or

$$(q_2 + q_1 + d_0)\left(\frac{1}{q_1} + \frac{1}{q_2} + \frac{1}{d_0}\right) = \frac{r_n k_{max} k_l r_1}{q_1 q_2 d_0} + 1.$$

Since all rate constants have units of 1/time, all terms in the preceding equation are dimensionless. We also note that it is symmetric in q_1, q_2 and d_0.

What is the minimum value of r_n for which this equation is satisfied? By symmetry, this minimum must have $q_1 = q_2 = d_0 \equiv c$. Thus, the minimum value of r_n is

$$r_n > 8 \frac{c^3}{k_{max} k_l r_1} \quad \text{or} \quad r_n > 8 \frac{q_1 q_2 d_0}{k_{max} k_l r_1}.$$

This is often called the secant condition (Tyson and Othmer 1978).

Now also note that around the fixed point $q_1 q_2 d_0 \bar{P}_n = k_{max} k_l r_1 f(\bar{P}_n)$. Let $\bar{r}_n = f(\bar{P}_n)/\bar{P}_n$.

We find that $r_n / \bar{r}_n > 8$. Since $\tilde{f}(P_n) \approx -r_n \tilde{P}_n$, it is more appropriate to consider $-(r_n / \bar{r}_n) < -8$. Now what is the meaning of $-r_n / \bar{r}_n$? For the following discussion, consider the case where we have linearized around a fixed point to find r_n. Near the fixed point $\bar{P}_n$, f is linear and $df/dP_n = -r_n$. So, $-(r_n/\bar{r}) = \left(df(P_n)/f(\bar{P}_n)\right)\left(\bar{P}_n/dP_n\right) < -8$.

To illustrate this, we simulate the Goodwin model with different values of m and plot the minimum and maximum of the rhythms (see figure 4.3). Rhythms appear (meaning the minimum and maximum are different) only when $m \geq 8$.

The quantity $\left(df(P_n)/f(\bar{P}_n)\right)\left(\bar{P}_n/dP_n\right)$ is typically called the sensitivity of f at $\bar{P}_n$; this result was first found by Thron (1991). Note that sensitivity is sometimes denoted $d\log f(P_n)/d\log P_n$. Finally, in our case, $-r_n/\bar{r}_n$ is the ratio of the amplification of the amplitude of the oscillation of P_n to the amplification of the steady-state value of P_n. To understand sensitivity, let us consider some examples:

Example 1: $f(x) = x$. Here the sensitivity is clearly 1.

Example 2: $f(x) = cx$. Here the sensitivity is also 1.

Example 3: $f(x) = x^m$. Here the sensitivity is m.

Example 4: $f(x) = cx^m$. Here the sensitivity is also m.

Example 5: $f(x) = 2 - x$. Here the sensitivity at $x = 1$ is -1.

Two further examples of the sensitivity of transcription regulation functions are shown in figure 2.3.

What is the biological meaning of the sensitivity of f? First note that, if k_f/k_r is large, transcription occurs at a rate $\approx \left(k_{max}k_r\right)/\left(k_f P_n^m\right)$. The sensitivity is then $-m$. Thus, for the Goodwin system to oscillate, $m \geq 8$. This implies that a complex of 8 molecules of the repressor must bind to the DNA. Having so many repressor molecules act together is likely to be rare in real life.

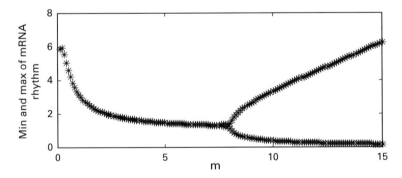

Figure 4.3
The minimum and maximum of the rhythms for the Goodwin model with different values of m, the Hill coefficient. When m is less than 8, oscillations do not appear, as shown by the minimum and maximum being the same.

What if k_f/k_r is not large? Let $f^0(P_n) \equiv (k_{max}k_r)/(k_f P_n^m)$, and recall that $f(P_n) = k_{max}/(1 + (k_f/k_r)P_n^m)$. We wish to show that the sensitivity of $f(P_n)$ is always less than the sensitivity of $f^0(P_n)$.

We may calculate directly the sensitivity of $f(P_n)$ and find that it is $-m(k_f/k_r)P_n^m/(1 + (k_f/k_r)P_n^m)$, which is greater than $-m$. Oscillations occur only when the sensitivity is < -8, so m must be > 8. Historical note: In Goodwin's original paper (1966), he shows oscillations for $m = 1$. This is clearly wrong!

We may summarize what the linearization does and does not tell us:

1. We found that only select biochemical mechanisms can yield oscillations in the Goodwin system (e.g., 8 or more repressor molecules forming a complex).

2. We found a simple relationship for the approximate frequency of the oscillator:

$$w = \sqrt{q_1 d_0 + q_1 q_2 + q_2 d_0} \,.$$

3. We have no clue about how some parameters (e.g., k_{max}) affect the period, other than that their effect might be small.

4. We have no estimates for the amplitude of the oscillations.

Biological observations help find a solution: Surprisingly, the Kim-Forger model (Kim and Forger 2012) does not need such strict requirements for transcription regulation mechanisms such as having 8 molecules bind together to achieve a high sensitivity and oscillations (see figure 2.3). The model oscillates, so long as K_d is low and the activators and repressors have similar concentrations. This may explain why the transcription regulation mechanism described by Kim and Forger has been previously found in many natural circadian clocks.

Finally, we note that different transcription regulation functions can give rise to different types of behaviors in a feedback loop. Figure 4.4 shows an example of the Goodwin oscillator with five elements and three possible transcription regulation functions. Each transcription regulation function contains two binding sites for a transcription factor. In the top panels, we see a case where oscillations can coexist with a steady state, depending on the initial conditions. In the middle panel, we see oscillations whose period changes drastically for a small change in the initial conditions. In the bottom panels we see chaotic oscillations. This further illustrates the role that transcription regulation can play in shaping rhythms.

4.4   Frontiers: Futile Cycles Diminish Oscillations, or Why Clocks Like Efficient Complex Formation

One biochemical mechanism that impedes efficient complex formation is futile cycles. To illustrate this, consider a protein, X, which can act as a repressor, as in the Goodwin model.

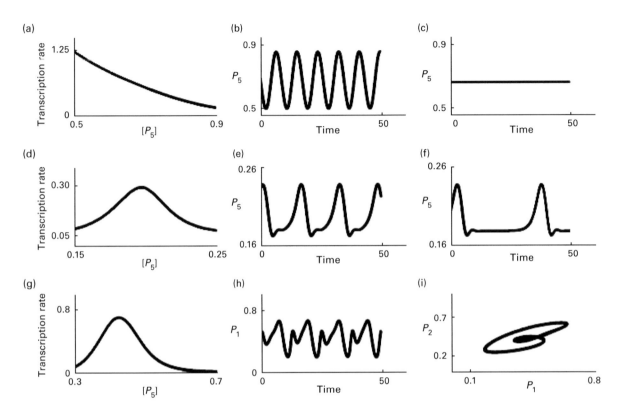

Figure 4.4
Example of behaviors that can be seen in the Goodwin model with 5 elements for different transcription regulation functions. Example transcription regulation functions are shown on the left and are derived from a simple
gene with two binding sites (DeWoskin et al. 2014), with the middle and right panels showing the behaviors of
the system. Graphs (*b*) and (*c*) show that oscillations or quiescence can be seen depending on the initial conditions. Graph (*e*) uses the same transcription regulation function, but with the overall transcription rate increased
by 5 percent. This results in a near doubling of the period (*f*), showing how the period can carefully depend on
individual parameters. Graphs (*h*) and (*i*) illustrate chaos. Taken from DeWoskin et al. (2014).

We assume that the protein can be converted to a form Y, which is inactive unless it is converted back to X. This can be represented by the following equations:

$$dX/dt = s(t) - dX - r_f X + r_r Y,$$

$$dY/dt = r_f X - r_r Y - qY,$$

where $s(t)$ is the production rate of X, d is the clearance rate of X, r_f is the conversion rate of
X to Y, r_r is the conversion rate of Y to X, and q is the clearance rate of Y. Taking the Fourier
transform of these equations, we have

$$iw\tilde{Y} = r_f \tilde{X} - (r_r + q)\tilde{Y},$$

and (after substitution of the preceding equation)

$$iw\tilde{X} = \tilde{s} - \tilde{X}(d + r_f) + \frac{r_f r_r}{iw + r_r + q}\tilde{X}$$

We let $\rho = r_f r_r$ and $l = r_r + q$, and we assume $d + r_f$ is 1 by scaling time. We then have

$$\frac{\tilde{s}}{\tilde{X}} = iw + 1 - \frac{\rho}{iw + l}$$

The steady state of this equation can be found where $w = 0$, which yields that the steady state of s divided by the steady state of X is $1 - \rho/l$, where we assume that $\rho/l < 1$. Scaling the preceding equation by this factor, we have

$$\left(\frac{\tilde{s}}{\tilde{X}}\right)\left(\frac{\overline{X}}{\overline{s}}\right) = \frac{iw + 1 - \dfrac{\rho}{iw + l}}{1 - \dfrac{\rho}{l}}$$

As before in figure 4.2, we next look at a phase/amplitude plot of this equation. This is shown in figure 4.5 for random choices of ρ and l. This curve is bounded by the curve of $\rho = 0$, which would happen if there was no futile cycle.

We can see this analytically by expressing the preceding numerator as

$$iw + 1 - \frac{\rho}{iw + l} = 1 - \frac{\rho l}{w^2 + l^2} + iw\left(1 + \frac{\rho}{w^2 + l^2}\right).$$

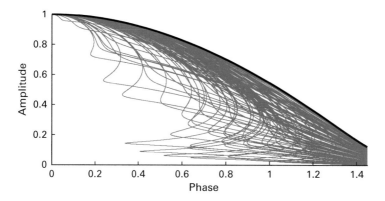

Figure 4.5
Phase-amplitude relationship for random choices of parameters of a futile cycle. The result of including a futile cycle, whose parameters were chosen randomly, is shown in blue. The case where no futile cycle is included is shown in black.

We now consider what happens to the phase and amplitude of this equation as w increases from 0. If $\rho = 0$, the imaginary part of this equation grows, while the real part is unaffected as w increases. However, with $\rho > 0$, the real part decreases as w increases, which causes the amplitude for any given phase to decrease and be bounded by the case of $\rho = 0$.

As discussed previously, oscillations require a proper phase relationship between the rhythms of the system variables. Including a futile cycle will decrease the likelihood of oscillations since, to achieve a given phase shift, the amplitude of the rhythm will attenuate more than if the futile cycle was not present. However, additional nonlinear feedback loops, particularly those capable of bistability, can yield oscillations, as seen in section 4.8.

Most biochemical clocks consist of feedback loops made up of protein complexes. While experimental techniques can identify which proteins are in these complexes, very little is known about how the complexes form. This detail is typically ignored in mathematical models, which usually use a rapid equilibrium assumption to reduce the complexity of the model. Thus, modelers often consider only the initial proteins and the final complex, without considering intermediates. This is understandable, since the number of possible intermediate complexes of n proteins is 2^n. Simulating 2^n complexes, where n can be five or more, can be computationally challenging, not only because of the 2^n differential equations but also, and more important, because of the growth in the number of terms in these equations as n gets larger. DeWoskin et al. (2014) sought to determine if anything was lost when the details of complex formation are not considered. Considering four types of models, they found that the details of protein complex formation are very important. Part of these results could be explained by the addition of futile cycles, as many of the reactions in protein complex formation form futile cycles.

4.5   Example: Case Study on Familial Advanced Sleep Phase Syndrome

In a well-studied sleep disorder, familial advanced sleep phase syndrome (or FASPS for short), patients wake up very early each morning (e.g., 4:30 AM) and cannot go back to sleep (Toh et al. 2001). If patients were allowed to live without time cues, they would have a short period (< 24 hours). We shall see later why a short period causes FASPS patients to wake up early in the morning. For the moment, we are interested in understanding what causes this short period.

The human circadian clock is a genetic transcription-translation feedback loop. The genetic mutation that causes FASPS affects how a kinase phosphorylates the PERIOD (PER) protein (Toh et al. 2001). Based on experiments, there are 3 possible effects of this mutation:

1.   It could cause PER to be more stable.
2.   It could cause PER to enter into the nucleus of the cell more slowly.
3.   It could decrease the rate at which PER is blocked from entering the nucleus of the cell.

Based on the techniques we have learned thus far, we wish to determine which of the three possible effects of this mutation cause a short period in FASPS patients.

We begin with the Goodwin oscillator. This is a great oversimplification of the full mammalian circadian clock, but it does have features that resemble the main feedback loop. At the heart of this clock are the PER proteins, which, with the CRY proteins, act as repressors to their own transcription. The details of this transcription regulation do not matter for our analysis here. Thus, we represent this PER feedback loop with the Goodwin system. Here P_1 is protein in the cytoplasm of the cell, P_2 is protein in the nucleus of the cell, and r_1 is the rate of transport of PER protein from the cytoplasm of the cell to the nucleus of the cell.

$$\frac{dM}{dt} = \frac{k_{max}}{1+\left(k_f/k_r\right)P_2^m} - d_0 M,$$

$$\frac{dP_1}{dt} = k_l M - d_1 P_1 - r_1 P_1,$$

$$\frac{dP_2}{dt} = r_1 P_1 - d_2 P_2.$$

Finally, we add a new state into the model, P_1^*, which is a state of the PER protein that cannot enter into the nucleus of the cell. Thus, our equations become:

$$\frac{dM}{dt} = \frac{k_{max}}{1+\left(k_f/k_r\right)P_2^m} - d_0 M,$$

$$\frac{dP_1}{dt} = k_l M - d_1 P_1 - r_1 P_1 - b_1 P_1,$$

$$\frac{dP_2}{dt} = r_1 P_1 - d_2 P_2,$$

$$\frac{dP_1^*}{dt} = b_1 P_1 - d_1 P_1^*.$$

We use the same techniques as before to analyze this system. First, we represent the system in the following form:

$$\left(\frac{d}{dt}+d_0\right)M = \frac{k_{max}}{1+\left(k_f/k_r\right)P_2^m},$$

$$\frac{1}{k_l}\left(\frac{d}{dt}+q_1\right)P_1 = M,$$

$$\frac{1}{r_1}\left(\frac{d}{dt}+d_2\right)P_2 = P_1,$$

$$\frac{1}{b_1}\left(\frac{d}{dt}+d_1\right)P_1^* = P_1,$$

with $q_1 = d_1 + r_1 + b_1$. We can take the Fourier transform and get

$$(iw+d_2)(iw+q_1)(iw+d_0)\tilde{P}_2 = k_{max}k_l r_1 \tilde{f}(P_2),$$

$$\frac{1}{b_1}(iw+d_1)\tilde{P}_1^* = \tilde{P}_1.$$

The second equation can be ignored. We have encountered the first equation before. It has a solution we found previous (4.3.1):

Consequences of CKI phosphorylation
  Degradation of PER1 and PER2
  Delayed nuclear entry of PER1
  Accelerated nuclear entry of PER1

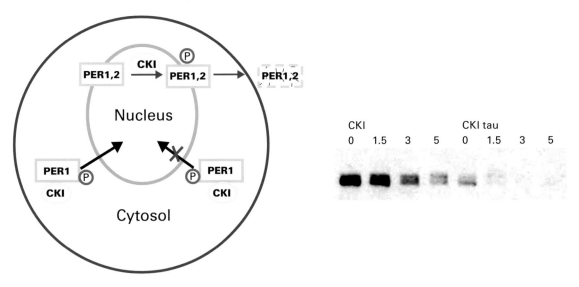

Figure 4.6
Determining the effect of the tau mutant. (Left) Possible effects of the tau mutant of PERIOD2 (PER2) protein levels. (Right) Modeling led to the prediction that tau increases, not decreases, the degradation rate of PER2 by CKI. The data shown are PER2 levels for the indicated hours after PER2 translation was blocked. Indeed PER2 degrades quicker in the presence of tau. See Gallego et al. (2006), Virshup et al. (2007), and Zhou et al. (2015) for more details. Data courtesy of David Virshup.

$$w = \sqrt{q_1 d_0 + q_1 d_2 + d_2 d_0}.$$

Let us substitute for q_1. We have:

$$w = \sqrt{(d_1 + r_1 + b_1)d_0 + (d_1 + r_1 + b_1)d_2 + d_2 d_0}.$$

Now, the FASPS mutation can either cause slower degradation of PER (d_1 or d_2 decreases), slower nuclear entry of PER (r_1 decreases) or slower blocking of the nuclear entry (b_1 decreases). In all cases, this causes w to decrease, or the period to increase. Thus, none of the proposed explanations for FASPS are valid—a situation that leads to the prediction that the mutation may be acting in an opposite way to what was thought. This prediction was experimentally validated for an animal model of FASP, the *tau* mutant hamster, which was the first mammalian circadian mutant discovered (Gallego et al. 2006) (see figure 4.6).

4.6 Frontiers: An Additional Fast Positive Feedback Loop

Many biochemical oscillators have a combination of two feedback loops, one fast, and one slow. The fast feedback loop typically contains an activator, and the slow feedback loop contains a repressor. In such a case, it is reasonable to assume that the dynamics of the mRNA of the activator and repressor are the same. However, we assume that the activator itself is short-lived (fast degradation) and thus is "fast." First, let us rethink our earlier model of transcription. The model developed in section 2.5 does not take into consideration the activator. Instead, it assumes a maximal rate of transcription (k_{max}), and takes the fraction of time the gene is "on" (or the probability that a repressor complex is not bound) to be $1/\left(1 + \left(k_f/k_r\right)P_2^m\right)$. Now assume that, in addition to no repressor complex being bound, an activator molecule must also be bound. Here, we write a differential equation for the probability that a particular binding site is *occupied* (G):

$$\frac{dG}{dt} = -k_{ra}G + k_{fa}(1 - G)A_2,$$

where A_2 is the activator that can bind to DNA. Note that this is the *opposite* of the equation derived in section 2.5 for the probability a site is free. As before, we can take this at equilibrium, and here find that

$$k_{ra}G = k_{fa}(1 - G)A_2,$$

$$G = \frac{A_2}{\dfrac{k_{ra}}{k_{fa}} + A_2}.$$

So the total rate of transcription here is

$$\frac{t_{max} A_2}{\left(1+\dfrac{k_f}{k_r} P_2^m\right)\left(\dfrac{k_{ra}}{k_{fa}} + A_2\right)}.$$

With the assumption that the mRNA of A and P have the same dynamics, we have

$$\frac{dM}{dt} = \frac{k_{max} A_2}{\left(1+\left(k_f / k_r\right) P_2^m\right)\left(\left(k_{ra} / k_{fa}\right) + A_2\right)} - d_0 M,$$

$$\frac{dP_1}{dt} = k_l M - d_1 P_1 - r_1 P_1,$$

$$\frac{dP_2}{dt} = r_1 P_1 - d_2 P_2,$$

$$\frac{dA_1}{dt} = k_{1A} M - d_{1A} A_1 - r_{1A} A_1,$$

$$\frac{dA_2}{dt} = r_{1A} A_1 - d_{2A} A_2.$$

Now assume that the activator is unstable (i.e., that $d_{1A}, d_{2A} \gg 0$). In this case, we can assume that A_1 and A_2 reach equilibrium quickly. At equilibrium, we have

$$\frac{k_{lA}}{d_{1A} + r_{1A}} M = A_1,$$

$$\frac{r_{1A}}{d_{2A}} A_1 = A_2,$$

and our equations become

$$\frac{dM}{dt} = \frac{\dfrac{k_{max} k_{lA} r_{1A}}{(d_{1A} + r_{1A}) d_{2A}} M}{\left(1+\dfrac{k_f}{k_r} P_2^m\right)\left(\dfrac{k_{ra}}{k_{fa}} + \dfrac{k_{lA} r_{1A}}{(d_{1A} + r_{1A}) d_{2A}} M\right)} - d_0 M,$$

$$\frac{dP_1}{dt} = k_l M - d_1 P_1 - r_1 P_1,$$

$$\frac{dP_2}{dt} = r_1 P_1 - d_2 P_2.$$

Now, since $d_{1A}, d_{2A} \gg 0$, it is reasonable to approximate the preceding as

$$\frac{dM}{dt} = \frac{\dfrac{k_{max}k_{1A}r_{1A}}{(d_{1A}+r_{1A})d_{2A}}M}{\left(1+\dfrac{k_f}{k_r}P_2^m\right)\left(\dfrac{k_{ra}}{k_{fa}}\right)} - d_0 M, \tag{4.6.1}$$

$$\frac{dP_1}{dt} = k_l M - d_1 P_1 - r_1 P_1,$$

$$\frac{dP_2}{dt} = r_1 P_1 - d_2 P_2.$$

Finally, let

$$\frac{\dfrac{k_{max}k_{1A}r_{1A}}{(d_{1A}+r_{1A})d_{2A}}}{\left(\dfrac{k_{ra}}{k_{fa}}\right)} \equiv k_t.$$

We now have

$$\frac{dM}{dt} = \frac{k_t M}{\left(1+\left(k_f/k_r\right)P_2^m\right)} - d_0 M,$$

$$\frac{dP_1}{dt} = k_l M - d_1 P_1 - r_1 P_1,$$

$$\frac{dP_2}{dt} = r_1 P_1 - d_2 P_2.$$

First, let us solve for the steady-state values of this equation:

$$\frac{k_t}{\left(1+\left(k_f/k_r\right)\bar{P}_2^m\right)} = d_0$$

$$k_l \bar{M} = d_1 \bar{P}_1 + r_1 \bar{P}_1$$

$$r_1 \bar{P}_1 = d_2 \bar{P}_2.$$

Note that this is a different structure than the previous equations. In particular, the steady-state value of P_2 does not depend on M. Let us now linearize the rate of transcription around the fixed point. Ignoring the constant terms (which are zero at the fixed point), we can approximate:

$$\frac{k_t M}{(1+(k_f / k_r)\mathrm{P}_2^m)} \approx M \frac{k_t}{(1+(k_f / k_r)\bar{\mathrm{P}}_2^m)} - r_n P_2 = d_0 M - r_n P_2.$$

Using this approximation, we find that

$$\frac{dM}{dt} = -r_n P_2 ,$$

$$\frac{dP_1}{dt} = k_t M - d_1 P_1 - r_1 P_1 ,$$

$$\frac{dP_2}{dt} = r_1 P_1 - d_2 P_2 .$$

Let $q_1 = d_1 + r_1$, and we now find by taking the Fourier transform,

$$(iw)(iw + q_1)(iw + d_2) = -r_n r_1 k_l ,$$

$$(iw)(-w^2 + iw(q_1 + d_2) + q_1 d_2) = -r_n r_1 k_l ,$$

$$(-w^2 (q_1 + d_2) + iw(q_1 d_2 - w^2)) = -r_n r_1 k_l ,$$

and setting the imaginary part equal to zero, we have $w = \sqrt{q_1 d_2}$. With this, we find the real part yields

$$q_1 d_2 (q_1 + d_2) = r_n r_1 k_l .$$

Note that this gives no restraint on the cooperativity. Thus, in this case, oscillations are much more likely.

Summarizing, if we assume linear dynamics, an additional positive feedback loop yields a possible oscillation, even with low m, with a fixed period and undetermined (at least by linear theory) amplitude.

4.7   Example: Increasing Activator Concentrations in Circadian Clocks

The *Drosophila* circadian clock contains an activator (*Pdp1*) and a repressor (PER-TIM) in a feedback loop similar to the preceding model. However, it has been noted that increasing the production of the activator (*Pdp1*) causes the period to decrease (Cyran et al. 2003). We now wish to understand this. First, note that the estimate of the period in section 4.6, $w = \sqrt{q_1 d_2}$, does not contain the rate of transcription of the activator, so an improved estimate is needed. One possibility is to reconsider our assumption (4.6.1) that

$$\frac{k_t M}{\left(1+\frac{k_f}{k_r}P_2^m\right)\frac{k_{ra}}{k_{fa}}} \approx \frac{k_t M}{\left(1+\frac{k_f}{k_r}P_2^m\right)\left(\frac{k_{ra}}{k_{fa}} + \frac{k_{1A}r_{1A}}{(d_{1A}+r_{1A})d_{2A}}M\right)}$$

In fact, without this assumption, we find that the activation by the activator saturates (i.e., there is a maximal transcription rate that occurs when the activator is bound 100 percent of the time). With the preceding assumption, we found

$$
\frac{k_t M}{\left(1+\frac{k_f}{k_r}P_2^m\right)\left(\frac{k_{ra}}{k_{fa}}+\frac{k_{lA}r_{1A}}{(d_{1A}+r_{1A})d_{2A}}M\right)} \approx M\frac{k_t M}{\left(1+\frac{k_f}{k_r}\bar{P}_2^m\right)}-r_n P_2 = d_0 M - r_n P_2.
$$

Without it, we find

$$
\frac{k_t M}{\left(1+\frac{k_f}{k_r}P_2^m\right)\left(\frac{k_{ra}}{k_{fa}}+\frac{k_{lA}r_{1A}}{(d_{1A}+r_{1A})d_{2A}}M\right)} \approx b_0 M - r_n P_2,
$$

where $b_0 < d_0$. We then find

$$
\frac{dM}{dt} = -r_n P_2 - (d_0 - b_0)M,
$$

$$
\frac{dP_1}{dt} = k_1 M - d_1 P_1 - r_1 P_1,
$$

$$
\frac{dP_2}{dt} = r_1 P_1 - d_2 P_2.
$$

We can then use our original formula (4.3.1) for the period to obtain

$$
w = \sqrt{(d_1 + r_1)(d_0 - b_0) + (d_1 + r_1)d_2 + d_2(d_0 - b_0)}.
$$

So, as more activator is present, transcription becomes more saturated, b_0 approaches 0, $d_0 - b_0$ becomes greater, w increases, and the period decreases.

4.8   Bistability and Relaxation Oscillations

With a fast positive feedback loop added to a slow negative feedback loop, the opposite behavior to what was described before can also occur: oscillations with a fixed amplitude and a period that varies depending on the clearance rates of the repressor. This clock design is based on the idea of genetic switches (Novak and Tyson 2008). Switches have been implicated as part of the oscillatory mechanism of many systems, most particularly the cell cycle (Novak and Tyson 2008). The cell cycle is designed such that the cell makes irreversible decisions on whether to replicate DNA or divide. We saw in section 4.6 that positive feedback can help generate oscillations. However, this positive feedback contained a Hill coefficient of 1 or a sensitivity of at most 1. If this positive feedback has higher sensitivity than 1, bistability, or having more than one stable steady state, can occur. We first study how bistability arises.

Consider a simple feedback loop where an activator activates itself. We assume that the activator forms a complex of m molecules of activator that activates the gene. We also assume that the gene is "leaky" in that a small (k_r) amount of transcription occurs regardless of whether the activator is bound. Thus, the transcription rate would be

$$k_r + \frac{A^m}{A^m + K^m}.$$

Next, assume that the mRNA is at rapid equilibrium with the activator, or that the activator protein is simply a constant times the concentration of its transcript. This yields the following equation:

$$\frac{dA}{dt} = k_r + \frac{A^m}{A^m + K^m} - \gamma A.$$

There are three key plots that can help understand the behavior of this simple system. The first is the rate balance plots where we plot both the production rate and clearance rate of A on the same plot. When the two curves intersect, a steady state can be found. Three examples, for different values of γ can be seen in figure 4.7.

In the leftmost plot, γ is high and there is only one steady state of A that occurs when A is low. For moderate values of γ there are three steady states. It is not difficult to determine that the low and high values are stable with an unstable steady state in between. In the rightmost plot, corresponding to low γ, only one steady state exists, when A is high.

The second way to visualize the dynamics is simply to plot the rate of change of A as a function of A. Steady states are shown for the same parameters as the middle of figure 4.7 in figure 4.8.

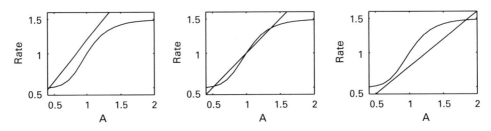

Figure 4.7
Production rate and degradation rate for a chemical species. When they intersect, steady states occur. Since the production rate has a sigmoidal shape, changing the degradation rate causes the number of steady states to change from 1 to 3, producing bistability. Here $k_r = 0.5$, $m = 5$, K $= 1$, and $\gamma = 1.5$ (left), 1 (middle) and 0.5 (right).

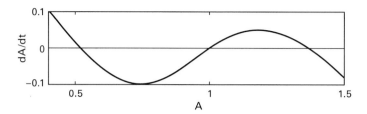

Figure 4.8
The rate of change of A for the middle plot shown in figure 4.7. Note the three steady states where the curve crosses $dA/dt = 0$. Parameters are as in figure 4.7 with $\gamma = 1$.

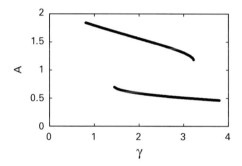

Figure 4.9
Bifurcation diagram showing stable steady states of A as a function of γ for a model related to that used in figure 4.8.

Again, this shows three values where $dA/dt = 0$. The highest and lowest values of A where $dA/dt = 0$ are stable in that if A is slightly higher than these values, $dA/dt < 0$, and if A is slightly lower, $dA/dt > 0$. By similar arguments, the middle steady state of A is unstable.

Figure 4.9 illustrates the behavior of A as a bifurcation diagram, which is a graph of all the stable steady states of the model as a parameter is varied. Here A is shown, with its steady states plotted as a function of γ.

Again, one can clearly see a region of bistability. This bistability exists for a range of parameter values. The range can be increased by having two interlocked positive feedback loops, a design seen in many cellular systems; see the work of Chang et al. (2010) for more details.

Now let us assume that a repressor binds to the activator and acts as a catalyst for its degradation (other assumptions also work, for example, having the repressor inactivate the activator). Let us also assume that the activator also activates the repressor. This yields the following equations:

$$\frac{dA}{dt} = k_r + \frac{A^m}{A^m + K^m} - (\gamma + R)A,$$

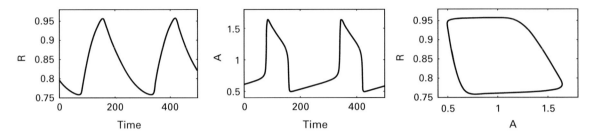

Figure 4.10
Simulations of the activator and repressor model. (Left) The repressor concentration and (middle) the activator concentration. (Right) The activator and repressor (scaled by 10) concentrations shown alongside the bifurcation diagram of figure 4.9. Here we used $k_r = 0.5$, $m = 6$, $K = 1$, $\gamma = 0.1$, and $b = c = 0.005$.

$$\frac{dR}{dt} = bA - cR.$$

Again, this has a double feedback loop structure with a positive feedback loop (activator activating itself) and a repression feedback loop (repressor degrading its activator). The solutions of this model are shown in figure 4.10.

This behavior is called relaxation oscillations. On the left and center of figure 4.10, we find the repressor and activator concentrations plotted with respect to time. We see that when the activator is high, we have an exponential rise in the repressor concentration, and when the activator is low, there is an exponential decay in its concentration. On the right plot of figure 4.10, we plot a parametric plot of the repressor concentration as well as the activator concentration. The activator concentration follows its steady states as indicated on the bifurcation diagram. When it is high (on the upper branch of the bifurcation diagram), the repressor concentration increases until it gets too high for the upper branch. At this point the activator concentration quickly decreases to its lower branch, and the repressor concentration decreases. This happens until the repressor concentration is too low for the lower branch, at which point the activator concentration quickly increases to the upper branch, and the cycle repeats.

4.9   Frontiers: Calculating the Period of Relaxation Oscillations

Here, we calculate the period of a simple negative feedback loop with an additional fast positive feedback loop, as described in section 4.8. It is important to note that similar analysis has been performed on neuronal models, e.g., the Morris-Lecar model (Ermentrout et al. 2010). Thus, relaxation oscillations occur in many systems, and the analysis presented here can easily be extended to other systems. Other estimates of the period of relaxation oscillations can be found (Pontryagin 1957).

We first consider the fast positive feedback loop, which describes the activator concentration:

$$\frac{dA}{dt} = f(A) - c(R)A$$

This is assumed to occur on a fast time scale and $df/dA > 0$. On a slower time scale, we assume the clearance rate of A changes in accordance with a slowly changing repressor concentration R. Note that this occurs for the model described in section 4.8:

$$\frac{dA}{dt} = k_r + \frac{A^m}{A^m + K^m} - (\gamma + c(R))A,$$

$$\frac{dR}{dt} = bA - cR.$$

As described earlier, there are three regions to consider: when R is low or high, there is just one steady state, but for intermediate R, both a high state and a low state exist for A. When A is increasing (because R is decreasing on a slow time scale), we assume that A tracks the lower steady state until it is lost at a critical point. Then A switches to the higher steady state, and R increases, lowering A until the high steady state is lost at another critical point. Steady states are lost when $f(A)$ and $c(R)A$ are tangent, which means that since $c(R)A$ is a locally linear function of A,

$d\log(c(R)A)/d\log A = 1$,

$d\log f/d\log A = 1$.

Also, at these points $f(A) = c(R)A$, which means that

$\log(f(A_{max})) = \log(c(R_{min})) + \log(A_{max})$

and

$\log(f(A_{min})) = \log(c(R_{max})) + \log(A_{min})$.

Subtracting the last two equations yields

$\log(f(A_{min})) - \log(f(A_{max})) - \log(A_{min}) + \log(A_{max}) = \log(c(R_{max})) - \log(c(R_{min}))$.

The left-hand side can be written as

$$-\int_{\log(A_{min})}^{\log(A_{max})} \left(\frac{d\log f}{d\log A} - 1 \right) d\log A,$$

which is equal to $\log(c(R_{max})/c(R_{min}))$. We can find the region where $(d \log f / d \log A) > 1$, determining A_{max} and A_{min}. The integral can then be used to determine $\log(c(R_{max})/c(R_{min}))$, which will tell us R_{max} and R_{min}.

Now assume the repressor obeys the equation

$$dR/dt = \beta A - \gamma R.$$

We can now calculate the period. Consider the case where $R = R_{min}$ and $A = A_{max}$. We then have

$$dR/dt = \beta A_{max} - \gamma R$$

and $R(0) = R_{min}$. With $R_{ss} = \beta A_{max}/\gamma$, the amount of time to get to R_{max} can be worked out to be

$$\tau_1 = (1/\gamma) \log((R_{min}-R_{ss})/(R_{max}-R_{ss})).$$

A switch then occurs, and we have

$$dR/dt = \beta A_{min} - \gamma R,$$

and with $R_{ss} = \beta A_{min}/\gamma$ now, this state lasts for

$$\tau_2 = (1/\gamma) \log((R_{max}-R_{ss})/(R_{min}-R_{ss})).$$

The period of the oscillation is thus $\tau_1 + \tau_2$.

4.10   Theory: The Global Secant Condition

Consider the model

$$\frac{dp_i}{dt} = f_{i-1}(p_{i-1}) - g_i(p_i), \quad 1 \leq i \leq n, \tag{4.10.1}$$

with $f_0(p_0) \equiv f_n(p_n)$ giving a feedback loop and f_i and g_i continuous.

Strong mathematical results called secant conditions, going back almost 50 years (Goodwin 1966; Sontag 2006; Tyson and Othmer 1978), show that many biochemical feedback loops of the form (4.10.1) also do not show rhythms. If the secant condition is violated, the system must go to a steady state. It was originally determined as a condition for where a steady state would become unstable. However, it is also possible to have oscillations and a stable steady state, meaning this local condition may not be sufficient to rule out oscillations. Several global secant conditions exist that, if violated, guarantee that oscillations will not be seen. The strongest is by Forger (2011), which will be described here.

The global secant condition: For systems of the form (4.10.1), oscillations occur only if

$$\Pi_i \frac{\|f_i\|^2}{\langle f_i, g_i \rangle} \geq \sec(\pi/n)^n,$$

with τ the period of the oscillations $\langle x \rangle \equiv (1/\tau) \int_t^{t+\tau} x \, dt'$ (i.e., the mean of x) and the $\|f_i\|^2 = \langle f_i^2 \rangle$ (i.e., the amplitude of f_i). We also let $\langle x, y \rangle \equiv (1/\tau) \int_t^{t+\tau} xy \, dt'$. Here, we give an intuitive explanation of the left-hand side of the global secant condition.

We begin by subtracting off the mean of the rates so that $\langle f_i \rangle = \langle g_i \rangle = 0$, and $\|f_i\|^2 = \langle f_i^2 \rangle$ has the interpretation of the variance of f_i in the following sense. If we have oscillations, they have a set period τ; let us draw a random number between t and $t + \tau$ and determine the rates in the system f_i and g_i at that random time. The variance of this random rate is then

$$\|f_i\|^2 = \langle f_i^2 \rangle.$$

An added benefit of this approach is it is inherently stochastic, so it applies even when the number of molecules of any particular species is not large.

Consider the system after it has settled to its steady state. Assume we have oscillations, and that $\|f_i\|^2 = c$. It is shown in Forger (2011) that

$$\|f_i\|^2 - \|g_{i+1}\|^2 = \|dp_{i+1}/dt\|^2.$$

From p_{i+1} we can calculate $\|g_{i+1}\|^2$ and then $\|f_{i+1}\|^2$. Working our way around the feedback loop, we need to recover the same value $\|f_i\|^2 = c$ when we return to the original step of the feedback loop. If we do not recover the same variance, we cannot have oscillations.

Here is why one might imagine this variance might go down. The rate of production of p_i could vary greatly, like the rate of water flowing into a bathtub; however, the overall number of molecules of the species, like the overall water level, varies less since it reflects the production rate not at a single time, but averaged over time, and a clearance rate dependent only on the total number of molecules of the current species.

There is only one way to combat this decrease in variance. That is if the rate of production of the next species, which depends on the number of molecules of the current species, adds more variance than is found in the rate of clearance of the current species. The way to measure this is by the covariance of the production rate of the next species and the clearance rate of the current species. This is $\langle f_i, g_i \rangle$.

So we can measure the increase in variance between g_i and f_i by considering $\langle f_i, f_i \rangle / \langle f_i, g_i \rangle$. However, note that the only variance of g_i that matters in the fraction is that which is held in common with f_i. Thus, it actually measures the ratio of the variance of f_i to the variance of f_i that is contributed by g_i. For example, if $f_i = g_i$, then this fraction would be 1. If f_i and g_i were independent, then this fraction would be infinite. The secant condition

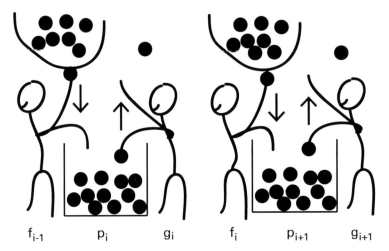

$$f_{i-1} \qquad p_i \qquad g_i \qquad f_i \qquad p_{i+1} \qquad g_{i+1}$$

Figure 4.11
An illustration of the principle behind the global secant condition.

gives a bound for how large this new variance needs to be in order to overcome the loss of variance by pooling the molecules created at different times.

The effects of pooling can be illustrated by the cartoon in figure 4.11. Here we animate f_i and g_i, by portraying them as individuals who are either creating molecules of the species (e.g., moving them into the volume of interest from a store) or clearing the molecules of the species. Note that the individuals who add molecules of a species to the volume look to the previous species to determine how many to add. The individuals who clear molecules of the species look only at the number of molecules of the species they clear. Normally, the system would go to a steady state. However, if the production rates are more "sensitive" than the degradation rates, an instability could be generated that would propagate through the feedback loop and create oscillations.

One final point about the secant condition is worth mentioning. As more steps are added to a feedback loop, oscillations are more likely to appear. We saw that delay equations are approximations of an infinite number of steps, so they artificially increase the likelihood of oscillations in feedback loops. Predictions from such models should be scrutinized.

Code 4.1 Effects of Feedback

(see figure 4.1)
```
%This code is provided to allow the reader to explore how phase and
%amplitude changes as one proceeds through a feedback loop
function z = drivenmammod(t, x)
global period
```

```
eps = 1;
z(1) = max(10 - x(20), 0) - x(1)+ eps*(1-sin(2*pi*t/period));
%z(1) = 0*max(10 - x(20), 0) - x(1)+ eps*(1-sin(2*pi*t/period));
%the above line is for when feedback is turned off
z(2) = x(1) - x(2);
z(3) = x(2) - x(3);
z(4) = x(3) - x(4);
z(5) = x(4) - x(5);
z(6) = x(5) - x(6);
z(7) = x(6) - x(7);
z(8) = x(7) - x(8);
z(9) = x(8) - x(9);
z(10) = x(9) - x(10);
z(11) = x(10) - x(11);
z(12) = x(11) - x(12);
z(13) = x(12) - x(13);
z(14) = x(13) - x(14);
z(15) = x(14) - x(15);
z(16) = x(15) - x(16);
z(17) = x(16) - x(17);
z(18) = x(17) - x(18);
z(19) = x(18) - x(19);
z(20) = x(19) - x(20);
z = z';
end
%The following lines should be entered at the command line.
%for ij = 1:500
%period = ij/10
%[T, X] = ode45(@drivenmammod, 0:0.01:1000, [1.1 1 1 1 1 1 1 1 1 1 1 1.1
%1 1 1 1 1 1 1 1]);
%amp(ij) = std(X(50001:100001, 20));
%meanp(ij) = mean(X(50001:100001, 20));
%end
```

Code 4.2 Effects of the Hill Coefficient on Rhythms in the Goodwin Model

(see figure 4.3)
```
%This code explores how the Hill coefficient determines when rhythms are
%present in the Goodwin model. It also gives a sample code for
%bifurcation diagrams that will be explored in chapter 5.
clear
global n % This will be the Hill coefficient
for ij = 1:150
n= ij/10
[T, X] = ode45(@good, 0:0.01:1000, [1 1 1]);
```

```
minm(ij) = min(X(90000:100000, 1)); %finds the min and max of the limit
%cycle
maxm(ij) = max(X(90000:100000, 1));
end
figure(1)
plot(.1:.1:15, maxm, '*k')
hold on
plot(.1:.1:15, minm, '*k')
% We also add the following line to Goodwin.m
%global n
% and change a later line to read
%Y(1) = 10/(.001^n + R^n) - M;
```

Exercises

General Questions
For these problems pick a biological model of interest, preferably one you study.

1. Determine the feedback loops in your model and relate them to the biological system you study. Does your system structure and behavior match any of the designs (e.g., double negative feedback loop) described in the chapter?

2. If your system shows relaxation oscillations, calculate (estimate) the period using methods shown in section 4.9. If your system does not show relaxation oscillations, calculate (estimate) the period of the system by the methods shown in section 4.3.

3. How are individual rates predicted to affect the period? Can you relate this to the biology of the underlying system?

4. Can you identify parts of the system that promote oscillations?

Specific Questions
1. Explain intuitively how a simple motif where a repressor binds to an activator can generate arbitrarily high sensitivity.

2. For the Goodwin model, what biochemical processes are the key determinants of the period?

3. For a hysteresis oscillator, what biochemical processes are the key determinants of the period?

4. Does a positive feedback loop need to show hysteresis to enhance the possibility of oscillation when added to a negative feedback loop?

5. Set the Hill coefficient of the Goodwin model to 1. Add in a positive feedback loop, and find parameters that allow for oscillations.

6. Consider the negative feedback loop with fixed f and g:

…

$$dP_i/dt = f(P_{i-1}) - g(P_i)$$

…

Using the global secant condition, show that this cannot generate feedback oscilla-tions. Show that adding an additional linear degradation term to this equation cannot cause oscillations.

7. Which of the following terms, when included in a biochemical feedback loop, promote oscillations?

   a.   A transcription term with a high Hill coefficient

   b.   A Michaelis-Menten degradation term

   c.   A fast positive feedback loop that creates bistability

   d.   A Michaelis-Menten translation term

II

BEHAVIORS

5

System-Level Modeling

Many biological clocks can be accurately modeled based on their mathematical structure. Here, we show how a bifurcation structure can be used to develop models. We focus on type 1 models, which arise from a saddle-node-on-invariant-circle bifurcation, and type 2 models, which arise from a Hopf bifurcation. We show how these two models lead to different predictions with regard to the oscillator's response to noise. Each prediction is tested using data from squid giant axon electrophysiology. We also develop a model of the human circadian and sleep-wake system illustrating how behaviors can be used to build models. Finally, we demonstrate how shared mathematical structures can be used to compare biochemical and electrophysiological models.

5.1 Introduction

Sometimes it is better to construct a high-level model without much biophysical detail. This is particularly useful if little is known about the mechanisms of timekeeping. In this chapter, we describe two general classifications of biological oscillators. From these principles, we generate high-level models that can be fit to data. This approach is different from that used in chapters 2–4, since we will not consider the detailed biophysical mechanisms in generating models.

Much of this chapter is devoted to understanding behaviors that are classified as type 1 or type 2 oscillators, a classification originally developed by Hodgkin (1948) for studying neuronal firing. He noted that these two types of systems showed different behaviors as parameters were changed at the onset of oscillations. Beyond just illustrating these principles, we aim to show how they can be used to develop models in practice. This is a different approach than taken in chapter 2.

As a way of illustrating this type of model building in some detail, and of introducing the van der Pol model in particular, we will build a model for the human circadian pacemaker in section 5.10–5.12. Since human circadian clocks are so important, we discuss light inputs to this model, as well as outputs that include the timing of sleep and being awake. We also show how these techniques can be applied to understand the firing of the squid giant axon.

5.2   General Remarks on Bifurcations

Bifurcations occur when a small change a system's parameters cause a large change in the system's behavior, for us typically values of parameters that are at the border of when oscillations occur and do not occur. This change in behavior is called a bifurcation. We will see that different types of bifurcations lead to different system behaviors. Without going into the advanced mathematical theory of bifurcations, we will present an overview of some of the simplest bifurcations, particularly those of importance to biological clocks. The reader is pointed to other texts for more mathematical details (Kuznetsov 2004). We consider bifurcations in two-dimensional models since, in a three-dimensional system, much more complicated bifurcations occur, and these can lead to complicated behaviors like chaos. No general theory of bifurcations in a three-dimensional system exists (Kuznetsov 2004).

Bifurcations are usually illustrated with bifurcation diagrams. In these diagrams, one of the parameters is plotted on the horizontal axis, and the steady states of the system are plotted on the vertical axis. An example of this is shown in figure 5.1, which is reproduced from (Barik et al. 2010). In this example, the total cell volume is changed, and the resulting behavior of the system is plotted. These diagrams could plot multiple steady states, or oscillatory states, where the peak and trough of the rhythm is plotted. Only sustained oscillations are plotted on this kind of graph. Sometimes bifurcation diagrams plot the period of an oscillator as the parameter of interest is varied. Both styles of diagram are illustrated in figure 5.1.

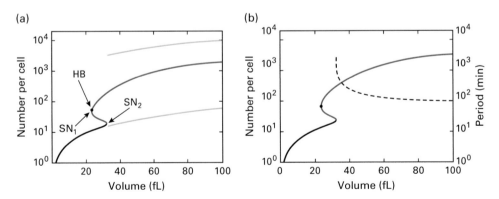

Figure 5.1
Example bifurcation diagrams. Two bifurcation diagrams showing the behavior of a mathematical model of the yeast cell cycle. (Left) Amount of model variable ClbM as the cell volume is increased. A Hopf bifurcation (HB) and Saddle-Node on Invariant Circle (SN) bifurcations are shown. These bifurcations are discussed later in the chapter. (Right) The period of the model. Taken from Barik et al. (2010).

It is somewhat surprising that bifurcations exist at all, for the following reason. Most models can be represented by differential equations. These differential equations are simply functions that take in the parameter values and the particular state of the system and determine how the state of the system changes with time. In mathematical terms,

$$\frac{dx}{dt} = \underline{f}(\underline{x}, \underline{c}).$$

In most models, this function is continuous (f is C^1 in mathematical parlance), in that a small change in the parameters will lead to a correspondingly small change in the rate of change of the variables. So imagine we are near the boundary between a region of parameters where oscillations occur and a region where they do not occur. Very close to this boundary, a small change in parameters can cause a very big difference in the overall system behavior, but, by continuity, it may alter the rate of change of any of the variables at a particular point by only a very small amount. However, as time progresses forward, the change in system state depends on the integral of the rate of change. Thus, a relatively small change in the rate of change with respect to time may add up to a large change over a large amount of time.

Typically, we are interested in the state of the system as time approaches ∞. These states may change drastically as parameter values change. For example, consider

$$dx/dt = x^2 - x^3 + \varepsilon.$$

We study this system in a similar way to the analysis presented in section 4.8. The plots of dx/dt vs. x for $\varepsilon = 0$ and $\varepsilon = 0.01$ are shown in figure 5.2.

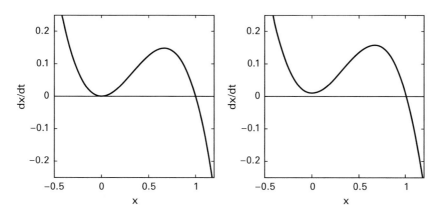

Figure 5.2
Plots of the dynamics of the system $dx/dt = x^2 - x^3 + \varepsilon$ for $\varepsilon = 0$ and $\varepsilon = 0.01$. The system evolves from having three steady states and the behavior of a switch to having just one steady state.

The steady state near $x = 1$ is stable since if x is slightly on the right of it, $dx/dt < 0$ (x will decay) and if x is slightly to the left of it, $dx/dt > 0$ (x will grow). When $\varepsilon < 0$, a second steady state appears just below $x = 0$. The system can act as a switch since it has two stable steady states, and transient perturbations can move the system from one to another. These switches are very important for many biochemical systems, and can determine whether cells choose one fate or another. As we saw in chapter 4, relaxation oscillations occur when the system switches from one state to another. Thus, switches can help generate oscillations.

The system has steady states at 0 and 1 for $\varepsilon = 0$. The state at 0 is neutrally stable meaning that if we are exactly at this state, we will remain there forever, but otherwise x will increase. For very small positive ε, the 0 steady state is lost. So, over time, the system will move toward the steady state near $x=1$. However, when ε is small, dx/dt is small around $x = 0$, and the system gets stuck there for a long time. This is sometimes called a "ghost," since the system still slows down near the steady state that has just disappeared as ε is increased. Although the steady state has officially gone, it still has some lingering effects, since removing it involved only a small change in the system's dynamics.

Because of these ghosts, studying bifurcations is particularly important. Bifurcations give an approximate description of the behavior of a system over a range of parameters, not just at the exact boundary between two states. Our focus will be more on the practical effects of these bifurcations, using a classification established by Rinzel and Ermentrout (1998), some elements of which were originally proposed by Hodgkin (1948).

5.3   SNIC or Type 1 Oscillators

Section 5.2 considered steady states and bifurcations in one dimension. The behavior of steady states in two dimensions is more complicated. In a two-dimensional situation, there are nodes that are analogous to the stable or unstable steady states in the one-dimensional situation. All initial conditions near a stable node approach the steady state, while all initial conditions near an unstable node are repelled from the steady state. Another possible type of node is a saddle, which is a steady state that in one direction attracts and repels in another direction. To understand how this is possible in the two-dimensional situation, consider the behavior of the system starting at states that lie on eigenvectors determined by the linear system and emanating from the steady state. We see in box 5.1 that if all the eigenvalues corresponding to these eigenvectors have negative real parts, then the fixed point will be a stable node. However, if some eigenvalues are positive real numbers and some are negative real numbers, some initial conditions will be attracted toward the steady state as $t \rightarrow \infty$, while others will be repelled from the steady state. Steady states that both attract and repel in this way are called saddles.

Just as a stable and unstable steady state can coalesce or emerge in one dimension, as described in section 5.2, a saddle and a node can also coalesce or emerge. In fact, in two

Box 5.1
Finding Stability of Fixed Points in Multidimensional Systems Using Eigenvalues

Consider the system

$$\frac{dx}{dt} = f(x).$$

Assume the system has a fixed point x_0 where $f(x_0) = 0$. We now construct a Taylor series around the point x_0,

$$\frac{dx}{dt} = f(x_0) + A(x - x_0) + \ldots,$$

where

$$A = \begin{bmatrix} \dfrac{\partial f_1}{\partial x_1} & \cdots & \dfrac{\partial f_1}{\partial x_n} \\ \cdots & \cdots & \cdots \\ \dfrac{\partial f_n}{\partial x_1} & \cdots & \dfrac{\partial f_n}{\partial x_n} \end{bmatrix}.$$

For x near x_0 we consider $x' \equiv x - x_0$, which has the following approximate behavior:

$$\frac{dx'}{dt} \approx A\, x'.$$

The solution of this equation can be constructed with the eigenvalues, λ_i, and eigenvectors, q_i, of A, which are defined by

$$A\, q_i = \lambda_i\, q_i.$$

If x' happens to be a multiple of an eigenvector, $x'(0) = c_i q_i$, then at future times

$$x'(t) = c_i e^{\lambda_i t} q_i.$$

We now express $x'(0)$ as a linear combination of the eigenvectors and find the following solution:

$$x'(t) = \sum_{i=1}^{n} c_i e^{\lambda_i t} q_i.$$

As $t \to \infty$, $x'(t)$ will $\to 0$ if the real part of all $\lambda_i < 0$, which means that solutions that are near x_0 will approach x_0. In this case, we say that x_0 is stable.

dimensions, they can do this in a way that can cause oscillations to appear. This bifurcation is called a saddle-node on invariant circle (or SNIC, so called because "invariant circle" is another name for a cycle). Before we explore this further, some historical details may be helpful.

Neurons that show this behavior are called type 1 in the nomenclature of Rinzel and Ermentrout (1998). Actually, these types of neurons were first considered by Hodgkin when he noticed that the period of repetitive firing of action potentials could show great variability over a small range of parameters (Hodgkin 1948). The classification is also applicable to genetic oscillators (Conrad et al. 2008; Guantes and Poyatos 2006). To help understand this bifurcation, let us consider a two-dimensional system in polar (r, ϕ) coordinates.

Assume that the dynamics in the radial direction do not change with the changing parameter, e.g.,

$$dr/dt = r(1 - r),$$

where the system tends to an amplitude of 1 unless $r = 0$. Looking at the ϕ direction, we can use a similar model to that described in section 5.2. Consider

$$d\phi/dt = (\phi - 0.5)^2 + \varepsilon. \tag{5.3.1}$$

We also assume that when ϕ reaches a threshold value (e.g., 1), it fires and resets to another value (e.g., 0). This model is illustrated in figure 5.3.

For ε positive, $d\phi/dt > 0$, and the model oscillates. For ε negative, no oscillations exist, and there is a steady state near 0.5 to which ϕ approaches.

For $\varepsilon = 0$, $d\phi/dt = 0$ at $\phi = 0.5$, and this point has neutral stability. If ε is close to, but just above, 0, the system will still be stuck by $\phi = 0.5$ for a long time. This is a "ghost" of the steady state as discussed in section 5.2. One can consider the $\varepsilon = 0$ case to be the case of infinite-period oscillations. For ε negative, we have one stable and one unstable state.

In this bifurcation, the fixed points collide in the phase space as ε increases. Both fixed points are attracting in the r direction, and one is attracting and one is repelling in the ϕ direction. In other words, one point is a node, and another is a saddle. Both of them lie on what will be the limit cycle, and this gives rise to the name of the bifurcation as a SNIC.

Oscillators that show SNIC bifurcations are called type 1 oscillators, as mentioned earlier. These oscillators are also called integrators, for reasons we will now see.

In the type 1 system, before the bifurcation occurs, the system will be attracted to a node and stop oscillations. Thus, in the phase direction, we have

$$\frac{d\phi}{dt} \approx -a\phi$$

near the node. If the system is forced by influence of an external signal, we have

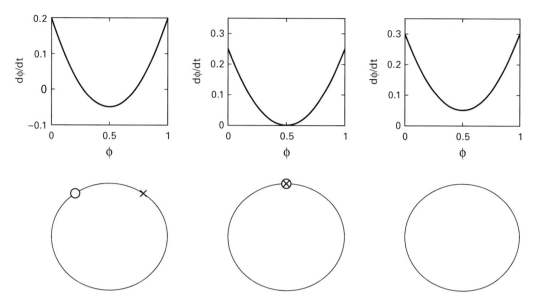

Figure 5.3
The dynamics of the model $d\phi/dt = (\phi - 0.5)^2 + \varepsilon$ in the vicinity of $\varepsilon = 0$. Note the steady states merge with $\varepsilon = 0$ (center) at a SNIC bifurcation. The top plots illustrate the cases of $\varepsilon = -0.05$ (left), 0 (center) and 0.05 (right). The bottom plots show the stable steady state (o) and unstable steady state (x).

$$\frac{d\phi}{dt} \approx f(t) - a\phi$$

or, using the solution of this system,

$$\phi(t) = \int_0^t f(t - t') e^{-at'} dt'.$$

Thus, the signal is integrated against a kernel e^{-at}. If the integrated signal passes the saddle, the system passes a threshold. The oscillator then travels around the cycle and approaches the node.

5.4 Examples of Type 1 Oscillators: Simplifications of the Hodgkin–Huxley Model

The original Hodgkin–Huxley neuron model is an example of a type 2 oscillator, which will be explained in section 5.5. One simplified version of the Hodgkin–Huxley model was proposed by Morris and Lecar (1981). Technically, this is a model of a different neuronal system (a nerve fiber in a barnacle, not a squid), but it follows the general formalism proposed by Hodgkin and Huxley. We will see that this model can act as a type 1 oscillator or a type 2 oscillator. This is a fairly common property (Conrad et al. 2008).

The following changes are made to simplify the model and apply it to the barnacle:

1. Na is replaced by Ca.
2. Ca gating is assumed to be instantaneous, so the m and h variables are not needed.

We also note that the gating variable is defined in terms of its steady-state value, w_∞, and time constant τ_w.

The model has the following equations:

$$C\, dV/dt = -g_{Ca}\, m_\infty\, (V - V_{Ca}) - g_L\, (V - V_L) - g_K\, w\, (V - V_K) + I, \qquad (5.4.1)$$

$$dw/dt = a\, (w_\infty(V) - w)/\tau_w(V), \qquad (5.4.2)$$

$$m_\infty(V) = 0.5(1 + \tanh((V - V_1)/V_2)), \qquad (5.4.3)$$

$$w_\infty(V) = 0.5(1 + \tanh((V - V_3)/V_4)), \qquad (5.4.4)$$

$$\tau_w(V) = 1/(\cosh((V - V_3)/(2 \cdot V_4))). \qquad (5.4.5)$$

The parameters are similar to those of the Hodgkin–Huxley model. For example, C is the neuronal capacitance, V_{Ca} is the equilibrium potential for calcium, w is a gating variable, $w_\infty(V)$ is the steady-state value of the gating variable, and $\tau_w(V)$ is the time scale. This model, like the Hodgkin–Huxley model, has both positive and negative feedback. As the voltage increases, calcium channels open and increase the voltage further (positive feedback). However, as the voltage increases, potassium channels also open, and these decrease the voltage (negative feedback).

For certain parameters, the model behaves as a type 1 neuron (see code 5.1). Figure 5.4 plots the Morris-Lecar model for parameters around the SNIC bifurcation. As the applied current is changed by a small amount (from 39.5 to 40.5), the period changes drastically.

Another example of a type 1 neuron is the "integrate-and-fire" neuron. We start with the following description of the voltage V and current I:

$$dV/dt = I - aV.$$

In addition to the continuous changes described by this differential equation, when the voltage reaches a threshold, V_{th}, we assume an action potential is fired, and the voltage resets to a set value V_{reset}.

The steady state of this model, if it exists, is at $V = I/a$. If $I/a = V_{th}$, we have the case of infinite period. As I is increased above aV_{th}, oscillations appear.

However, a better simplification of the Hodgkin–Huxley model is the "quadratic integrate-and-fire" model, given by

$$dV/dt = B + V^2 + I,$$

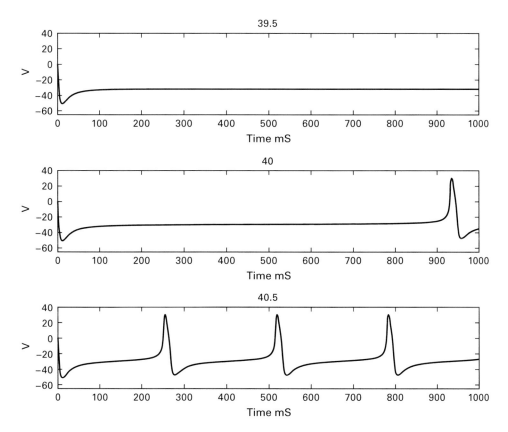

Figure 5.4
Simulations of the Morris-Lecar model around the SNIC bifurcation point. Small changes in the applied current (listed above the plots) show a dramatic change in the period, characteristic of a SNIC bifurcation that occurs around an applied current of approximately 39.97. See code 5.1 for more details.

since this preserves the saddle and the node discussed earlier (Ermentrout et al. 2010). There are other possibilities. One example is the model used in section 5.3. Another is

$$dV/dt = 1 - b\sin(V - V_0).$$

In these examples, we consider just one variable, which takes the role of the phase, and the other variables (e.g., amplitude) are assumed to be fixed. This assumption is not made with two-dimensional models like the Morris-Lecar model.

We also note that the van der Pol model has been proposed as a simplification of the Hodgkin–Huxley model (Fitzhugh 1961; van der Pol and van der Mark 1928). This is a type 2 oscillator, and it will be explored in sections 5.5 and 5.6.

5.5   Hopf or Type 2 Oscillators

The behavior of saddles and nodes are determined by eigenvalues that are real numbers. Eigenvalues can also be complex numbers. The behavior of the system near steady states with corresponding complex eigenvalues can be described by solutions of the form ae^{bt+ict}. These are solutions that spiral into or out of the steady state. What determines whether oscillations will persist is whether these solutions are attracted toward or repelled from the steady state. This in turn is determined by whether the real part of these eigenvalues is positive or negative.

A Hopf bifurcation occurs when a system switches from spiraling in toward a steady state to spiraling away from a steady state. Here, in contrast to the type 1 case, the ϕ direction is not affected by the bifurcation parameter, so we can assume

$d\phi/dt = c$.

Thus, the period of the oscillations remains constant. Additionally, a stable fixed point loses stability and becomes unstable and oscillates. Mathematically, what happens is that the real part of a pair of eigenvalues associated with that fixed point changes from negative to positive. Letting r denote the amplitude, we assume that this stable fixed point occurs at $r = 0$.

There are two main cases. The first is the supercritical Hopf bifurcation, where a stable limit cycle is born and the point at $r = 0$ is unstable. Consider the following example:

$dr/dt = r(\varepsilon - r)$.

For $\varepsilon > 0$, we have a limit cycle at $r = \varepsilon$. For $\varepsilon = 0$, we have neutral stability at the fixed point. For $\varepsilon < 0$, we have a stable fixed point. Thus, the system goes from a fixed point to oscillations with arbitrarily small amplitude (radius ε).

The second case is the subcritical Hopf bifurcation. In this case, the born limit cycle is unstable and the point at $r = 0$ is stable. Consider the following example:

$dr/dt = -r(\varepsilon - r)$.

For $\varepsilon > 0$, we have an unstable limit cycle at $r = \varepsilon$. For $\varepsilon = 0$, we have neutral stability at the fixed point. For $\varepsilon < 0$, we have a unstable fixed point. Thus, the system goes from a stable fixed point to an unstable fixed point. What the system does after this bifurcation depends on the far-field behavior. The amplitude of the oscillation may grow without bound in some other models. There may also be another limit cycle that the system approaches, where, for $\varepsilon > 0$, the system is bistable. Within the unstable limit cycle, the system tends toward the origin. Outside of the unstable limit cycle, the system tends toward another state (e.g., another limit cycle).

Actually, there is a third type of Hopf bifurcation, called the degenerate Hopf bifurcation. In this case, there are many possible steady states at the bifurcation. For example, consider

$dr/dt = -\varepsilon r(\varepsilon - r)$.

At $\varepsilon = 0$, we have $dr/dt = 0$, and there are no limit cycles (every initial point is on a cycle, and cycles of any amplitude exist).

In the SNIC bifurcation, we saw what happens when two fixed points merge. In the Hopf bifurcation, we saw what happens when a limit cycle and a fixed point merge. Another type of bifurcation is the "stable-unstable limit cycles" bifurcation, in which two limit cycles merge. Again, we can consider dynamics in the ϕ direction as being uninfluenced by changes in the bifurcation parameter, so we can assume

$d\phi/dt = c$.

Next, consider a system where there are two limit cycles:

$dr/dt = -r + \varepsilon r^2 - r^3$.

For $\varepsilon = 2$, there is just one limit cycle, at $r = 1$. For $\varepsilon < 2$, there are no limit cycles. For $\varepsilon > 2$, there are two limit cycles, one of which is stable while the other is not stable. This is shown in figure 5.5.

An example of type 2 behavior is shown in figure 5.6 for the Morris-Lecar model. Oscillations appear with a fixed period.

Simulation trick: In cases of bistability with a fixed point and a limit cycle, we are often interested in two regions: the basin of attraction of the fixed point, which is the region of the phase plane that tends toward the fixed point, and the basin of attraction of the limit cycle, which is the region that tends toward the limit cycle. The boundary between these two regions is an unstable limit cycle. To determine this boundary, we can run simulations backward in time. When we do this, the unstable limit cycle becomes a stable limit cycle, and the system will evolve, with backward time, toward the boundary. This gives a very simple way of calculating the boundary.

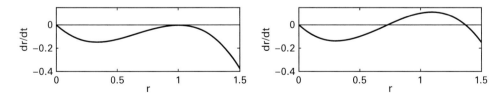

Figure 5.5
A bifurcation of limit cycles shown by the dynamics of the amplitude r. The left plot has $\varepsilon = 2$. The right plot has $\varepsilon = 2.1$.

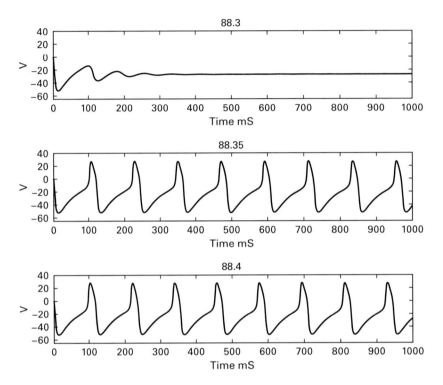

Figure 5.6
Simulations of the Morris-Lecar model around the Hopf bifurcation point. Small changes in the applied current (listed above the plots) show oscillations being created with a finite period, a situation characteristic of Hopf bifurcations. Here the bifurcation occurs around an applied current of 88.3. See code 5.1 for more details.

This method works well in two-dimensional (planar) systems. In higher-dimensional systems, the boundary will likely be of dimension greater than one, and not an unstable limit cycle. It can also have a complicated geometry. Thus, this method only works reliably for two-dimensional systems.

Type 2 oscillators are often thought of as resonators. They "color" an input signal by shifting it toward a preferred frequency b. To see this, consider a simple type 2 oscillator consisting of one complex variable forced by a signal with frequency w. With this forcing, our equations become

$$dz/dt = e^{iwt} + \lambda z.$$

Looking at the Fourier transform of this equation at w (noting the transform of a variable by a tilde, e.g., $\tilde{z}$), it can be shown that all other Fourier coefficients of x are zero in the attractor, so we have

$$iw\tilde{z} = 1 + \lambda\tilde{z}$$

or

$$\tilde{z} = \frac{1}{iw - \lambda}$$

Let $\lambda = -a + bi$. We choose the real part of λ to be negative so that the system will not oscillate on its own. We then have

$$\tilde{z} = \frac{1}{a + i(w - b)}.$$

The solution of the equation is

$$z(t) = \tilde{z}e^{iwt}$$

Note that the amplitude of the solution is given by the amplitude of the Fourier coefficient at w, $\tilde{z}$. As $w \to b$, this amplitude increases. This is shown in figure 5.7.

This is why this oscillator is called a "resonator." It will show the greatest response when subjected to an input frequency equal to b.

5.6 Examples of Type 2 Oscillators: The Van der Pol Oscillator and the Resonate-and-Fire Model

A good example of a type 2 oscillator is the van der Pol oscillator. This limit-cycle oscillator has been studied in various engineering applications since the early twentieth century (van der Pol and van der Mark 1928). It is, in one sense, the simplest possible limit-cycle model, since it has just one nonlinear term. It also is arguably the best studied. We start with the simple harmonic oscillator, for example with x representing the position of a pendulum and x_c the velocity of the pendulum:

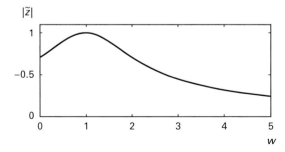

Figure 5.7
Resonance in a type 2 model where the amplitude $|\tilde{z}|$ increases around the resonant frequency $b = 1$.

$$\frac{dx}{dt} = \mu x_c$$

$$\frac{dx_c}{dt} = -\mu x$$

The problem with this oscillator is that it has no preferred amplitude, in that different initial conditions will lead to different amplitudes. To see this, consider the amplitude, defined by

$$r^2 = x^2 + x_c^2.$$

We note that

$$2r dr/dt = 2x(dx/dt) + 2x_c(dx_c/dt) = 2\mu x x_c - 2\mu x x_c = 0.$$

The equations need to be modified to allow for a limit cycle (where oscillations tend toward a preferred amplitude). We will see in chapter 10 that the simplest possible model is the following:

$$\frac{dx}{dt} = \mu\left(x_c + \varepsilon\left(x - \frac{4}{3}x^3 \right) \right)$$

$$\frac{dx_c}{dt} = -\mu x$$

We will also see in chapter 10 that the constant 4/3 is chosen to give the limit cycle an amplitude of around 1. Sometimes the van der Pol oscillator is presented as a differential equation in just one variable (typically x). To get it into this form, take the derivative of the dx/dt equation. We then have

$$\frac{d^2x}{dt^2} = \mu\left(\frac{dx_c}{dt} + \varepsilon\frac{dx}{dt}(1 - 4x^2) \right).$$

We can now substitute the second equation into this and find

$$\frac{d^2x}{dt^2} = \mu\left(-\mu x + \varepsilon\frac{dx}{dt}(1 - 4x^2) \right).$$

There are two limits where this equation is studied. In the limit of $\varepsilon \to 0$, the system is not very different from the harmonic oscillator. This is the "quasilinear" case, and the system behaves sinusoidally. Most of our discussions will focus on this case. The other case is where ε is large. In this case,

$$\frac{dx}{dt} \approx \mu\varepsilon\left(x - \frac{4}{3}x^3\right).$$

The steady states of this approximate system are $x = 0$ and $x = \pm\sqrt{3/4}$. Looking at the stability of this approximate equation, $x = 0$ is unstable, and the other steady states are stable. Since ε is large, x quickly approaches $\pm\sqrt{3/4}$. Starting with the positive choice, we find

$$\frac{dx_c}{dt} = -\mu x \approx -\mu\sqrt{\frac{3}{4}}.$$

Thus, on a slower time scale, x_c decreases. After enough time we find that the x_c term in the dx/dt equation starts to have an effect. In fact, eventually the stability of the positive steady state of x will be lost. Then x switches to the negative steady state, and the process continues. This is illustrated in figure 5.8.

Because of the fast-slow dynamics, the system is often called a relaxation oscillator because x_c "relaxes" and changes slowly after a quick jump in x. It is similar to the behavior studied in section 4.9. See box 5.2.

The resonate-and-fire model is often associated with resonating neurons, and it is valid just before the supercritical Hopf bifurcation, or just after the subcritical Hopf bifurcation (Izhikevich 2001). In either case, the fixed point is stable. Consider a perturbation in a Cartesian coordinate x:

$dx/dt = -wy + bx + \varepsilon$,

$dy/dt = wx + by$.

We also assume that if $x = 1$, x changes to x_{reset}. This model can be thought of as a linearization around the fixed point, combined with a threshold. However, similarly to the integrate-and-fire neuron model, this model does not fully capture the bifurcation. To see this, let $x = x' + c_x$ and $y = y' + c_y$. Substituting, we find that

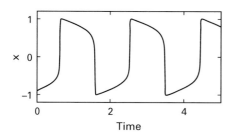
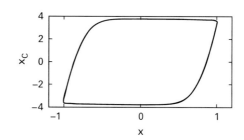

Figure 5.8
Simulations of the van der Pol model with $\varepsilon = 10$ and $\mu = 10$. The left plot shows x over time, whereas the right plot shows both x and x_c.

Box 5.2
Phase Plane Analysis of a Cardiac Electrophysiology Model

A version of the van der Pol model has been proposed as a simple model of cardiac electrophysiology. While other, much more detailed models exist of this system, this model
captures many of the key properties of the electrophysiology of cardiac cells, including their
typically very long action potential (at least as compared with neurons). This model takes the
form:

$$\frac{dx}{dt} = \frac{1}{\mu}\left(x_c - \frac{x^3}{3} + x\right),$$

$$\frac{dx_c}{dt} = -\mu x.$$

The parameter μ is chosen to be small (0.1 in our simulations). Our analysis of this model
follows the text of Kaplan and Glass (1995), which provides more details. A simulation of
the model is shown in the figure below.

(a)

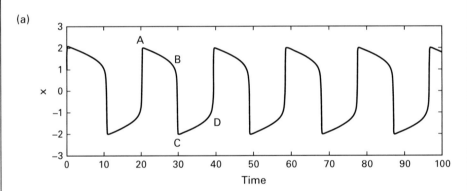

(b)

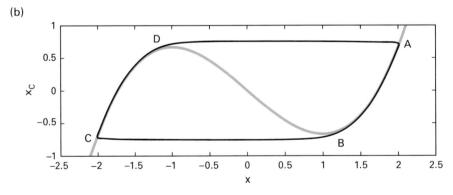

Box 5.2 (continued)

We note that the equation for x_c is much slower since it is multiplied by μ (0.1) rather than $1/\mu$ (10). Based on this, we can consider the dynamics of x and x_c separately. Because of the quick x dynamics, we can assume that the x variable quickly reaches an equilibrium. When $dx/dt = 0$, solutions follow the curve $x_c = -x + x^3/3$, called an isocline. This is shown in the bottom panel of the figure as the blue trace passing through points C, D, A, and B. Starting at point A, solutions begin by following this curve. On a slower timescale, we have $dx_c/dt = -\mu x$, so that the value of x decreases as x_c does. The two variables follow the blue curve until point B, where the system cannot follow the isocline any further. The x variable then quickly changes (see top panel of the figure) to the other branch of the isocline (point C). Now that we are on this isocline, x is negative and x slowly increases until point D. At point D, the solution can no longer follow the isocline, and another transition occurs. This simple example shows how taking account of isoclines, and studying the phase space of the model in general, can help explain the behavior of the model.

$$dx'/dt = -wy' + bx' + \varepsilon - wc_y + bc_x,$$

$$dy'/dt = wx' + by' + wc_x + bc_y.$$

Now let us choose c_x and c_y so that

$$\varepsilon = wc_y - bc_x,$$

$$wc_x = -bc_y.$$

We find that

$$\varepsilon/(w+b^2/w) = c_y,$$

$$-(b\varepsilon/(w^2 + b^2)) = c_x.$$

So, changing ε simply changes the location of the fixed point with respect to the threshold. If oscillations emerge, they will have finite amplitude, so the model does not accurately reflect the supercritical Hopf bifurcation. Unlike the subcritical Hopf bifurcation, there are no unstable limit cycles. What this model does capture is the resonance that appears in models that undergo Hopf bifurcations.

This model is an example of mixed-mode oscillations, where multiple types of rhythms coexist. In this case, there is a rhythm of resonance, where the system oscillates around the fixed point. There is also a rhythm of firing, where once the system passes a threshold, the model fires and resets. Both rhythms affect each other, and the overall oscillation depends on the combination of the two rhythms.

5.7 Summary of Oscillator Classification

Summarizing, we have two general classifications of oscillators.

Resonator	Integrator
Type 2	Type 1
Hopf	SNIC

The first row of names gives an intuitive explanation of the behavior of these oscillators. The next row is a classification that owes to Hodgkin, who studied neural oscillators (Hodgkin 1948). The third row gives the name of the mathematical structure for the oscillators.

So how can we tell these types of oscillators apart? In summary, type 1 oscillators integrate, and have their period sensitive to incoming signals. Type 2 oscillators resonate and have a fixed period that they wish to oscillate.

Mathematically, the best way to determine how the bifurcation occurs is to look at the steady states of the system and their eigenvalues. If changing a parameter causes the real part of an eigenvalue to be zero, then we have a Hopf bifurcation (type 2). If changing a parameter causes a saddle and node to be born, we would have a SNIC bifurcation if this occurs on an invariant circle (type 1).

SNIC and Hopf bifurcations can be thought of as opposites. In the Hopf bifurcation, changing a parameter causes minimal or no change to the period of the oscillation. However, the amplitude of the oscillation can change greatly. In the SNIC bifurcation, the dynamics of the amplitude are relatively unchanged. However, at the bifurcation point, $d\phi/dt \approx 0$ at some ϕ. At this point the period approaches ∞.

In figure 5.9 we plot the frequency (2π/period) and amplitude for the two types as a parameter, a, is changed.

In chapter 6 we will explore how the phase response curves of type 1 and type 2 oscillators are different.

5.8 Frontiers: Noise in Type 1 and Type 2 Oscillators

The behavior of a system near a bifurcation in the presence of noise is a very rich and interesting topic. Noise can not only affect the period of oscillation from one cycle to the next, it can also start or stop oscillations. Increasing the noise, in some situations, can actually make rhythms more regular. We will study the two bifurcations analyzed in the previous section. Analysis of the first, the SNIC bifurcation, will follow Lindner et al. (2003). Analysis of the Hopf bifurcation will follow Bodova et al. (2015). These analyses focus on calculating the coefficient of variation, which is a measure of scaled standard deviation of

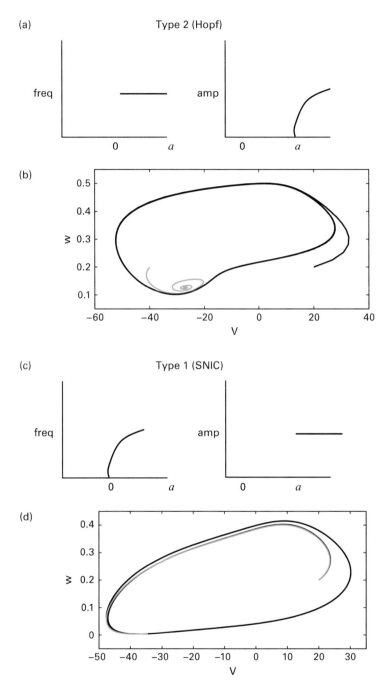

Figure 5.9
Summary of type 1 and type 2 oscillator behavior. For each behavior we show how the frequency and amplitude change before and after the bifurcation in parameter a (which is assumed to occur at $a = 0$). Below these plots, we show the behavior of the Morris-Lecar system just before (blue) and after (black) a bifurcation. In all cases, we use initial conditions $V = 20$, $w = 0.2$, except for the black curve for the type 2 behavior where initial conditions $V = -40$, $w = 0.2$ are used. We use an applied current of 38 (blue) and 40 (black) for the type 1 behavior and an applied current of 88.3 for the type 2 behavior.

the period of one cycle to the next. Both will use examples from neuronal firing. Our goal is to show that noise can increase neuronal firing, and can also cause either more regular or less regular rhythms.

5.8.1 Type 1

Following Lindner et al. (2003), we consider the following system, which is the quadratic integrate-and-fire model with noise:

$$dx / dt = B + x^2 + \sqrt{2D}\xi(t).$$

We will show that changing B and changing D have similar effects.

Without the last term, this equation is similar to the system (5.3.1) studied in section 5.3. Here, the critical point (i.e., the steady state) has been moved from 0.5 to 0 and the term in ε is replaced by B. Additionally, using the standard Wiener process, $\xi(t)$, we have added a noise term, which consists of a scaling $\sqrt{2D}$ combined with $\xi(t)$. The process $\xi(t)$ represents additive noise that is uncorrelated, so

$$<\xi(t), \xi(t+a)> = \delta(a),$$

where $\delta(a)$ is the Dirac delta function. Perhaps the easiest way to understand this system is by its numerical approximation in terms of successive time steps:

$$X_{j+1} = X_j + (B + X_j^2)\Delta t + \sqrt{2D\Delta t}\, \xi_i,$$

where ξ_i is a mean zero random number drawn from a Gaussian distribution with variance 1. Additionally, we assume that, similarly to the integrate-and-fire model, once the model reaches the point x_+ it resets to x_-. We can simulate the model and find a series of $T = (T_1, T_2, ..., T_i, ...)$, which is the set of times of interspike intervals (times between firings). Thus, the firing rate of the model is

$$1/<T>,$$

where $<T>$ is the average interspike interval. The coefficient of variation (CV) is the standard deviation of T scaled by its mean,

$$CV(B, D) = \frac{\sqrt{\langle \Delta T^2 \rangle}}{\langle T \rangle},$$

and $\Delta T_i = T_i - T_{i-1}$. Our goal is not to perform a full stochastic analysis of this model, but instead, to discover what results can be found by simple scaling.

Let us first consider the case where $B = 0$. This gives us

$$dx/dt = x^2 + \sqrt{2D}\xi(t),$$

and now we ask if we can scale time (and x) to remove the D term. Let $y = x/a$ and $t' = at$. This yields

$$dy/dt' = (x/a)^2 + (1/a^2)\sqrt{2Da}\ \xi(t'),$$

where the substitution $\sqrt{a}\,\xi(t') = \xi(t)$ comes from the properties of the Wiener process. If we let $a = D^{1/3}$, then we have

$$dy/dt' = y^2 + \sqrt{2}\,\xi(t').$$

We see that the scaling has indeed removed D from the model equation. Thus, we can solve the preceding equation once and then scale time and y to find x and t. This indicates $t \sim t'D^{-1/3}$ and that $<T>$ and $\sqrt{\langle \Delta T^2 \rangle}$ are scaled in the same way (the second expression scales as a^2 under the square root). Thus, the coefficient of variation remains the same regardless of the value of D. This means that, as D increases, the average $<T>$ increases, but the standard deviation of T increases by the same factor. In this way, increasing the noise in the case does not cause the firing to be more noisy. Instead, it changes the mean rate of firing.

Now consider the case where B is not zero. To allow for B to be negative or positive, we now let $y = x/a$, $t' = at$ and $a = |B|^{0.5}$. This then yields the following equations:

$$dy/dt = B/|B| + y^2 + (2D/|B|^{3/2})^{1/2}\ \xi(t')$$

or

$$dy/dt = \pm 1 + y^2 + (2D/|B|^{3/2})^{1/2}\ \xi(t'),$$

so

$$CV(B, D) = CV(\pm 1, |B|^{-3/2}D).$$

There is only one parameter here, $|B|^{-3/2}D$, which also determines the firing rate. The idea here is that changes in B can always be made up by changes in D. Also,

$$\lim_{D \to \infty} CV(B, D) = \lim_{D \to \infty} CV(\pm 1, |B|^{-3/2}D) = \lim_{B \to 0} CV(\pm 1, |B|^{-3/2}D) = CV(0, D),$$

with $CV(0, D)$ being the $B = 0$ case solved for the preceding arguments. So in the limit of large noise, we again see that further increases in the noise do not affect the CV, but instead affect the firing rate.

The values of $<T>$ and $<\Delta T^2>$ can be solved for explicitly. For example, in the case of $B=0$, Lindner et al. (2003) find that the firing rate $r(0, D) \sim 0.201D^{1/3}$ and $CV \sim 3^{-1/2}$.

In the case of weak noise and $B > 0$, we have $r = B^{1/2}/\pi$ and $CV = (3D/(t\pi))^{1/2}B^{-3/4}$ when $B >> D^{2/3}$. In this case, the interspike distribution looks like an inverse Gaussian. In the case of weak noise and $B < 0$, $r = (|B|^{1/2}/\pi)\exp(-4|B|^{3/2}/(3D))$ for $|B| >> D^{2/3}$, which looks like an exponential distribution. In the first case, increasing noise does not affect the firing rate. In the second case, noise causes the firing rate to increase.

5.8.2 Type 2

A completely different picture is needed in the type 2 case. Here we assume that two states coexist. First, there is a stable steady state, with solutions near the steady state spiraling toward the steady state with a period μ_2 and a standard deviation of σ_2. Secondly, there is a stable oscillating state with period μ_1 and a standard deviation of σ_1. The system switches between these states. Simulations of type 2 models often exhibit geometry like that shown in figure 5.10; see also (Bodova et al. 2015).

In figure 5.10, we see part of the trajectory of the limit cycle. At a particular phase, it becomes very close to the stable fixed point. At this point, noise can cause the system to enter the region of stability of the fixed point. After oscillating around the fixed point, it can transition back to the limit cycle. This can be rephrased as a simple probabilistic model that can be easily analyzed (Bodova et al. 2015).

We assume two states: a spiking state (S), where the system stays on the limit cycle, and a quiescent state (Q), where the system spirals toward the fixed point. Each time the system reaches the phase where the limit cycle and fixed point are close, a choice needs to be made. If starting in the spiking state, the system can enter Q with probability p_{SQ} or stay spiking with probability $p_{SS} = (1 - p_{SQ})$. If starting in Q, the system can enter the spiking state with probability p_{QS} or stay in Q with probability $p_{QQ} = (1 - p_{QS})$. In S, going around the

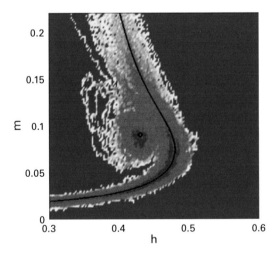

Figure 5.10
A limit cycle and fixed point close together, both shown in black. The density of stochastic trajectories for the Hodgkin–Huxley model is shown in yellow and red. Taken from Bodova et al. (2015).

limit cycle takes a time μ_1, whereas going around the fixed point in Q takes a time μ_2. These times can also be drawn from Gaussian distributions with a chosen variance for each.

In the case of large noise, we can assume that $p_{SQ}=1-p_{QS}$. With this assumption, the previous state does not matter.

As an example of how one can perform calculations with this model, we calculate the expected value of times between spikes:

$$<T> = \mu_1 p_{SS} + (\mu_1 + \mu_2)p_{SQ}p_{QS} + (\mu_1 + 2\mu_2)p_{SQ}p_{QQ}p_{QS} + (\mu_1 + 3\mu_2)p_{SQ}p^2_{QQ}p_{QS} + \cdots$$
$$= \mu_1(p_{SS} + p_{SQ}p_{QS} + p_{SQ}p_{QQ}p_{QS} + p_{SQ}p^2_{QQ}p_{QS} + \cdots) + \mu_2 p_{SQ}p_{QS}(1 + 2p_{QQ} + 3p^2_{QQ} + \cdots).$$

However, note that

$$(1 + p_{QQ} + p^2_{QQ} + \cdots) = 1/(1 - p_{QQ}) = 1/p_{QS},$$

so the first term becomes $\mu_1(p_{SS} + p_{SQ}p_{QS}/p_{QS}) = \mu_1$.

Also note that

$$(1 + 2p_{QQ} + 3p^2_{QQ} + \cdots) = 1/(1 - p_{QQ})^2 = 1/p_{QS}^2$$

since $0 \le p_{QQ} \le 1$.

This yields

$$<T> = \mu_1 + \mu_2 p_{SQ}/p_{QS}.$$

When p_{SQ} is small and p_{QS} is large, for example when there is a very stable limit cycle and an unstable fixed point, this formula shows that we approach a mean firing rate of μ_1.

One can also calculate the coefficient of variation. To do this, note that the variance is

$$\sigma_1^2 + (p_{SQ}/p_{QS})(\sigma_2^2 + \mu_2^2(p_{QQ} + p_{SS})/p_{QS}).$$

Now, as the noise increases, transitions from the quiescent state or from the spiking state increase, so $(p_{QQ} + p_{SS})/p_{QS}$ should decrease. For this reason, increasing the noise typically *decreases* the variance. So long as (p_{SQ}/p_{QS}) does not decrease as quickly, the coefficient of variation will then drop as the noise decreases. The CV cannot drop below σ_1/μ_1, which would be its value if the neuron never left the spiking state.

Another case of interest is where p_{SQ} is large and p_{QS} is small. Here, $<T>$ becomes very large and one waits a long time between firings. The variance in this case will look like

$$\sim(p_{SQ}/p_{QS})(p_{QQ}/p_{QS})\mu_2^2,$$

and if $p_{SQ} \sim p_{QQ}$, the CV will approach 1, which is the coefficient of variation of the Poisson distribution. However, we note that there are many possible scenarios where the CV could be greater than 1.

We also note that this model typically predicts inter-spike-interval (ISI) histograms that have multiple peaks corresponding to whether the system revolves around the fixed point

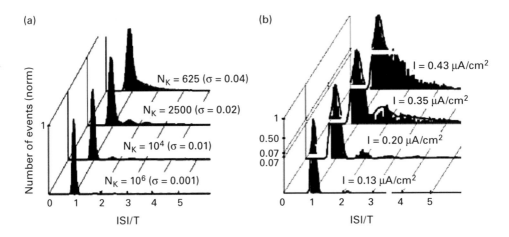

Figure 5.11
Interspike-interval histograms from simulations of a stochastic version of the Hodgkin–Huxley model (A) and experimental data from squid giant axon (B). Curves shown in (B) are fits to these data using the type 2 noise model described in section 5.8. Taken from Bodova et al. (2015).

one time, two times, etc. This has been found experimentally as well, and is shown in figure 5.11.

5.9   Frontiers: Experimentally Testing the Effects of Noise in Squid Giant Axon

Here, we test the theory presented in section 5.8. Figure 5.12 shows experimental data collected by the author and colleagues (Forger et al. 2011; Paydarfar et al. 2006) from squid giant axon. The experimental setup was similar to that used by Hodgkin and Huxley. Additionally, a noisy current was generated, simulating the input a neuron might receive in a large noisy network, and this noisy current was applied as an input into the squid giant axon. Figure 5.12 shows three different examples of the CV: low, around 1, and high.

We next seek to determine if the relationships derived in section 5.8 hold in these data.

We first consider the case where p_{SQ} is large and p_{QS} is small. In the standard squid giant axon preparation, p_{SQ} is large, basically 1, as the axon does not repetitively fire. Additionally, we can choose p_{QS} to be small by having a very small input current. This was the protocol used in (Forger et al. 2011). We note in figure 5.13 (top) that the CV tends to be around 1, as predicted, regardless of the number of spikes that were caused.

We next test to see if increasing the noise could make the firing more regular. To test this, we need p_{SQ} not to be as large as previously, and this can be done by alkalinizing the axon (Paydarfar et al. 2006). In this setup, increasing the amplitude of the input noise caused the mean ISI to decrease, which corresponds to a decrease in p_{SQ}/p_{QS} in terms of our formula.

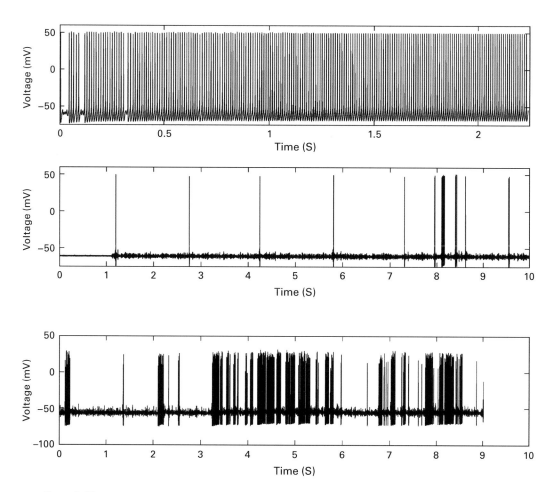

Figure 5.12
Three examples of neuronal firing with different coefficients of variation. The top and middle plots come from the same neuron. When low-amplitude noise was applied to the neuron, rare spiking occurred (middle). However, after about 8 seconds, the neuron transitioned into repetitive firing (top). The coefficient of variation (CV) of the top plot is 0.25 and that of the middle plot is 1.36. We also illustrate another axon with a large CV (3.18), as shown on the bottom.

Nevertheless, as the noise amplitude increased, even though $<T>$ decreased, the CV also decreased. Thus, more regular firing was seen in the presence of more noise.

5.10 Example: The Van der Pol Model and Modeling Human Circadian Rhythms

Here, in sections 5.10–5.12, we illustrate how mathematical models can be constructed based on known behaviors of biological systems and the known behaviors of systems of

(a)

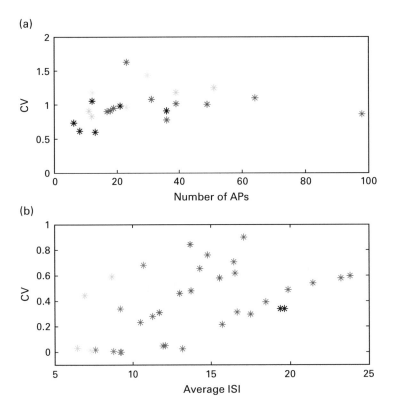

(b)

Figure 5.13
(a) Calculation of the coefficient of variation for squid giant axon with a noisy input. Color corresponds to different neurons. Different points of the same color correspond to changes in the noisy input current. As the noise was very small, and the quiescent state stable, the CV was around 1. (b) Plots of the CV for axons that were alkalinized in a separate experiment, so that both the spiking state and the quiescent state were stable. Additionally, larger noise levels were used than in the top plot. As the noise increases, the average ISI ($<T>$) decreases, and the CV decreases. This is a case of noise leading to more regular firing. Numbering of the axons in the paper by Paydarfar et al. (2006) is as follows: 17b (magenta), 21a (yellow), 14b (black), 15a (green), 21c (red), 16c (blue), and 17a (cyan).

equations described earlier. The biological system we study is the human circadian clock. Our goal is not just to understand the human circadian clock, but to systematically show how a model for it can be constructed, as an example of modeling building based on behaviors when detailed physiology, for example as used in chapter 2, is unknown. Thus, reader should not only pay attention to the application but the method of model building.

There is much evidence to suggest that circadian clocks are limit cycles. Clocks must be cycles of some kind. Experiments show that circadian clocks reset to a fixed amplitude after perturbations. This leads us to propose several types of limit cycle oscillators for the human circadian clock. R. Wever (1973) suggested that the van der Pol oscillator could be used as a model. Other models have been proposed. One common example is the radial isochron clock, which is given in polar coordinates:

$dr/dt = r(1 - r^2),$

$d\phi/dt = w.$

Note that this is sometimes also defined by $dr/dt = r(1 - r)$, e.g., (Kaplan and Glass 1995). Rather than looking at the amplitude and phase of the rhythms, we can look at the Cartesian coordinates x and x_c. Another model for the human circadian clock was proposed by Pavlidis (1978):

$dx/dt = x_c - ax,$

$dx_c/dt = Rx_c - cx - bx^2.$

The differences between these models will be discussed in chapter 10.

Environmental stimuli must affect the oscillator in some way. Returning to the example of the van der Pol oscillator, we can incorporate these stimuli in the following way:

$$\frac{dx}{dt} = \mu \left(x_c + \varepsilon \left(x - \frac{4}{3} x^3 \right) \right) + \eta_x B,$$

$$\frac{dx_c}{dt} = -\mu x + \eta_{x_c} B,$$

where the unit vector (η_x, η_{xc}) determines in which phase-space direction the stimulus is applied, and B is the stimulus strength. To proceed, though, we need more information on circadian timekeeping before describing how this class of models was fit (Forger et al. 1999; Jewett et al. 1999; Kronauer et al. 1999).

5.11   Example: Refining the Human Circadian Model

We now show how a simple model for the human circadian clock can be refined. Light is the most important stimulus that affects the human circadian clock (Duffy et al. 1996). Other factors, such as activity, hormones (e.g., melatonin), and food may affect the mammalian circadian clock, as well. However, if one changes the light schedule then regardless of change in activity or any other stimulus, the human circadian pacemaker will often follow the light stimulus and not be affected by the presence or absence of these other "zeitgebers" (time-givers). We wish to model the effect of light on the circadian pacemaker.

The pathway from external light to the pacemaker in the brain has received much recent attention from researchers. Originally, it was thought that only rods and cones in the retina could send light signals to the circadian clock in the brain. However, mice that lack rods or cones can still be entrained to light and dark signals. Even some "blind" people can entrain to such signals. These observations led to a search for some other means of phototransduction. The result was an amazing discovery that a subset of retinal ganglion cells originally presumed to process information from rods and cones were actually photoreceptor cells in their own right. These cells, at the outermost level of the retina, and called intrinsically photosensitive retinal ganglion cells have a photopigment, melanopsin, which can sense light. Light causes these ganglion cells to depolarize and fire action potentials.

These retinal ganglion cells project from the retina to the central pacemaker in the brain through the retinal hypothalamic tract (RHT). Projections come from both optic nerves (one from each eye). The pacemaker sits just about where the two optic nerves cross, in the suprachiasmatic nuclei (SCN) (figure 5.14).

The SCN are composed of two groups of neurons, and in mice contain approximately 10,000 neurons. The ventral part of the SCN receives input from the RHT. The dorsal SCN contains cells that have a primary timekeeping function. Within these cells are the proteins and genetic transcription loops that form the molecular clock.

For the simplest mathematical model, consider the following:

light $\rightarrow$ photoreceptors $\rightarrow$ core pacemaker.

We can model the pacemakers by the van der Pol equation with B as the input from the RHT. This is a huge simplification of the biochemistry and electrophysiology of the clock, and the heterogeneity of the SCN. Nonetheless, we need a place to start.

We now need a model of the input through the retina. We propose the following model. Light causes photoreceptive molecules to go from a ready state to a used state. This process causes a signaling cascade that leads to phase shifts of the pacemaker. The total amount of ready or used photopigments only indirectly affects the pacemaker. Rather, it is the act of conversion that triggers the signaling process in this model. Assume some rate $\alpha(I)$ for the conversion of ready molecules to used molecules, where I is the intensity of light. Let n be the probability that a photopigment molecule is in the used state. Let β be the recycle rate. This system is depicted in figure 5.15.

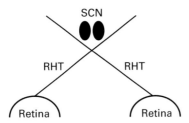

Figure 5.14
Structure of the visual circadian system.

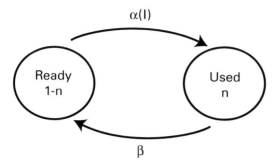

Figure 5.15
Dynamics of the gating variable in the human circadian model.

The equations for this system are similar to those used for the gating variables in the Hodgkin–Huxley model and follow from figure 5.15:

$$\frac{dn}{dt} = \alpha(I)(1-n) - \beta n.$$

The dynamics of this process are often on the time scale of minutes, while the time scale of the circadian clock is typically hours, so if we initially define α and β in terms of minutes, α and β should be divided by 60 before being incorporated into the circadian model.

We assume that as photoreceptor molecules are activated, they send a signal to the circadian clock. The signal to the circadian clock should be proportional to the rate at which photoreceptor molecules are activated:

$$B \sim \alpha(I)(1-n).$$

The circadian pacemaker is more sensitive to light at certain times of the day. In fact, the whole visual system is more sensitive at certain times rather than others (Knoerchen and Hildebrandt 1976). One way of incorporating this overall sensitivity into our model would be to directly build it into α (e.g., having photoreceptors more active at certain times of the day). However, let us assume instead that it is the pacemaker itself that is more or less sensitive. One possible reason for this is that some mediating molecule in this input to the pacemaker is part of the circadian clock and, as such, it could oscillate. The most general statement of this is

$$B \sim \alpha(I)(1-n)m(x, x_c).$$

We assume that light acts only in the x direction:

$$\frac{dx}{dt} = \mu\left(x_c + \varepsilon\left(x - \frac{4}{3}x^3\right)\right) + B,$$

$$\frac{dx_c}{dt} = -\mu x.$$

Let $x'_c = x_c + B/\mu$. We now have

$$\frac{dx}{dt} = \mu\left(x'_c + \varepsilon\left(x - \frac{4}{3}x^3\right)\right),$$

$$\frac{dx'_c}{dt} = -\mu x,$$

which does not contain B. The period of this system (where x_c is shifted) is the same as the period of the unforced system, which shows that B does not change the period of this model. However, we wish to account for "Aschoff's Rule," which states that diurnal animals shorten their period with increasing intensity of constant light, whereas nocturnal animals lengthen their period with increasing intensity of constant light (Aschoff 1960). Thus, we need to add an extra term to account for the effect of light on the circadian period. One possibility is

$$\frac{dx'_c}{dt} = -\mu x(1 + aB),$$

where a is simply a scaling factor for the strength of the effect of light on the diurnal human circadian period. Fits of the model to human phase response curve data are shown in figure 5.16.

More information on phase response curves is given in chapter 6. The final model proposed is that of Forger et al. (1999):

$$\frac{dn}{dt} = 60(\alpha(1-n) - 0.0075n),$$

$$\frac{dx}{dt} = \frac{\pi}{12}(x_c + B),$$

$$\frac{dx_c}{dt} = \frac{\pi}{12}\left(0.23\left(x_c - \frac{4}{3}x_c^3\right) - x\left(\left(\frac{24}{0.99669\tau_x}\right)^2 + 0.55B\right)\right),$$

where $B = 33.75\alpha(1 - 0.4x)(1 - 0.4x_c)$ and $\alpha = 0.05(I/9500)$, with I being the light intensity in lux and τ_x the period of the human circadian clock (typically 24.2 hours). This model reflects several changes from those discussed earlier.

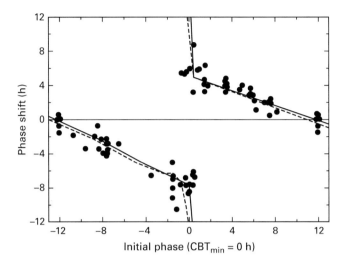

Figure 5.16
Fits of a model for the human circadian clock to experimental data on the phase shifts caused by a light signal given at different times of day. CBT_{min} is the minimum of the core body temperature, a common marker of the circadian clock, which occurs a few hours before the normal waking time. See Forger et al. (1999) for more details.

5.12   Example: A Simple Model of Sleep, Alertness, and Performance

We developed a model of the human circadian clock, which can predict how well a person's central circadian pacemaker entrains to external signals. A person's alertness and performance is of more relevance to many applications. This is influenced by the circadian clock, but also is affected by other factors (e.g., time since last sleep event). These factors can be modeled using a systems-level approach such as the one proposed by Jewett and Kronauer (1999). In this approach, there are three components to alertness (A) and performance (T):

A or $T = C + H + W.$

The term C is the circadian component of alertness. The component H is the homeostatic drive, which is basically a measure of how long one has been awake and how much sleep one has gotten. Finally, W is the sleep inertia (to be defined later). While these three parts exist for both alertness and performance, the parameter values for each component will be different. All rate constants are given in inverse hours (h^{-1}).

We now describe the model, starting with the homeostatic drive, a measure of how tired a person is, which will affect alertness, performance, etc. From this, a simple sleep-wake model can be generated in that, when H is high enough, sleep is triggered, and when H is low enough, waking is triggered, another simple oscillator.

First, consider sleep. Based on experimental data, it is assumed that a person recovers exponentially during sleep. So we have

$$dH/dt = r_{Hsl}(u_H - H)$$

where $u_H = 0.9949$. It is known that subjective alertness recovers at a slower rate than cognitive throughput. If we are keeping track of cognitive throughput, $r_{Hsl} = 1/2.14$, while for subjective alertness, $r_{Hsl} = 1/9.09$.

It is known that a person can sleep better at certain times of the day than others. This necessitates some circadian modulation of the sleep recovery. The model is modified to:

$$dH/dt = (1 - 0.1x)r_{Hsl}(u_H - H),$$

where x is a variable in the human circadian clock model.

Finally, during the waking day, it is assumed that a person's homeostatic drive decreases exponentially. The longer a person is awake, the more tired they become. When they first wake up, their subjective alertness (corrected for any circadian variations) decreases at a faster rate. This is modeled by

$$dH/dt = (-t^2/(t + t_0))r_{Hw}(H - u_c).$$

The parameters used were $t_0 = 18.88$, $r_{Hw} = -0.006$, and $u_c = 0.21$. However, cognitive throughput decreases in a different way. We find

$$dH/dt = -2t(r_{Hw})^2(H - u_c)$$

with $r_{Hw} = 1/32$.

Next, just after waking, a person typically is still very tired and cannot perform well. To account for this, a "sleep inertia" component is included in the model (Jewett and Kronauer 1999). Upon waking from sleep, we assume that the sleep inertia, W, is some constant value, $W_0 = -0.5346$ for subjective alertness and -0.2868 for cognitive throughput. The only problem is that the prediction of alertness and throughput can never be less than zero. If

$$H + C < |W_0|, W = -(H + C).$$

Once a person is awake, they recover exponentially from sleep inertia. We find

$$dW/dt = -r_w W$$

and $r_w = 1/0.79$ for subjective alertness and $1/0.86$ for cognitive throughput.

Finally, we need to model the circadian component of subjective alertness and cognitive throughput. It was experimentally found that as the homeostatic drive increases, the circadian component of subjective alertness and cognitive throughput increases. Thus, if you have been awake for a long time, the variation in your performance based on the circadian clock will increase. This is why, after an all-nighter, one gets a second wind during the morning of the next day, but is really exhausted by the evening of that day.

First we consider the circadian component as a function of the variables x and x_c of the circadian model,

$$C = A_c(0.91x - 0.29x_c).$$

A_c, varies according to the following equation:

$$A_c = u_c - ae^{H/h_{Ac}},$$

where a is 1.8×10^{-5} and $h_{Ac} = 0.11$. This completes the model and allows us to simulate a person's performance.

These models have been used in simulation software to predict human performance, as is shown in figure 5.17. More details on models of human sleep-wake patterns can be found in (Enright 1980; Gleit et al. 2013; Pavlidis 1973; Strogatz 1986).

5.13   Frontiers: Equivalence of Neuronal and Biochemical Models

We next show how neuronal and biochemical models can have the same mathematical structure and be capable of both type 1 and type 2 behavior. Consider a biochemical feedback loop model proposed by Guantes and Poyatos (2006), which contains both positive and negative feedback:

$$dA/dt = f(A, R) - aA - bRA, \tag{5.13.1}$$

$$dR/dt = g(A) - R. \tag{5.13.2}$$

In this model, A is the concentration of the activator A, and R is the concentration of the repressor R. The clearance of the activator A has rate constant a, while R binds with A to cause its degradation with rate constant b. It is assumed that the clearance of A by R does not remove R, so the term $-bRA$ does not appear in the dR/dt equation. We also assume that $dg/dA > 0$, $df/dA > 0$, and $df/dR < 0$, which gives the structure of a negative feedback loop (A activates R, which represses the production of A), as well as a positive feedback loop (A activates itself). The key question is whether the Morris-Lecar model can be put in this form. It is easy to see that dw/dt in (5.4.2) and dR/dt in (5.13.2) have the same form by replacing A with V and R with w, if $\tau_w(V)$ is approximately constant. Now consider (5.4.1), which we may rearrange as

$$C\,dV/dt = -g_{Ca}\,m_\infty\,(V - V_{Ca}) - g_L\,(V - V_L) - g_K\,w\,(V - V_K) + I$$
$$= I + g_L\,V_L - g_{Ca}\,m_\infty\,(V - V_{Ca}) + g_K\,w\,V_K - g_L\,V - g_K\,w\,V.$$

Here g_L replaces a and g_K replaces b in matching (5.13.1) to this equation from the Morris-Lecar model. Let us equate $f(A, R)$ with $I + g_L\,V_L - g_{Ca}\,m_\infty\,(V - V_{Ca}) + g_K\,w\,V_K$, by

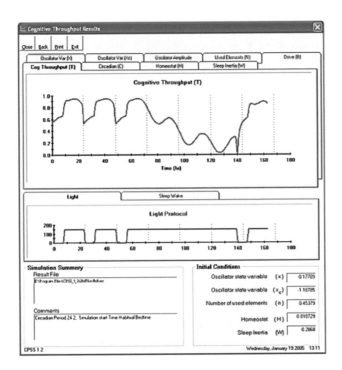

Figure 5.17
Simulations of a software package, CPSS, predicting performance as measured by cognitive throughput after no sleep for 3 days. (Image kindly provided by Dennis Dean.)

noting that $(V - V_{Ca})$ is always negative for physiological values of the voltage and that V_K is negative. We note that $f(A, R)$ and $(I + g_L \, V_L - g_{Ca} \, m_\infty \, (V - V_{Ca}) + g_K \, w \, V_K)$ have similar forms, as well as $g(A)$ and $w_\infty(V)$ (see figure 5.18).

This explains why developing and studying models based on their mathematical structure can be so powerful. One structure can apply to many systems. It is also interesting to note that biochemical models proposed by Guantes and Poyatos (2006) also show both type 1 and type 2 behaviors like the Morris-Lecar model. See Conrad et al. (2008) for more details of this analysis. Interesting work has also been conducted on linking molecular and electrophysiological models (Diekman et al. 2013; LeMasson et al. 1993; Marder and Goaillard 2006; Siegel et al. 1994). But the take-home message is that many models can show either type 1 or type 2 behavior depending on the choice of model parameters.

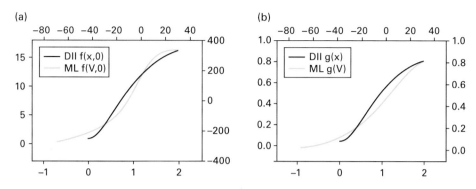

Figure 5.18
Similarity between biochemical and neuronal models. (A) Plots of $f(A, 0)$ from the Guantes-Poyatos (2006) (GP) Design II (DII) model of a biochemical clock and $f(V, 0) = (I + g_L V_L - g_{Ca} m_∞ (V - V_{Ca}))$ from the Morris-Lecar (ML) model. (B) Plots of $g(x) = g(A)$ from the GP model and $g(V) = w_∞(V)$ from the ML model. This shows how the dynamics of the two models are similar, despite the fact that they describe a biochemical (GP) and an electrophysiological (ML) system. From (Conrad et al. 2008), where more details can be found.

Code 5.1 Simulation of Type 1 and Type 2 Behavior in the Morris-Lecar Model

(see figures 5.4, 5.6)

```
% The following code simulates the Morris Lecar and human circadian clock
%models
function y = ml(t, x)
% first type global II in your MATLAB command line.
%Then define II to be whatever value you wish
% The bifurcation for this parameter set is around II = 88.3
%This is for the type II parameters
global II;
V = x(1); w = x(2); Vone = -1.2; Vtwo = 18; Vthree = 2; Vfour = 30;
gca = 4.4; gk = 8.0; gl = 2; Vk = -84; VL = -60; Vca = 120;
phi = 0.04;
% to try the other parameter set uncomment the following line, the
% bifurcation occurs around II = 40 and shows Type 1 behavior
%Vthree = 12; Vfour = 17.4; gca = 4.0; phi = (1/15.0);
minf = 0.5*(1+tanh((V-Vone)/Vtwo));
winf = 0.5*(1+tanh((V-Vthree)/Vfour));
tw = 1/(cosh((V-Vthree)/(2*Vfour)));
y(1) = (1/20.0)*(-gca*minf*(V—Vca) - gl*(V—VL)- gk*w*(V—Vk) + II);
y(2) = phi*(winf—w)/tw;
y = y';
% Enter the following in the command line.
```

```
%II = 40.5;
%[T, Y] = ode45(@ml, 0:0.1:1000, [.5 .5]);
%plot(T, Y(:,1), 'k')
%title('40.5')
%axis([0 1000-65 40])
%xlabel('Time mSec')
%ylabel('V')
function z = circmodel(t, X)
alp = 0.05*LI(t); % The function LI should be defined to determine the
%light levels
pa = 0.2618; c = 33.75; mu = 0.23; k = 0.55; cm = 0.4; bet = 0.0075;
a = (24.0/(0.99669*24.2))^2;
B = c*alp*(1-X(1))*(1-cm*X(2))*(1-cm*X(3));
y(1) = 60*(alp*(1-X(1)) - bet*X(1));
y(2) = pa*(X(3) + B);
y(3) = pa*(mu*(X(3) - (4/3)*X(3)^3) - X(2)*(a+k*B));
z = y';
end
```

Exercises

General Problems
For these problems pick a biological model of interest, preferably one you study.

1. Vary the parameters in your model. What types of bifurcations can you see?

2. Does your system act as an integrator or a resonator? Verify this by adding in appropriate (e.g., sinusoidal) perturbations, or check the system's behavior in the presence of noise.

Specific Problems
1. Consider a positive feedback loop that shows bistability. Draw an example of a bifurcation diagram and a rate balance plot that could represent the behavior of this system.

2. Discuss the differences between the Morris-Lecar and Hodgkin–Huxley models. What are the differences in model assumptions.

3. Simulate the Morris-Lecar model for the type 1 and type 2 parameter set. Change the applied current I, and see how the period changes as oscillations are gained or lost. Explain your results through the bifurcation structure of type 1 and type 2 oscillators.

4. Using the methods of section 4.3, calculate the period of the Morris-Lecar model.

5. Start the model in both the type 1 and type 2 regions before oscillations begin. Demonstrate that, in one region, the model behaves as a resonator, and in another region it behaves as an integrator.

6. Start the model in both regions where oscillations are present. Develop a phase response curve to a brief pulse of current (I) and explain why the differences between these phase response curves can be explained by the model being in the type 1 or type 2 region.

7. Simulate the model with the two parameter sets listed earlier and with additive noise of varying magnitude in the dw/dt equation using the Euler method as described in section 3.4. Generate a histogram for the time between spikes, and explain why the differences are accounted for by the type 1 or type 2 behavior. You may need to simulate many spikes (i.e., > 200) to see the difference.

8. Consider the human model of the human circadian clock (code 5.2) with $\tau_x = 24.2$:

$$\alpha(I) = 0.05 \frac{I^{0.6}}{9500^{0.6}}$$

$$B(n, x, x_c) = 33.85(1-0.4x)(1-0.4x_c)\alpha(I)(1-n)$$

$$\frac{dn}{dt} = 60(\alpha(1-n)-0.0075n),$$

$$\frac{dx}{dt} = \frac{\pi}{12}(x_c + B),$$

$$\frac{dx_c}{dt} = \frac{\pi}{12}\left(0.23\left(x_c - \frac{4}{3}x_c^3\right) - x\left(\left(\frac{24}{0.99669\tau_x}\right)^2 + 0.55B\right)\right),$$

a. What is the predicted period of the human circadian clock during darkness? What about when the light level, I, is 10, 100, 1000, or 10000 lux?

b. Assume that a person leads a simple life with 8 hours of sleep each night ($I=0$) and 16 hours of day ($I=100$). In four days the person will go from Michigan to France. Find the best schedule so that, when in France, the person will continue to live a simple French life with 8 hours of sleep each night ($I=0$) and 16 hours of day ($I=100$), and so that, upon arrival, the person's circadian cycle will be as close as possible to French time.

9. By choosing the following equation for dx_c/dt,

$$\frac{dx_c}{dt} = \frac{\pi}{12}\left(-x\left(\left(\frac{24}{0.9969\tau_x}\right)^2 + 0.55B(n,x,x_c)\right) + 0.23\left(x_c - \frac{4}{3}x_c^3\right)\right),$$

we can set the natural period, τ_x, of a person's circadian clock. In hours, τ_x is, on average, 24.2, but it can vary between 23.5 and 24.5. The Martian day is 24.66 hours with, by assumption, 8.22 hours of sleeping ($I{=}0$) and 16.44 hours of being awake. Assume that, while awake, an astronaut receives a low amount of light ($I{=}5$).

a.   What can the natural period of the astronaut's circadian clock be to entrain exactly to the Martian day (i.e., each day at lights on, his clock is at the same state)?

b.   What values of the natural period of the astronaut's circadian clock would not be compatible with his living, on average, a 24.66 h day?

c.   What behaviors are seen in between the values of the period considered in a and b?

6

Phase Response Curves

Many biological clocks can be characterized by their response to stimuli. Here, we consider models of biological clocks and analytically derive phase response curves (PRCs) for them. These derivations show how PRCs can be calculated for short or long stimuli and models of different numbers of variables. We show how the geometry of the phase space (e.g., the amplitude of the rhythms or the shape of its rhythms) of the model affects the phase response and summarize Winfree's theory of phase resetting. We end with a discussion of how to experimentally determined PRCs and the behaviors that can be seen when a clock is entrained to a period stimulus.

6.1   Introduction and General Properties of Phase Response Curves

Clocks need to integrate signals from the external world to function accurately. The period of circadian clocks is around 24 hours, but it is not exactly 24 hours. Each day, these clocks need to be adjusted to 24 hours by signals from the external world. When we cross time zones, a new phase of the clock must be established so that signals occurring at a particular time of day match the new time zone rather than the old time zone. Such signals are called zeitgebers, and we will now study their effects.

The effect of zeitgebers is often biologically measured by a phase response curve (PRC). Here a stimulus is applied at different times (keeping in mind that phase is defined as time from a marker of the rhythm), and the resulting change in phase is measured. Calling ϕ_{old} the phase before a stimulus is applied and ϕ_{new} the phase of the oscillator after the stimulus, PRCs plot $\phi_{new} - \phi_{old}$ against ϕ_{old}. These plots show how the effect of a stimulus depends on the phase at which it is applied.

Another related curve is the phase transition curve (PTC). Here the initial phase ϕ_{old} is plotted against ϕ_{new} rather than $\phi_{new} - \phi_{old}$. If $\phi_{new} > \phi_{old}$, then the phase is said to be advanced. On the other hand, if $\phi_{new} < \phi_{old}$, the phase is said to be delayed (note: some authors use the opposite convention). Finally, if $\phi_{new} \approx \phi_{old}$, we are said to be at a dead zone of the PRC. An in-depth discussion of this is provided by Winfree (1980). We also introduce a convention. If a brief stimulus is applied at phase ϕ, we denote ϕ_{old} as ϕ and ϕ_{new} as ϕ'.

At a particular phase, one can apply stimuli of different strengths and get a dose response curve that measures how the strength of the stimulus affects the shift in phase.

PRCs are typically found in an assay-stimulus-assay protocol. First, we must determine the phase of the clock. Stimuli are then timed with respect to clock phase, and the response is recorded. It is important to remember that clocks may take a long time to recover from stimuli. So it is best to wait several cycles to determine the changed phase unless the system is so noisy that waiting several cycles may lead to larger errors. Thus, we normally calculate the phase response curve after several cycles.

A way to test how long phase shifting takes is to do a "two-pulse PRC." We give a first stimulus, followed by a second stimulus when we think the phase shifting has finished. If the phase shifting from the first stimulus has indeed finished, we should see a phase response to the second stimulus that we would predict after the full effect of the first stimulus. If this is not what we see, the system has not adjusted yet.

Phase shifts are usually listed modulo the period of the clock. However, in some cases, a phase shift larger than the period of the clock is actually an appropriate result to report. For example, consider the integrate-and-fire model, which goes from a reset value of $V = 0$ to a firing value of 1. If a signal sets $V \ll 0$, it may take more than one period to recover.

Sometimes, a periodic signal is presented to an oscillator with a period different from the oscillator's natural period. If there is a phase where the phase shift caused by the signal equals the difference between the two periods, then entrainment can occur, and the clock will change its period to match that of the signal. To see the effect of the periodic stimulus, we can subtract the difference between the signal's period and the oscillator's natural period from the PRC. The result will tell us the relative phase in the oscillator's original period where the next stimulus will occur if the clock's dynamics are well described by the PRC.

As we will see throughout this chapter, PRCs depend on the geometry of the model's solutions (e.g., amplitude of the rhythm, or the location of a fixed point). We can still determine these effects even if we do not have a good measurement of all the state variables of the oscillation, and may not be able to accurately measure the geometry from all variables. For instance, we may be able to measure the end product of some biochemical reactions that are controlled by a cellular clock, such as one described by the Goodwin model. Such a measurement can be used as a reporter of the clock's rhythm, but the measured product is not involved in producing that rhythm. Although we can get a measurement of the clock's period this way, the phase of the reporter may be different than the phase of any state variable of the clock. For example, if the clock output acts as a repressor or an activator of the reporter, these two possibilities will determine two opposite phase relationships. Nevertheless, a phase change, after any transients, should reflect the underlying oscillator, since the old and new phases will be offset by the same amount. Other properties of the phase response should also be preserved. One such property, as we shall see later, is that a lower clock amplitude will yield larger phase shifts.

This chapter is organized to show how geometric properties of clocks can affect phase response curves. Our general methodology is to consider a model, a stimulus, and then derive a PRC. Some of the details of these derivations could be skipped by some readers without harming the overall conclusions. These details are included for readers who want to see where results come from.

6.2   Type 1 Response to Brief Stimuli in Phase-Only Oscillators

In this section, we derive three phase response curves to one-variable clock models. This illustrates the different shapes that PRCs can take.

An oscillator that has only one variable that evolves along a cycle may be called a phase-only oscillator (and the coordinate that measures how far it has proceeded along the cycle is its phase). Consider the following clock (case 1):

$d\phi/dt = \omega + s$, where ϕ is defined modulo 1.

Now assume that s is a very brief, but powerful, stimulus ($c\delta(t_{stim} - t)$, a Dirac delta function). There are two interesting properties of this stimulus: (1) After this stimulus, the new phase is instantaneously reset to $\phi' = \mathrm{mod}(\phi + c, 1)$; and (2) The signal will always achieve the same phase shift regardless of when it is applied during a cycle. Such signals are characteristic of phase-only clocks.

A slightly more complicated case (case 2) occurs when the effect of signals depends on when they are applied. This is, by far, the most common case. Consider the integrate-and-fire model, which is similar to the previous model. In this case, V determines the phase. Such a model is often used to describe neuronal oscillations; however, in this case (as in the following case), we assume exponential growth for V. Again, we have

$dV/dt = -aV + I,$

and, if $V > V_{th}$, this is changed to $V = V_{reset}$.

After applying a stimulus, $\varepsilon\delta(t_{stim} - t)$, we can calculate how much time there will be till the next time the neuron fires. First, note that the time needed to go from $V_{th} - \varepsilon$ to V_{th} is more than the time from $V_{th} - c - \varepsilon$ to $V_{th} - \varepsilon$ because of the exponential approach to the steady state. We assume that $V_{reset} = 0$ by scaling V. This model will be discussed more in chapter 7.

We wish to find out how much time it takes for the system to go from V_0 to $V_0 + \varepsilon$, since this is the amount of time the stimulus advances the phase. The solution of the model, assuming we do not hit the threshold, and regarding phase zero as the point where $V = V_{reset} = 0$, is

$V(t) = (1 - \exp(-a(t - t_a)))I/a,$

where we define phase ϕ as the time t from the last firing ($\phi = t - t_a$, $V(t_a) = 0$). The stimulus increases V by ε, so we wish to determine ϕ', the phase after the stimulus, from ϕ, the phase before the stimulus. This gives

$$V(t) + \varepsilon = (1 - \exp(-a\phi'))I/a,$$

$$(1 - \exp(-a\phi))I/a + \varepsilon = (1 - \exp(-a\phi'))I/a,$$

$$\exp(-a\phi) - \varepsilon a/I = \exp(-a\phi'),$$

or

$$\phi' = \log(\exp(-a\phi) - \varepsilon a/I)/(-a).$$

We also note that the phase shift can be calculated as

$$\phi' - \phi = \log(1 - \exp(a\phi)\varepsilon a/I)/(-a).$$

The one exception to this is that, if $V + \varepsilon > V_{th}$, then we have

$$\phi' = \log(\exp(-a\phi)) - \varepsilon' a/I)/(-a)$$

where $\varepsilon' = V_{th} - V$.

A slightly more complex model (case 3) is the quadratic integrate-and-fire model, which takes several forms (Latham et al. 2000), including

$$dV/dt = -aV^2$$

and

$$d\phi/dt = 1 - \cos\phi,$$

which are shown to be equivalent by Ermentrout and Kopell (1986). More generally, one can write

$$d\phi/dt = 1 - b\sin(\phi - \phi_0).$$

Working with this general form of the model, and considering it as arising from a two-dimensional limit cycle, we will see in section 6.3 that when a stimulus pushes the oscillator in a particular direction, the component of this stimulus along the phase direction is the stimulus magnitude multiplied by $\sin(\phi)$. Thus, we scale the stimulus strength by $\sin(\phi)$. This is different than the scaling used in the original quadratic integrate-and-fire model, e.g., $dV/dt = V^2 + \delta(t - t_0)$. With our choice of stimulus scaling, we now have

$$d\phi/dt = 1 - b\sin(\phi - \phi_0) + \varepsilon\sin(\phi)\delta(t - t_0).$$

We can divide by $1 - b\sin(\phi - \phi_0)$. We then have

$$\frac{d\phi}{dt} \frac{1}{1-b\sin(\phi-\phi_0)} = 1 + \frac{\varepsilon\delta(t-t_0)\sin(\phi)}{1-b\sin(\phi-\phi_0)}.$$

We can simplify by scaling time such that

$$dt' = dt(1-\varepsilon b\sin(\phi-\phi_0)).$$

We now have

$$\frac{d\phi}{dt'} = 1 + \frac{\varepsilon\delta(t(t')-t_0)\sin(\phi)}{1-b\sin(\phi-\phi_0)}.$$

Note that the signal has the strongest effect on the phase when $\phi - \phi_0$ is any multiple of π.

We plot the PRCs for the three cases we studied in figure 6.1. A common notion about PRCs for type 1 oscillators is that they tend to favor phase advances over phase delays, but it is important to remember that this is not always true (Ermentrout et al. 2012). The quadratic integrate-and-fire model in figure 6.1, for example, shows both advances and delays.

6.3 Perturbations to Type 2 Oscillators

A much more common example of phase resetting is that of a stimulus in a Cartesian direction forcing a limit-cycle oscillator. For simplicity, we can choose the radial isochron clock, whose dynamics can be represented as

$$\frac{dx}{dt} = c(x_c + n_x B(t)),$$

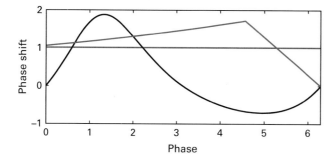

Figure 6.1
Plots of the three PRCs derived for type 1 models. (Red) A phase-only oscillator. (Blue) The integrate-and-fire model. (Black) The quadratic integrate-and-fire model.

$$\frac{dx_c}{dt} = c\left(-x + n_{x_c} B(t)\right).$$

Here we require that $\sqrt{n_x^2 + n_{x_c}^2} = 1$, and for reasons to be explained later, we assume that the total phase shift is less than one quarter of the cycle. Letting $t' = t + a$ is equivalent to rotating the phase portrait. We then find that

$$\frac{dx}{dt} = c\left(x_c + B(t - a)\right),$$

$$\frac{dx_c}{dt} = -cx.$$

Next, we can scale time so that

$$\frac{dx}{dt} = x_c + B(t),$$

$$\frac{dx_c}{dt} = -x.$$

We compare this with the undriven system,

$$\frac{dx}{dt} = x_c,$$

$$\frac{dx_c}{dt} = -x,$$

in order to find the phase shift. Let us look at these systems in the complex plane. Let $z = x + ix_c$. (For this approach, the author acknowledges a suggestion by Richard Kronauer.) We then have for the unforced system $dz/dt = -iz$, or $z = z_0 e^{-i(t-t_0)}$, where $z(t_0) = z_0$. For the driven solution, we have

$$dz/dt = -iz + B(t).$$

Let $B(t)$ be brief, say a delta function, and small in magnitude. Then convert to polar coordinates. We have

$$d\phi/dt = b + \varepsilon \sin(\phi)\delta(t - t_0).$$

In this case, the PRC looks like $\sin(\phi)$.

More generally, the solution of the driven system becomes

$$z(t) = z_0 e^{-i(t-t_0)} + \int_{t_0}^{t} e^{-i(t-\rho)} B(\rho) d\rho = z_0 e^{-i(t-t_0)} + e^{-it} \int_{t_0}^{t} e^{i\rho} B(\rho) d\rho.$$

This takes into consideration the fact that the equations are linear, and therefore have the superposition property. The change in state between this case and the case where $B(t) = 0$ is:

$$\Delta z = e^{-it} \int_{t_0}^{t} e^{i\rho} B(\rho) d\rho.$$

This is a complex number that can take any phase. However, by the same arguments as before, we can scale time ($t' = t + a$) so that Δz is real. Another way to see this is that there exists a time, somewhere during the cycle, when an instantaneous perturbation in the x direction of the same magnitude as Δz can be applied to give the same phase shifting as in the original system. We can also scale time so that it is equivalent to giving an instantaneous pulse at the center of the stimulus ($t/2$).

The drive can also vary with time. However, for simplicity, let us assume that it is constant and, without loss of generality, that $t_0 = 0$. We then calculate the strength of the drive on the system:

$$|\Delta z| = B \left| \int_0^t e^{i\rho} d\rho \right| = B \left| \frac{e^{it} - 1}{i} \right| = B |\cos(t) - 1 + i \sin(t)|$$

$$= B \sqrt{(\cos(t) - 1)^2 - \sin^2(t)} = 2B \sin\left(\frac{t}{2}\right).$$

The relationship between the amount of time the stimulus is presented and the phase shift is plotted in figure 6.2, with $B = 0.5$.

6.4 Instantaneous Perturbations to the Radial Isochron Clock

Now consider an oscillator where the isochrons (lines of constant phase) are radial lines emerging from the origin. The harmonic oscillator (radial isochron clock) described before is one such example (see figure 6.3). We consider the effect of a stimulus that moves the

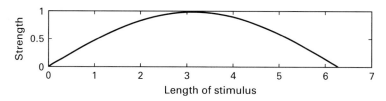

Figure 6.2
The strength of a stimulus as a function of its length in time. The stimulus strength increases until it reaches half the period, and the maximum possible strength is achieved.

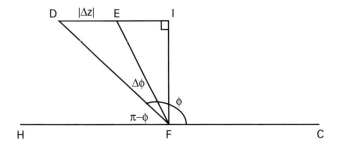

Figure 6.3
Geometry of a phase shift to the radial isochron clock.

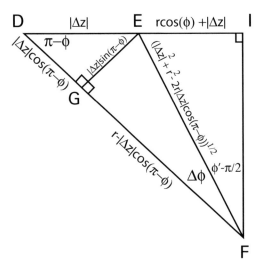

Figure 6.4
This shows a more detailed geometry than that of figure 6.3.

system from point D to point E and derive a formula for the phase shift. Point F is the position of the center of the cycle. The phase before the stimulus is ϕ. The phase after the stimulus is ϕ'. The phase shift is $\Delta\phi$. The point D is the state before the stimulus, whose coordinates are $(r\cos(\phi), r\sin(\phi))$. After the stimulus, we move to point E, whose coordinates are $(r\cos(\phi) + |\Delta z|, r\sin(\phi))$. Point I has coordinates $(0, r\sin(\phi))$ and is used in the derivation of the phase shift. This is illustrated in figure 6.3.

Figure 6.4 fills in the lengths of many of the line segments in figure 6.3. We note that angle FDE is the same as angle HFD in figure 6.3 since DI and HF are parallel. Drawing a perpendicular line from E to DF, we have a right triangle EGD in figure 6.4. We then find the lengths of EG and DG from the definitions of sine and cosine. Since DF has length r, GF has length $r - |\Delta z|\cos(\pi-\phi)$. The length of EF follows from the law of cosines applied to triangle FDE. The length of EI is the first coordinate of the system after the stimulus, or $r\cos(\phi) + |\Delta z|$. Finally, we note that the final phase, ϕ', is the angle CFE in figure 6.3. Since CFI is a right angle, we have that angle IFE is $\phi' - \pi/2$.

With $\sin(\phi' - \pi/2) = \cos(\phi')$, and $\cos(\pi-\phi) = -\cos(\phi)$ by the standard trigonometric identities and also $\sin(\phi' - \pi/2) = $ (the length of EI)/(the length of EF) from figure 6.4, we have

$$\cos(\phi') = \frac{r\cos(\phi) + |\Delta z|}{\left(|\Delta z|^2 + r^2 + 2r|\Delta z|\cos(\phi)\right)^{1/2}} = \frac{\cos(\phi) + \dfrac{|\Delta z|}{r}}{\left(1 + \left(\dfrac{|\Delta z|}{r}\right)^2 + 2\dfrac{|\Delta z|}{r}\cos(\phi)\right)^{1/2}}.$$

This gives a formula for the phase transition curve found in the treatment of this system by Keener and Glass (1984).

In figure 6.5 (top), we plot the phase transition curve for $|\Delta z|/r = 1/3, 1/1.5, 1, 1.5$ and 3. For small $|\Delta z|/r$, we have that $\phi' \approx \phi$ (blue curve). However, if $|\Delta z|/r = 1$, we have ϕ' undefined for $\phi = \pi$, since the denominator of the preceding expression is zero. For large $|\Delta z|/r$, we have $\cos(\phi') \approx 1$, or that $\phi' \approx 0$ or 2π (yellow curve).

For the phase response curve we note that $\tan(\Delta\phi) = $ (the length of EG)/(the length of GF). This gives

$$\tan(\Delta\phi) = \frac{|\Delta z|\sin(\pi - \phi)}{r - |\Delta z|\cos(\pi - \phi)} = \frac{|\Delta z|\sin(\phi)}{r + |\Delta z|\cos(\phi)} = \frac{\sin(\phi)}{\dfrac{r}{|\Delta z|} + \cos(\phi)}.$$

A similar formula is given by Forger and Paydarfar (2004).

In figure 6.5 (bottom), we plot the phase response curve for $|\Delta z|/r$ as $1/3, 1/1.5, 1, 1.5$, and 3. For small $|\Delta z|/r$, the formula for the phase shift looks like a sinusoid, $\pm(|\Delta z|/r)\sin(\phi)$, where after examining several values, one can see that the correct choice of sign is $-$ rather than $+$. However as $r = |\Delta z|$, the formula becomes undefined for $\phi = \pi$. In fact, at this point we can use the fact that:

$$\tan\left(\frac{\phi}{2}\right) = \frac{\sin(\phi)}{1 + \cos(\phi)}$$

to see that $\Delta\phi = \pm\phi/2$. Substitution of several values shows that the correct choice of sign is $-$ (see red curve). Around $\phi = \pi$, the PRC switches from $-\pi$ to π. In the limit of large $|\Delta z|$, we have that $\Delta\phi$ approaches $-\phi$ (see yellow curve).

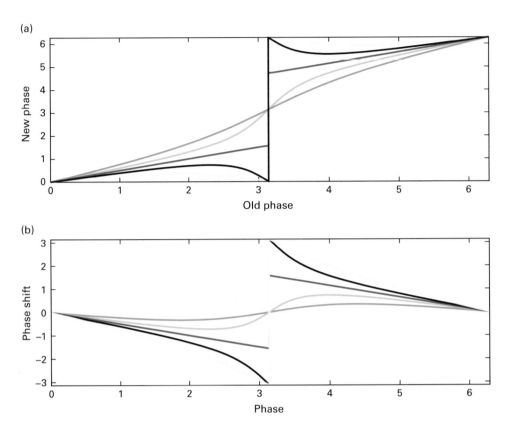

Figure 6.5
Calculated phase transition curve (a) and phase response curve (b) for stimuli of different strengths. These are coded as $|\Delta z|/r = 3$ (yellow), 1.5 (black), 1 (red), 1/1.5 (green) and 1/3 (blue).

This shows that as the strength of stimulus increases from small to large, the phase transition curve moves from an average slope of 1 to 0. A critical point is where $|\Delta z|/r = 1$ where this transition occurs (see figure 6.5). This is further discussed in section 6.8.

This formula also gives the predicted dose response curve at a fixed ϕ. Note the nonlinear relationship between the strength of the drive and the phase shift it causes. For instance, suppose $r = 1$ and $\phi = 3\pi/4$. We then have

$$\phi = \arctan\left[\frac{|\Delta z|}{\sqrt{2} - |\Delta z|}\right].$$

The plot of this function, which is close to linear, but not exactly, is shown in figure 6.6.

The phase response curve depends on the geometry of the limit-cycle model. For instance, let us consider moving the position of the fixed point within a limit cycle. More-

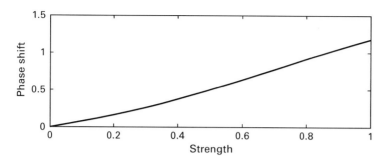

Figure 6.6
Calculated dose response curve.

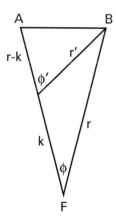

Figure 6.7
Geometry of an oscillator with the center moved.

over, let us move the fixed point without moving the limit cycle. The distance to the limit cycle can be calculated by the law of cosines. In particular, let r be the original radius, and k the distance the fixed point was moved. See figure 6.7 for details.

The new radius is given by

$$r' = \sqrt{r^2 + k^2 - 2rk\cos\phi}$$

We then find that

$$\tan(\phi) = \frac{|\Delta z|\sin\phi}{\sqrt{r^2 + k^2 - 2rk\cos\phi} + |\Delta z|\cos\phi}.$$

This is plotted in figure 6.8 for $k = 0.2$.

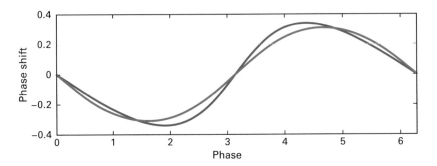

Figure 6.8
(Blue) Phase shift as shown in figure 6.6. (Magenta) Phase shift with the center moved ($k = 0.2$).

6.5   Frontiers: Phase Resetting with Pathological Isochrons

Phase resetting becomes more complex when isochrons (lines of constant phase) are not radial. However, if the isochrons can be calculated, then phase responses can easily be calculated. Simply determine which isochron the system ends on after an instantaneous stimulus. The phase of the isochron is the new phase, and the difference between it and the original phase is the phase shift.

This is further complicated since, in certain cases, isochrons cannot even be defined. For instance, consider the Birkhoff equations (Strogatz 2000) that are sometimes referred to as the radial isochron clock:

$$\frac{dr}{dt} = r\left(1 - r^2\right),$$

$$\frac{d\phi}{dt} = 1.$$

Around the limit cycle, the approach will be exponential (e.g., $r(t) = 1 - be^{-at}$). Now we construct a stimulus with a small amplitude, but with an interesting property:

$$\frac{dr}{dt} = r\left(1 - r^2\right),$$

$$\frac{d\phi}{dt} = 1 + \frac{\varepsilon}{-\log\left(\dfrac{1-r}{b}\right)}.$$

Using our solution for $r(t)$ ($r(t) = 1 - be^{-at}$), we find

$$\frac{dr}{dt} = r\left(1 - r^2\right),$$

$$\frac{d\phi}{dt} = 1 + \frac{\varepsilon}{at}.$$

So, at any r, we can calculate the approximate period and find a relatively nice answer. As we approach the limit cycle, the period speeds up. However, as we approach the limit cycle, the change in phase between the approaching trajectory and the limit cycle changes too much. As time approaches infinity, the change in phase due to this small term (make it as small as you like) approaches infinity. In this case, no isochrons can be defined. However, this case is somewhat rare in that, as shown by Guckenheimer (1975), so long as the model's equations are smooth, and solutions approach or recede from the limit cycle exponentially fast, isochrons exist.

Our definition assumes that all points are eventually attracted to a limit cycle. However, all two-dimensional or higher oscillators contain at least one point that is not attracted to the limit cycle. This phaseless region need not be just a point. One case is a system that is hard excited in that it has a region of attraction for the fixed point as well as another region of attraction for the limit cycle. Such a system is given by

$dr/dt = -r(r - 1)(r - 0.5),$

$d\phi/dt = 1.$

Here the region $0 < r < 0.5$ is the region of attraction of the fixed point. Within this domain, phase cannot be defined, since solutions of this system never attract to the limit cycle. Thus, for a point that does not tend toward a limit cycle, isochrons do not exist. Other interesting examples of isochrons come from Best (1979) as well as Glass and Winfree (1984) and more modern treatments, such as that shown in figure 1.12 (Langfield et al. 2014).

6.6   Phase Shifts for Weak Stimuli

In some cases, as we will see later, amplitude must be taken into consideration in the study of oscillators. However, Art Winfree, in his early work, championed consideration of oscillators that never deviate far from their limit cycle (Winfree 1967). For those oscillators that essentially stay on the limit cycle, the system has, in essence, only one direction in which it evolves, namely, that along the limit cycle. Although biological oscillators typically contain two or more variables, often we are interested only in the phase, the one variable an oscillator cannot do without. Ignoring other variables, we can construct a phase-only oscillator.

To illustrate this phase-only oscillator, consider the harmonic oscillator or the radial isochron clock:

$$\frac{dx}{dt} = x_c,$$

$$\frac{dx_c}{dt} = -x.$$

Let us now consider a small perturbation in the x direction that does not move us far from the limit cycle. Converting to polar coordinates, we find that

$$x = r\sin\phi,$$

$$x_c = r\cos\phi,$$

and

$$\frac{dx}{dt} = \frac{dr}{dt}\sin\phi + r\cos\phi\frac{d\phi}{dt},$$

$$\frac{dx_c}{dt} = \frac{dr}{dt}\cos\phi - r\sin\phi\frac{d\phi}{dt}.$$

Now multiply the first equation by $\cos\phi$ and the second equation by $-\sin\phi$. Then, adding the two equations, we find

$$r\frac{d\phi}{dt} = \cos\phi\frac{dx}{dt} - \sin\phi\frac{dx_c}{dt}.$$

So for a perturbation $\varepsilon(t)$ in the x direction, we have

$$\frac{dx}{dt} = x_c + \varepsilon(t) = r\cos\phi + \varepsilon(t),$$

$$\frac{dx_c}{dt} = -x = -r\sin\phi.$$

We then find that

$$\frac{d\phi}{dt} = \frac{\varepsilon(t)}{r}\cos\phi + 1.$$

The idea here is that $r(t)$ is known on the limit cycle. Since the perturbations of the system do not change the amplitude much, we remain essentially on the limit cycle. Thus, the known function $r(t)$ can be used for r in the previous equation.

Another possibility is that oscillations do not proceed evenly through the limit cycle (suppose they speed up or slow down at different phases). In the unperturbed harmonic oscillator, we have $d\phi/dt = 1$. However, we can convert this model to another model where time progresses unevenly by defining

$(1/f(\phi'))d\phi'/dt = d\phi/dt.$

We then find

$d\phi'/dt = f(\phi'),$

and perturbations become

$$\frac{d\phi'}{dt} = f(\phi') + \frac{\varepsilon(t)}{r(t)}\cos\phi.$$

6.7 Frontiers: Phase Shifting in Models with More Than Two Dimensions

Next, consider the case where we have an oscillation in many dimensions. If the limit cycle lies approximately on a plane, we can redefine the space into two components: one component with two coordinates r and ϕ that represent the oscillation's amplitude and phase on this plane, and other component with multiple coordinates $l, m, n, \ldots$ that represent variables that are "perpendicular" to the oscillation. Our full set of coordinates is thus $(r, \phi, l, m, n, \ldots)$. Suppose we perturb the system in some particular direction. Let the angle between this direction and the (r, ϕ) plane be χ. Small perturbations from the limit cycle become

$$\frac{d\phi}{dt} = \frac{\varepsilon}{r}\cos\phi\cos\chi + 1.$$

In some biochemical applications, we can think about stimuli as perturbations in a Cartesian direction. For instance, in the Goodwin equations, increasing the rate of transcription could be thought of in this way:

$$\frac{dM}{dt} = \varepsilon(t) + \frac{k_{max}}{1 + \frac{k_f}{k_r}P_n^m} - d_0 M,$$

and for small perturbations, we find phase response curves similar to those predicted before. An example where the perturbation is to increase for 1 hour the rate of transcription by 0.1 is shown in figure 6.9.

However, for larger stimuli, nonlinear effects are seen. An example like that in figure 6.9, but with the increase in transcription rate raised to 0.5, is shown in figure 6.10.

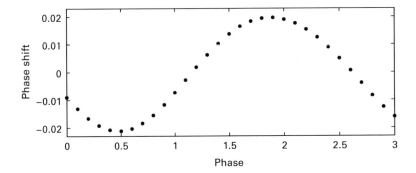

Figure 6.9
Small perturbations to the Goodwin model.

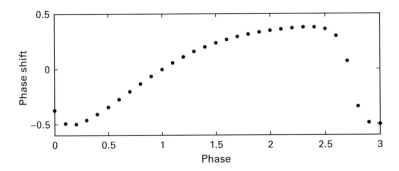

Figure 6.10
Larger perturbations to the Goodwin model.

For instance, in the Goodwin model, shown in figure 6.10, the period of the system is shortened as it returns to the limit cycle favoring phase advances. In fact, the integral of the phase response curve gives an estimate of the effect of the stimulus on the period of the oscillation.

6.8   Winfree's Theory of Phase Resetting

We began our discussions of phase response curves by using geometric arguments to consider the phase shift due to moderate stimuli. We then considered the case of very small stimuli in section 6.6. We now proceed to the case of moderate and large stimuli. Before we proceed, we need one more definition. A phase-amplitude resetting map (PARM) is a plot of all the initial phases and amplitudes just before a stimulus is applied, and the phases and amplitudes just after the stimulus is applied (Jewett et al. 1994). This, along with the phase response curve and the phase transition curve, gives a complete description of the effect of a stimulus if the system has only two variables (Winfree 1980).

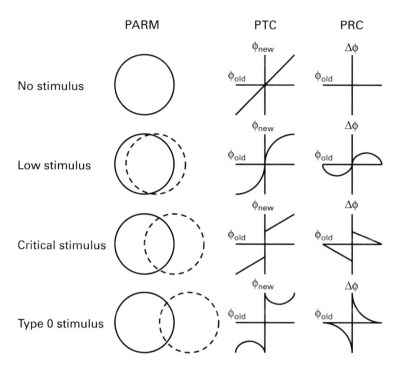

Figure 6.11
Winfree's classification of phase response curves. Shown on the left are the original points on the oscillator (solid) and the new points after a stimulus has been applied at each phase (dashed). The resulting phase transition curve (center) and phase response curve (right) are plotted. All plots have a scale of 0 to 2π on the horizontal axis. The phase response curves have a vertical axis from $-\pi$ to π and the phase transition curves have an axis from 0 to 2π. These curves are schematics of the curves calculated in figure 6.5.

With the PARM, we can now see the effect of stimuli of different magnitudes illustrated in figure 6.11. For small stimuli, the PTC does not deviate much from a line of slope 1, and the PRC does not deviate much from a line of slope 0. Art Winfree called this type 1 resetting (Winfree 1980). For a stimulus that can reach the singularity, we have certain points of the phase response curve (e.g., if the initial phase is at the leftmost point) that are undefined after the stimulus is applied. This is called critical resetting. For even stronger stimuli, the PRC is discontinuous, with a slope of -1, and the PTC becomes close to a line of slope 0. Art Winfree called this type 0 resetting (Winfree 1980). (Note that Winfree's PRC classification *has nothing to do with* the classification of oscillators into type 1 and type 2 that we saw in chapter 5.)

6.9 Experimental PRCs

So far we have mathematically derived PRCs. Experimentally calculating PRCs can be challenging because experiments need to be run many times at different phases. In fact, it is a rite of passage for chronobiologists to spend a night in the lab collecting data points. I was quite lucky to live just a few blocks from the lab of my postdoc so I could catch an hour or two of sleep in between collecting data points. Other considerations include properly building into the study a control, which often means running twice as many experiments as one would do without a control. Additionally, one needs to plan how to measure the state of the rhythm before and after the stimulus. This could easily require several cycles of time since stimuli could take several cycles before their effects are fully seen (and this could be confounded by the fact that noise or measurement error could build during this time). Thus, designing a good experiment is an art. Johnson (1999) offers some further guidelines.

Figure 6.12 illustrates a phase response curve calculated for a model of the human circadian clock. Code 6.1 illustrates how this curve is calculated, including the controls and times when the rhythm is measured.

6.10   Entrainment

Many biological clocks live in environments that change periodically with a period different from the clock's intrinsic period. Our circadian clocks, whose period is not exactly 24 hours, must adjust to a 24-hour period. Other examples are the many biochemical clocks that live within growing and dividing cells and must adjust to the overall cell cycle. The process of a clock adjusting its period so that it better matches that of its environment is called entrainment. As a clock adjusts its period, it can also adjust its phase to have an advantageous phase relationship with the cycle of its environment (e.g., fruit flies waking just before sunrise to search for food).

However, the process of entrainment can be quite complex. In particular, a clock may match its period to a multiple of the environmental period, or to a fraction of it. For example,

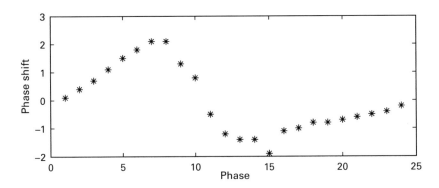

Figure 6.12
A predicted human phase response curve to a 5-hour light stimulus. See code 6.1 for more details.

a cellular biochemical network with a period of 11 hours may change its period to 12 hours so that every two periods, its rhythm coincides with a 24-hour environmental period. We now illustrate this with a example by Glass and Keener, following work by Guevara and Glass (1982). The model in this example was motivated by studies of periodic stimulation of heart cells.

We consider the model of a radial isochron clock with a brief stimulus applied every τ time units. Assuming that the clock resets to its preferred amplitude after each stimulus, the formula we determined in section 6.4 gives the change in the phase of this clock due to the stimulus, i.e., the PRC. Using this formula, we can determine what happens to this system in the presence of a periodic input in the following way. We advance time by τ time units, apply the formula, and then repeat. A simple code for doing this is provided as code 6.2.

In figure 6.13 we show the average new period of the radial isochron clock after repeated stimulation with a period τ ranging from 0 to the natural period of this clock and with the stimulus instantaneously moving the state of the system from 0 to two times the preferred amplitude of the clock (r).

Many possible behaviors can be seen in this model, some of which are illustrated. The black region corresponds to strong stimuli that are constantly applied. In this region, the clock stops oscillating since the stimuli never let it proceed around the cycle. The blue region corresponds to the clock matching the period of the stimulus. Most of this region corresponds to the clock having a constant phase relationship with the stimulus, but sometimes (particularly near the border with other regions) the clock can periodically change its phase relationship and regain the same phase relationship after going through 2, 3, … cycles. An experimental example of this behavior for human circadian rhythms is shown in the next chapter, in figure 7.10.

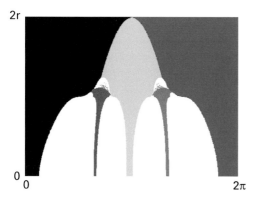

Figure 6.13
Average new period of the radial isochron clock under periodic stimulus. The horizontal axis is the stimulus period τ, and the vertical axis is the measure of the stimulus amplitude. See text and code 6.2 for more details.

Many other regions of entrainment also exist. For example, we plot those corresponding to the clock having a period of twice the stimulus period (green) 1.5 times the stimulus period (magenta), and 3 times the stimulus period (red). This example illustrates how iterating PRCs can explain entrainment, and shows some of the many possible behaviors that can be seen in entrainment.

Code 6.1 Calculating a Predicted Human PRC

(See figure 6.12)
```
clear
% The following code simulates the effect of a 5 hour light pulse
%mimicking human PRC experiments. A subject has a night of 8 hours plus
% however many additional hours are needed to time the stimulus. A 5
%hour light pulse is then presented, and an additional 40 hours of
%darkness to see how much the phase has changed.
for ij = 1:24
%Note [0.0240    -0.8332    0.7558] is an initial condition one would
%find for the model entrained to a normal sleep wake cycle just before
%night
[T, Y]= ode15s(@circmodeloff, 0:0.1:(8+ij), [0.0240-0.8332 0.7558]);
[T, Y]= ode15s(@circmodelon, 0:0.1:5, Y((8+ij)*10 + 1,:));
[T, Y]= ode15s(@circmodeloff, 0:0.1:40, Y(51,:));
[aa, bb] = min(Y(:,2));
prc(ij) = bb;
end
%The following code is a control where no light is given
for ij = 1:24
[T, Y]= ode15s(@circmodeloff, 0:0.1:(8+ij), [0.0240-0.8332 0.7558]);
[T, Y]= ode15s(@circmodeloff, 0:0.1:5, Y((8+ij)*10 + 1,:));
[T, Y]= ode15s(@circmodeloff, 0:0.1:40, Y(51,:));
[aa, bb] = min(Y(:,2));
prcc(ij) = bb;
end
prc = prc-prcc;% This calculated the phase shift
% The following assumes that the phase shifts will not be greater than
%12 hours
for ij = 1:24
if prc(ij) > 120
prc(ij) = prc(ij) - 242;
end
end
prc = prc/10;
% Note timepoints are taken by ode15s every 1/10th of an hour
figure(2)
plot(prc, '*k')
```

```
function z = circmodelon(t, X)
%This simulates the model of Forger et al. JRB 1999
% circmodeloff.m is the same as this code except with alp = 0;
alp = 0.05; pa = 0.2618; c = 33.75; mu = 0.23; k = 0.55; cm = 0.4;
bet = 0.0075; a = (24.0/(0.99669*24.2))^2;
B = c*alp*(1-X(1))*(1-cm*X(2))*(1-cm*X(3));
y(1) = 60*(alp*(1-X(1)) - bet*X(1));
y(2) = pa*(X(3) + B);
y(3) = pa*(mu*(X(3) - (4/3)*X(3)^3) - X(2)*(a+k*B));
z = y';
end
```

Code 6.2 Iterating PRCs

(See figure 6.13)

```
%The following code explores how PRCs can be iterated to understand
%entrainment. The reader is encouraged to comment the code to understand
%how entrainment can easily be calculated through PRCs.
clear
for ij = 1:800
for ik = 1:999
plock(ij, ik) = calplock(ij/400, (ik*2*pi/1000));
end
end
figure(14)
spy(abs(plock) > 15, 'k')
hold on
spy((.98 < plock) & (plock < 1.1))
spy((1.98 < plock) & (plock < 2.1), 'g')
spy((2.98 < plock) & (plock < 3.1), 'r')
spy((1.48 < plock) & (plock < 1.51), 'm')
axis off
function y = calplock(b, tau)
x = rand*2*pi;
phadif = 0;
for ij = 1:250
  if x > pi
      x = 2*pi-x;
      phin = 2*pi-acos((b+cos(x))./(1+2*b*cos(x)+b^2).^(1/2));
      x = 2*pi-x;
  else
      phin = acos((b+cos(x))./(1+2*b*cos(x)+b^2).^(1/2));
end
x = mod(phin + tau, 2*pi);
end
for ij = 1:50
```

```
  if x > pix = 2*pi-x;
      phin = 2*pi-acos((b+cos(x))./(1+2*b*cos(x)+b^2).^(1/2));
      x = 2*pi-x;
  else
      phin = acos((b+cos(x))./(1+2*b*cos(x)+b^2).^(1/2));
end
phadif = phin-x + phadif + tau;
x = mod(phin + tau, 2*pi);
end
y = tau*50/(phadif*tau/(2*pi));
```

Exercises

General Problems
For these problems pick a biological model of interest, preferably one you study.

1. Simulate perturbations to the system that could occur in the wild to determine PRCs to low-amplitude and high-amplitude stimuli.

2. Can one get both type 1 and type 0 PRCs?

3. Find similarities that the PRCs you generate have to those studied in this chapter.

4. How does the geometry of the phase space affect the PRC?

Specific Problems
1. Consider the following model for a biological oscillator that considers only its phase, $d\phi/dt = 1 + I(t)\sin(2\pi t)$.

   a. Develop a phase response curve (either through simulation or analysis) to a stimulus where $I(t) = 1$ for $t_a < t < t_a + 0.1$.

   b. Are there any stimuli that show critical resetting? Explain.

   c. Now assume we apply the stimulus in part (a) periodically. What is the range of entrainment?

   d. Assume the stimulus is applied with a period of 2π. What are the phases where the oscillator stably entrains?

2. Define the following terms: zeitgeber, entrained, phase response curve, dose response curve.

3. Simulate the Hodgkin–Huxley model with an applied current of 8. Choose initial conditions where a limit cycle is found. Apply a pulse of current to the model for at different phases and amplitudes. Determine which causes critical phase resetting.

4. Consider the harmonic oscillator $dx/dt = x_c + I(t)$, $dx_c/dt = -x$, and let $x(0) = 1$ and $x_c(0) = 0$.

a.   Assume $I(t)$ is 0 except when a stimulus is applied. During the stimulus, assume it takes a value of 10. Find when and how long the stimulus must be applied to bring the system to its phaseless region.

b.   Now consider a stimulus applied to two harmonic oscillators with half the strength of that found in (a), and applied for the same length of time: (i.e., $dx/dt = x_c + I(t)/2$, $dx_c/dt = -x$, $dx'/dt = x'_c + I(t)/2$, $dx'_c/dt = -x'$). These two oscillators need not be in phase (i.e., $x(0)$ need not equal $x'(0)$, and the same for $x_c(0)$ and $x'_c(0)$). This stimulus is not a critical stimulus on either of the harmonic oscillators. Can it be a critical stimulus when considering the output of both oscillators (i.e., $(x + x', x_c + x'_c)$)? Demonstrate.

5. Consider the model of firefly flashing described in equation 1.1 of Mirollo and Strogatz, which is the same as our integrate-and-fire model with $I = 2$, $a = 1$, and $\varepsilon = 0.1$, which is the signal from one firefly to the next.

a.   What is the period of an isolated firefly?

b.   Consider a system of two fireflies. Determine the change in phase when the two fireflies flash once. Compare it with the formula we derived for a pulse signal to the integrate-and-fire model.

c.   Using your answer to (b), find the phases that the fireflies never synchronize regardless of the number of flashes. (Hint: the formula $\Delta_n \phi = (-1)^{n-1} f((-1)^{n-1} \phi_{n-1})$.)

7

Eighteen Principles of Synchrony

Many biological systems are composed of coupled oscillators. After defining terminology related to oscillator coupling, we consider the Peskin model, a model of all-to-all coupling of integrate-and-fire model oscillators. Using this model, we demonstrate many of the key behaviors that are seen in coupled oscillator systems, with excitatory or inhibitory coupling and possible oscillator heterogeneity, including noise-induced oscillations, and pattern formation. Some generalizations to multidimensional oscillators are presented. Our results are summarized as eighteen principles of synchrony. We end with a simple proof of the result by Mirollo and Strogatz that pulse coupled oscillators almost always synchronize.

7.1 Basics and Definitions of Synchrony

Synchrony is a huge topic and the focus of thousands of papers (Pikovsky et al. 2003) covering everything from firefly flashing (Mirollo and Strogatz 1990b) to embryonic development (Ay et al. 2013; Ay et al. 2014). Our goal here is to introduce key ideas in the topic of synchronization, some very recent or novel, which should be widely applicable.

We will illustrate eighteen key properties of synchronization with simple examples. Some hold generally, while others indicate special behaviors that can be seen. We initially consider oscillators of just one variable, and, following the conventions of Winfree and Kuramoto (Kuramoto 1984; Winfree 1967), we denote the number of oscillators as N. As an alternative to the many other reviews of synchronization that focus more on the modeling formalism proposed by Kuramoto (Kuramoto 1984; Pikovsky et al. 2003), we focus mainly on pulse-coupled oscillators. Models that have considered higher-dimensional oscillators, such as the work of Pavlidis (1969), received less attention at first, perhaps because they are more difficult to study. At the end of this chapter, in sections 7.7–7.8, we present recent techniques to study higher-dimensional oscillators.

7.1.1   What Are Coupled Oscillators?

We often assume that each oscillator can easily be identified in the equations and that the coupling can be singled out as well. Sometimes the physiology makes these things clear. For example, individual neurons are often physically distinct, and communication between them is limited to synapses. On the other hand, coupling of biochemical networks remains complex. For example, consider figure 7.1, where two feedback loops share a common element. We may be tempted to consider them one system, but, as the example shown in the figure demonstrates, they can also show properties like those of coupled oscillators.

Principle of Synchrony 1: *Determining individual oscillators from a coupled system can be complex.*

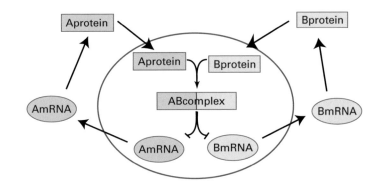

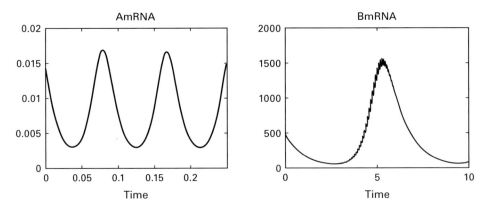

Figure 7.1
Example of two coupled molecular feedback loops that share the same element, an AB complex. This shared element could also be a signal (e.g., calcium) triggered by both loops. The dynamics of the B loop are much slower than those of the A loop. In this case, the primary oscillations in the A and B loops have different periods. See code 7.1 for more details. In the top plot, the circle represents the nucleus of the cell.

7.1.2 What Types of Coupling Can Be Found?

One way to classify coupling is by charting which oscillators affect which other oscillators. Five interaction schemes are typically found:

All-to-all (mean field) coupling. In this case, every oscillator affects every other oscillator.

Nearest-neighbor coupling. Here each of the oscillators can be ordered with oscillator j affecting only oscillator $j-1$ or oscillator $j+1$. The oscillators can form a ring, where the last oscillator affects the first oscillator, or a line, where the end oscillators do not affect each other.

Random coupling. Here, each oscillator has a probability, p, of affecting another oscillator. So, each oscillator will, on average, affect Np other oscillators, where N is the number of oscillators in the system.

Small-world network coupling (Watts and Strogatz 1998). Here, most interactions are local, meaning an oscillator mainly affects only its nearest neighbors or a small group (clique) of other oscillators that have a good chance of also affecting that oscillator. However, in addition to these local interactions, some interactions between cliques are allowed, where an oscillator from one clique affects an oscillator from another clique.

Scale-free networks coupling. Here, the probability of being connected to n nodes of the network is $n^{-\gamma}$ for some constant γ.

Another way to classify coupling is by what the coupling does. Excitatory coupling means that the coupling typically increases the state variable of the receiving oscillator. Inhibitory coupling decreases the state variables. Often it is assumed that excitatory coupling leads to more phase advances and inhibitory coupling leads to phase delays, but, as can be seen by our previous discussion of phase response curves, both excitatory and inhibitory stimuli often lead to advances as well as delays.

Yet another way to classify coupling is by when the coupling is active. Continuous coupling refers to oscillators continuously sending signals to other oscillators. This would occur, for example, when they share a component. Pulse coupling refers to oscillators that affect each other only at a specified phase during the cycle. A good example of this is coupling among neurons. Communication typically occurs only when an action potential is fired, at least according to conventional models.

7.1.3 How to Define Synchrony?

The definition of synchrony is complicated. For example, one might assume that synchronous oscillators have the same phase. However, this assumption is often too restrictive. For example, coupling can allow oscillators with different intrinsic periods to agree on a common period. When they do this, nonidentical oscillators rarely have the same phase. Thus, we call synchronized oscillators those that have agreed on a common fixed period and a fixed phase relationship. (We will see later, when discussing phase trapping in section 7.7, why it is important to specify that these are fixed.) Phase-synchronized oscillators are those that, in addition to being synchronized oscillators in the sense just defined, also have the same phase. In a possible abuse of terminology, we consider oscillators phase-desynchronized when they are approximately equally distributed over all phases, and we consider them desynchronized, for example, when oscillators with different periods are coupled, but they have not agreed on a common period.

7.1.4   How Can We Measure Synchrony?

Principle of Synchrony 2: *Synchrony can be measured via synchrony indexes.*

Assume we have N oscillators with phases $0 \leq \phi_j \leq 2\pi$, $1 \leq j \leq N$. The synchrony index provides a quantitative way to measure the synchrony of these oscillators. It is measured by the same measure of the standard deviation discussed in chapter 1, section 1.12. Here we express the magnitude v of the complex mean as an absolute value,

$$v = (1/N)|\Sigma\exp(i\phi_j)| = (1/N)|\Sigma\cos(\phi_j) + i\sin(\phi_j)|,$$

so v is a real number and $v \geq 0$. If $v = 1$, then $\phi_j = \phi_k$ for all j and k. If $v = 0$, then the phases of the oscillators are evenly spread from 0 to 2π. However, the condition $v = 0$ could be achieved in many ways, and we mention two contrasting examples. For the first example, $\phi_j = 0$ for $j \leq N/3$, $\phi_j = 2\pi/3$ for $N/3 < j \leq 2N/3$, and $\phi_j = 4\pi/3$ for $2N/3 < j \leq N$. In this case, there are three groups of oscillators, and each oscillator within a group shares the same phase as the other oscillators in that group. For the second example of how the condition $v = 0$ can be achieved, we choose the phases in such a way that no oscillator has the same phase as any other oscillator, and $\phi_{j+1} = \phi_j + 2\pi/N$.

Clustering is when oscillators form groups, but the groups are out of phase with each other. Clustering as in the first example yields $v = 0$, but it can nonetheless be detected by a higher-order synchrony index (Diekman and Forger 2009):

$$v_k = (1/N)|\Sigma\exp(ik\phi_j)| = (1/N)|\Sigma\cos(k\phi_j) + i\sin(k\phi_j)|.$$

Like v above, v_k is a real number and $v_k \geq 0$. If there are k groups of oscillators, then $v_k = 1$ even if $v = 0$.

Another measure we can use is the distance between the phases of oscillators. So the sum of the distances between one oscillator and all others is

$$\Sigma_j \text{mod}(\phi_i - \phi_j, 2\pi).$$

However, we could also have used

$$\Sigma_j f(\text{mod}(\phi_i - \phi_j, 2\pi)),$$

where f is a monotonic function. This could give more weight to oscillators that are closer or further apart in phase in the calculation of phase difference. A variation of this will be used in the proofs that follow at the end of the chapter.

7.2   Synchrony in Pulse-Coupled Oscillators

Principle of Synchrony 3: *Homogeneous oscillators typically show either phase synchrony or phase desynchrony, perhaps with clusters of oscillators evenly spread over all phases.*

We now define the types of oscillators that we will study for most of this chapter. These oscillators are a generalization of the integrate-and-fire model studied in section 6.2. Consider oscillators with phase $0 < \phi_i \leq 1$ (1 is chosen rather than 2π to match the convention of Mirollo and Strogatz). When $\phi_i = 1$, oscillator i "fires" and the phase of each oscillator j other than the firing oscillator is increased from ϕ_j to $\phi_j + p(\phi_j)$, which, if >1, causes the other oscillator to fire and synchronize. After firing, an oscillator's phase is reset to 0 even if other oscillators fire with it. In between firings, all oscillator phases are increased at the same rate until the next firing.

Consider the case of two oscillators. Thinking carefully about these oscillators, one might expect that eventually the oscillator's phase would be close enough so that when one oscillator fired, the other synchronized. However, another case could occur. Imagine if each time one oscillator fired, the other oscillators had the same phase distribution. This would also be a steady state, and one where the oscillators would never synchronize.

A key question in this field has been to determine when oscillators synchronize and when they do not. Under general conditions, it turns out they will synchronize. To illustrate this process, we include a simulation in figure 7.2. A proof of this fact is offered in section 7.10.

For inhibitory coupling, the opposite occurs. The oscillators typically reach a stable distribution of phases that are spread over all phases. Some may synchronize and form clusters, but they will be clusters firing out of phase with oscillators that are not in the cluster (Golomb et al. 1992). This is illustrated in figure 7.3 and proved in section 7.10.

We note that in our earlier example, oscillators fired when they reached a state of phase 1. If this was increased to a larger level (equivalent to increasing the amplitude of oscillators), the coupling signal would be effectively weakened.

Principle of Synchrony 4: *The effect of the coupling signal inversely scales with the amplitude of an oscillator.*

7.3   Heterogeneous Oscillators

It is the rule rather than the exception for coupled oscillators to be heterogeneous. One possible form of heterogeneity would be for the oscillators to produce coupling signals of different strengths. This turns out not to affect the previous arguments and the proof in 7.10, as we could imagine multiple identical oscillators synchronized with the same phase. These would then fire together, which would be equivalent to one oscillator with a coupling signal strength equal to the sum of those of the individual oscillators. By

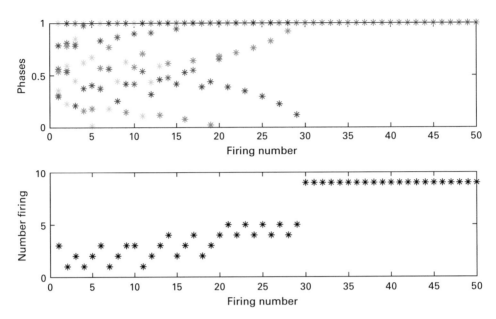

Figure 7.2
Simulations of 9 coupled oscillations as proposed by Mirollo and Strogatz. The top plot shows the phases of
the oscillators (each color-coded) at each firing, and the bottom plot shows how many oscillators fired at each
firing. As oscillators synchronize, their combined phase is shown by a single point with the color of one of the
oscillators. Random initial conditions are used. Note that the initial firing causes two other firings. See code 7.2
for more details.

appropriately increasing the number of oscillators, we could then form an identical problem to that of oscillators with different coupling strengths.

While there are many possible other forms of heterogeneity one can consider, we consider here one of the most common: that the oscillators have different periods but are
otherwise identical. If anything, the heterogeneity in period helps them synchronize, in
that it can increase the difference between the speeds of the fastest and slowest oscillators,
making their phases change more and more likely for a synchronization. In fact, even when
uncoupled, heterogeneous oscillators will come arbitrarily close to each other in phase
unless their periods are some multiple or fraction of each other's.

The preceding argument seems to suggest that heterogeneous oscillators are even more
likely to synchronize than homogeneous oscillators. However, the opposite is often true,
since we do not know if oscillators that are initially synchronized will remain synchronized. After the oscillators are synchronized, they will continue with their own periods
until the next pulse-coupled event, and thus desynchronize. So the question we address
here is: will heterogeneous oscillators that start synchronized, e.g., because of a synchro-
nizing event, remain synchronized?

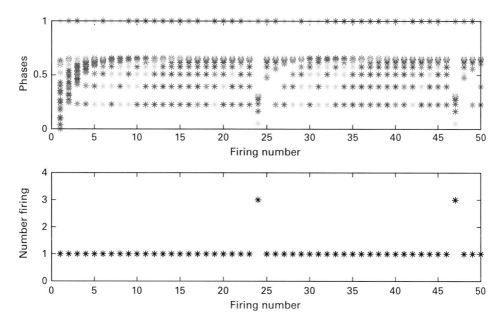

Figure 7.3
Simulations of 25 coupled oscillations as proposed by Mirollo and Strogatz. The top plot shows the phases of the oscillators (each color-coded) at each firing, and the bottom plot shows how many oscillators fired at each firing. Random initial conditions are used. Having epsilon be negative ($\varepsilon < 0$) in Mirollo and Strogatz's model causes inhibitory coupling. After firing, we set all phases >1 to be 1 and all phases <0 to be 0. See code 7.3 for more details.

This question may initially seem to be a simple function of the coupling strength, and how different the periods are. What matters is the ratio of the difference in period to the coupling strength.

Principle of Synchrony 5: *In the limit of strong coupling and small period differences, synchrony can persist. In the limit of weak coupling and large period differences, synchrony will not persist.*

An illustration of intermediate coupling is shown in figure 7.4. In fact, in such simulations there may be some oscillators that always remain synchronized and others that always remain desynchronized, even if the oscillators are identical except for their initial states. Such a situation is called a chimera.

Principle of Synchrony 6: *In between these limits, surprisingly, the question of whether two oscillators stay synchronized depends very much on the state of the other oscillators in the network.*

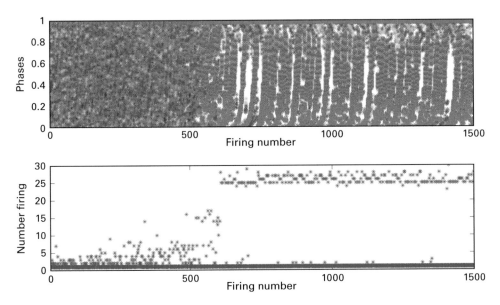

Figure 7.4
Intermediate coupling strengths lead to networks that transition into and out of synchrony. This simulation is of 30 coupled oscillators with different frequencies. A denser plot is seen since we consider a longer time interval. As can be seen for later firings, the network alternates between states where more or fewer oscillators fire in phase. The plot uses code 7.2 with So = 1.5, gamma = 2, eps = .003, and frq = ones(1, numosc) + 0.5*(1:numosc)/numosc.

Here is an example:

Consider three oscillators, which all fire together initially. First, consider the case where they are uncoupled. A is the fastest, and when it fires, we denote B's phase as $1 - \varepsilon_B$. When B fires, we denote C's phase as $1 - \varepsilon_C$. Now if they are coupled, A's firing could increase C and B's phase by more than ε_B and B's firing could increase C's phase by more than ε_C. In this case, all three oscillators will synchronize once A fires, as this will trigger B to fire, which will trigger C to fire. This is called a cascade, and it helps with synchrony, since the fastest oscillators fire to help the slower oscillators fire. Here B's firing helps A synchronize with C. If A's firing would not be enough to bring C to fire since, alone, it would need to increase C's phase by $\varepsilon_B + \varepsilon_C$, the intervention of B would allow them to synchronize.

However, B's firing can also desynchronize A and C. Consider the case that A and C start synchronized, but have different periods and B is ahead of them so that it will fire before A and C. B will increase both A's phase and C's phase. However, since A is faster, and the PRC has a positive slope, it will increase A's phase more than C's phase, and make it harder for A to synchronize C. Thus, we see two potentially opposite effects. Previously, we saw that B could help C fire with A, and now B could prevent C from firing with A.

This is a useful place to interject a side note about the cascade to illustrate general principles of coupled oscillations. First, when an oscillator fires in a cascade, it does not affect the oscillators that have just fired. If it did, it would push the phase of the oscillators that have just fired away from those that are about to fire.

Principle of Synchrony 7: *Having a refractory period, e.g., having the oscillators insensitive to firing after they have just fired, helps in synchrony.*

Next, consider the case where many oscillators in the network are not rhythmic. For example, this may occur if their state approaches some state below the threshold for firing. If they are coupled as part of the network, the coupling can push them over the threshold. This has been explored by many authors (e.g., Enright 1980; Ko et al. 2010; Smolen et al. 1993).

Principle of Synchrony 8: *Coupling can cause nonrhythmic oscillators to become rhythmic.*

This is illustrated in figure 7.5.

There are even examples where no oscillator in the network would be rhythmic when isolated, but the network could become rhythmic when coupled.

In general, intermediate levels of coupling are difficult to study, as there are a potentially infinite number of configurations of the oscillators. However, one key principle holds for most coupled oscillator problems with intermediate levels of coupling:

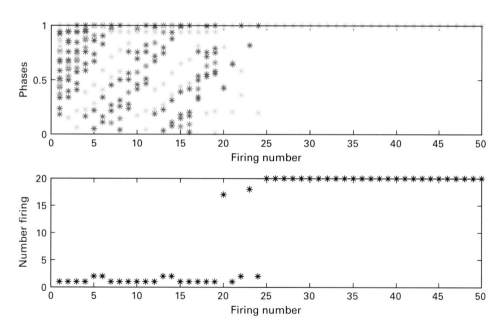

Figure 7.5
Nonrhythmic oscillators can become rhythmic when coupled. This is similar to figure 7.2 with 20 oscillators; however, the last 10 oscillators stop at phase 0.9 before the effects of the coupling are accounted for. All oscillators in the network nonetheless synchronize. This is done by replacing the line pha = frq*min((1-pha)./frq)+pha; with the following two lines: pha = frq*min((1-pha(1:10))./frq(1:10)) +pha; pha(11:20) = min(pha(11:20), 0.9).

Principle of Synchrony 9: *As more oscillators fire synchronously, they create a larger signal that can synchronize even more oscillators.*

This is reminiscent of positive feedback loops, but a better analogy is to phase transitions in physics. Triggering one firing could trigger another, which could trigger another, like a row of dominos falling, except that in the case of dominos, each domino depends only on the previous one, whereas here, each oscillator feels all others.

We now illustrate this key principle. Assume that at least one oscillator has fired, and calculate how many others are brought past threshold to fire. This can be easily seen by considering the cumulative distribution function $\vartheta(x)$, defined as

$\vartheta(x) \equiv$ the fraction of the oscillators whose phase lies between $1-x$ and 1, where 1 is the firing state.

As x goes from 0 to 1, $\vartheta(x)$ takes account of the fraction of oscillators in states ranging from 1 down to 0 along the phase coordinate, incorporating into its count those oscillators that are within a distance x of the firing threshold. Note that $\vartheta(0) > 0$, since some oscillator has fired.

Assume that all oscillators with phase $1 - x$ to 1 have fired. Now we wish to determine if more will fire. The collective signal from these oscillators is $\varepsilon\vartheta(x)N$, where N again is the number of oscillators. More oscillators will be brought to synchrony unless

$$\varepsilon\vartheta(x)N = x,$$

and this can be visualized by plotting $\varepsilon\vartheta(x)N$ and x, similar in spirit to the rate balance plots described in chapter 4, section 4.8 for positive genetic feedback loops. We can also solve $\varepsilon\vartheta(x)N - x = 0$ to determine this number.

Note that small changes in $\vartheta(x)$ can lead to very large changes in the number of oscillators that synchronize. This makes synchrony very dependent on the initial phase of the oscillators.

This property does not hold for uncoupled oscillators. Uncoupled oscillators ignore each other. There is a slight chance that two of them will have the same phase; however, there is a much lower chance that three uncoupled oscillators will have the same phase, and so on for four, etc. As we have seen earlier, in a coupled case, the states in which just two oscillators have the same phase can be less likely than states in which three or more oscillators have the same phase. Thus, in the coupled case, having more oscillators synchronized increases the likelihood that additional oscillators will join in the synchronization, while in the uncoupled case the opposite happens.

Now consider the case where the oscillators have similar periods, but not quite close enough for all oscillators to synchronize. Again, synchronization events must occur, and these could lead to the network being almost synchronized. However, the coupling is not quite strong enough, and at one point synchronization fails. The key here is that it can fail in an epic way. Imagine that a cascade is happening and one oscillator falls out of place. Now the oscillators that just fired are desynchronized from the oscillators that will next fire in the cascade. This illustrates another key point about intermediate coupling.

Principle of Synchrony 10: *For coupling strengths strong enough to cause some synchronization, but not enough to fully synchronize the network, the network can show large changes in the synchrony index over time, alternating between states of near total synchrony and near desynchrony.*

This is illustrated in figure 7.4. Such changes in the synchronization of the network are much larger than what one would find occurring in an uncoupled network.

7.4   Subharmonic and Superharmonic Synchrony

Principle of Synchrony 11: *Oscillators can synchronize to other oscillators if their periods are approximately related by simple fractions.*

This is illustrated in figure 7.6 and discussed by many authors (e.g., Jordan and Smith 1987).

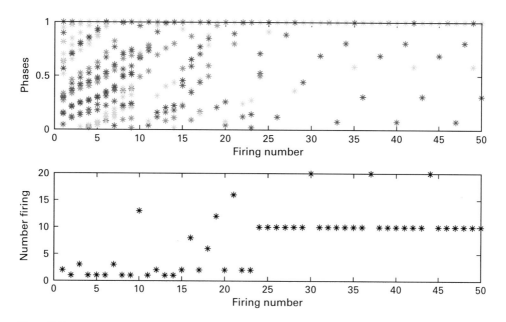

Figure 7.6
Simulations as in figure 7.2, but with 10 oscillators with a frequency of 1 and 10 oscillators with a frequency of 2.1. The oscillators form two populations that synchronize with each other.

Here we consider the synchronization of two oscillators, noting that the same principles apply for larger populations of oscillators. Even if oscillators that start out synchronized do not synchronize at the next firing, a later synchronization could occur. For example, if one oscillator's period is nearly double the other's, then the faster oscillator could fire twice, and at its second firing, the firing of the slower oscillator could be close enough for synchronization to occur. Such synchronizations are called subharmonic or superharmonic, coming from the terminology of entrainment (Jordan and Smith 1987): Subharmonic/superharmonic entrainment means that an oscillator synchronizes to a signal that has a longer (shorter) multiple of its own period.

Now consider any two oscillators with different periods. Initially, we regard them as uncoupled. The ratio of their periods can be approximated by

Period of oscillator A/Period of oscillator B $\approx a/b$,

where a and b are integers. Thus, starting with the oscillators synchronized, after a cycles of oscillator A and b cycles of oscillator B (which we denote as a-b synchronization) their phases will again be close, ignoring for the sake of simplicity, the resetting that occurs during the a cycles of oscillator A and b cycles of oscillator B. In fact, for appropriate choices of a and b, we can make this phase difference closer and closer. The strength of the coupling needed to a-b synchronize these two oscillators can be made as small as one would like. By this logic, any two oscillators should eventually synchronize. While this is correct mathematically, experimentally it is not particularly useful, since

1. the oscillators may synchronize first in a 1–1, 2–1, or some other subharmonic or superharmonic manner on the way to a-b synchronization, and
2. the amount that the oscillator's average period will be changed by this a-b synchronization could be minimal, particularly if the ratio of the oscillators' periods is close to a/b, perhaps even so small that it is impossible to experimentally measure.

7.5 Frontiers: The Counterintuitive Interplay between Noise and Coupling

Intuitively, one would expect that noise internal to oscillators could desynchronize a network, and that coupling would fight against noise, mitigating its effects by averaging over many oscillators (Liu et al. 1997). While noise can desynchronize a network and coupling can overcome noise, these general principles have exceptions that are important to consider.

First, recall that the more oscillators that are synchronized, the stronger the signal to synchronize other oscillators. So while noise may bring some oscillators' phases apart, it can also bring phases together (Goldobin et al. 2010). If enough phases happen to synchronize, a cascade can synchronize the entire network.

Principle of Synchrony 12: *If the effect of the cascade is strong enough, noise can synchronize a network of oscillators. Oscillators can also synchronize to a noisy external signal.*

In this section, we therefore consider the same pulse-coupled oscillator model as we discussed previously, however, we change the model to assume that the time needed to go from the reset state to threshold is drawn from a probability distribution. We also study different events that can trigger a cascade, for example, if the first element to fire triggers a cascade, or if it is the third element that does so. Our goal is to determine how accurately the population of oscillators fires.

The standard rule when coupling noisy elements together is that the ensemble noise should decrease by $1/N^{1/2}$ where N is the number of elements (Liu et al. 1997). However, this assumes a simple averaging of the state of the elements, which often does not occur. For example, assume that one oscillator in the network is significantly faster than all the other oscillators. Once this oscillator fires, also assume that the cascade is strong enough to cause all other elements of the network to fire. In this case, the network firing will be just as noisy as the firing of the fastest oscillator.

A slightly less noisy case is where every oscillator has the same average period, but again, it is the first oscillator to fire that synchronizes all the others. As pointed out by Enright (1980), this is much more noisy than, say, having the third-, fourth-, or fifth-firing oscillator, etc., synchronize the network. This property comes simply from probability. The largest variance occurs in the first (or last) element of independent and identically distributed random elements, the next largest variance occurs in the second (or next to last) element, and so on, until the variance of the middle element, with ordinal rank $N/2$ (or $N/2 + 1/2$ for N odd), which has the least variance. How a biological oscillator could implement synchronization by these less noisy elements is interesting to ponder: It seems that there would need to be some thresholding with the coupling, and that the effects of the coupling would need to be deferred rather than instantaneous, so that no coupling signal would be sent unless a certain number of oscillators had fired.

To further test this enhanced noise of the first oscillator to fire, we generate 100 random numbers from a normal distribution 10,000 times, representing 10,000 realizations of the phase of 100 oscillators in a noisy coupled oscillator system (see figure 7.7 and code 7.4). Assume that each oscillator has some large fixed mean period, so that the difference in period between the fastest and slowest oscillator is small with respect to the mean period. We then find the standard deviation of the phase of the oscillator that was the largest of the 100, then the standard deviation of the next largest oscillator out of the 100, and so on, until the smallest. As seen in figure 7.7, top left, the middle oscillators have the lowest standard deviation, suggesting, as before, that these are the least noisy. Now we average the top two phases, then the next two, and so on, until the bottom two, and repeat this for averaging three, four, all the way to a hundred, averaging all the oscillators. This is shown in figure 7.7, top right. Although the average of all the oscillators has the lowest standard deviation, it is not much less than the standard deviation of the middle oscillator. The opposite results are seen for phases taken from a uniform distribution, where the largest or smallest oscillators are the least noisy; see figure 7.7, bottom.

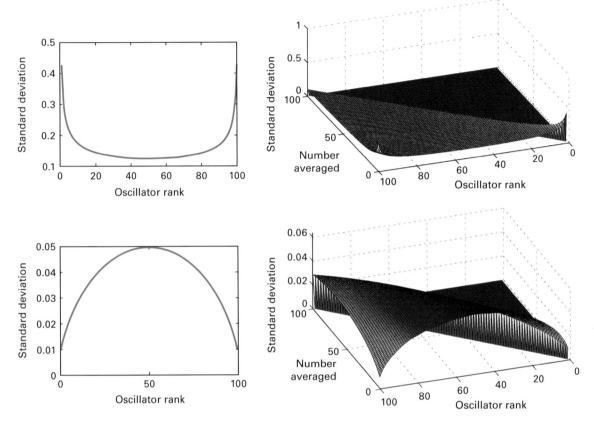

Figure 7.7
Determining the most accurate signal from noisy oscillators. Top left, the standard deviations of 100 Gaussian random variables (representing the phases of oscillators in a noisy system), rank ordered from the largest to the smallest. Top right repeats this plot, but it also shows the standard deviations of the averages of two oscillators, averages of three oscillators, and so on until all the oscillators are averaged. Bottom, the most accurate oscillators are first or last when phases of the oscillators are taken from a uniform distribution. See code 7.4 for details.

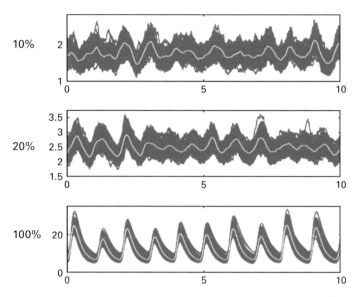

Figure 7.8
Simulations of a network of circadian oscillators. Coupled oscillations occur when BMAL levels are reduced to 10 percent of their original values. At this level, representing the removal of BMAL1, isolated cells are not rhythmic, nor is the network rhythmic without noise. Individual oscillators are shown in blue, with the average shown in yellow. Taken from Ko et al. (2010).

Finally, noise can cause oscillations in single cells, and coupling can propagate this information to cause an entire network to be rhythmic.

Principle of Synchrony 13: *Noise and coupling can synergistically cause rhythms.*

However, noise-induced network oscillations can be noisy on the population level, especially if one oscillator could have the effect of triggering a cascade that causes the rest of the network to fire. Ko et al. (2010) present a biological example of this, in which neurons in the SCN are made nonrhythmic by removing the BMAL1 protein. These authors show that rhythms can occur with cooperative effects of noise and coupling. See figure 7.8.

7.6   Nearest-Neighbor Coupling

Previously, we considered coupling that was all-to-all. When this is not the case, interesting spatial patterns can occur. Although this subject could easily take up a whole book, we will illustrate it with a simple example, inspired by Ermentrout and Kopell (1984), who drew upon work on coupled oscillators in the gut done by several authors (e.g., Linkens 1976). The result we will find here, with oscillators forming clusters with different frequencies, is a special case of the frequency plateaus discussed by Ermentrout and Kopell (1984). We consider a one-dimensional chain of oscillators, while noting that more interesting patterns can occur in two or more dimensions.

Principle of Synchrony 14: *Nearest-neighbor coupling can cause spatial patterns.*

Consider a chain of integrate-and-fire oscillators where each oscillator is coupled only to the previous oscillator. In other words, when one oscillator fires, it affects only the next oscillator in the chain. We also assume that as we proceed down the chain, the frequency of the oscillation decreases. An example of this for 20 oscillators is shown in figure 7.9. The first group of oscillators all fire about 1,300 times. After this, four more pairs of oscillators also fire the same number of times.

This can be understood heuristically by considering the following simple model. Assume oscillators synchronize if their periods have a difference less than ε. In this case, start the oscillators all at a particular phase. When the first oscillator fires, it can cause the next oscillator to fire, and so on. However, when the coupling is not strong enough to bring an oscillator to the same frequency as the first oscillator, that oscillator breaks free. Nevertheless, it can synchronize other oscillators, forming another cluster. An interesting property of this is that it depends on initial conditions. If we now assume the first oscillator does not

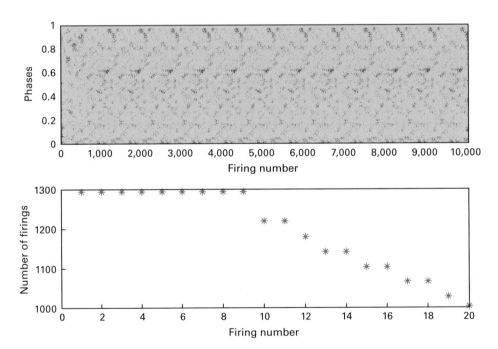

Figure 7.9
Simulations with nearest-neighbor coupling. As one proceeds down the chain of oscillators, the autonomous frequency of individual oscillators decreases. See code 7.5 for more details.

start with phase zero, it may be the second oscillator that forms the cluster. Such dependence on initial conditions is a property of Turing patterns (Turing 1990), which generate interesting spatial dynamics in coupled oscillators, e.g., the Belousov-Zhabotinskii reaction (Jahnke et al. 1989).

7.7   Frontiers: What Do We Gain by Looking at Limit-Cycle Oscillators?

Consider the case of two coupled van der Pol oscillators. The van der Pol model allows us to explore the effects of having the oscillators described by both phase and amplitude, and having a nonmonotonic PRC as well. This analysis is based largely on course notes by Richard Kronauer (1998), as well as on Rand and Holmes (1980). Here, we adopt one further convention to make clear where assumptions are being made. When an approximation is first made, we indicate it with $\approx$ but later with $=$ (otherwise most equations would have $\approx$ and it would be less clear what approximations are being considered).

The specific system we study is

$$d^2x_1/dt^2 + \varepsilon(-1+x_1^2)\,dx_1/dt + (1-\varepsilon\upsilon_1)x_1 + \varepsilon a_{12}x_2 = 0,$$

$$d^2x_2/dt^2 + \varepsilon\beta(-1+x_2^2)\,dx_2/dt + (1-\varepsilon\upsilon_2)x_2 + \varepsilon a_{21}x_1 = 0,$$

where the oscillators have an uncoupled limit cycle (i.e., $a_{12}=a_{21}=0$) with amplitude 2. The parameters υ_1 and υ_2 measure the differences between the periods of the oscillators, while ε is a parameter that measures the nonlinearity in the system. The coefficient β determines the relaxation rate to the limit cycle of oscillator 2 in relation to that of oscillator 1, while the cross-coefficients a_{12} and a_{21} tell us the strengths of the coupling of oscillator 2 to oscillator 1 and the coupling of oscillator 1 to oscillator 2, respectively.

Letting $x_1 = r_1\cos(\tau+\theta_1)$ and $x_2 = r_2\cos(\tau+\theta_2)$, we convert to polar coordinates. Using the method of averaging, which will be described in detail in chapter 10, one can show that this system is well approximated by

$$(1/\varepsilon)dr_1/dt = r_1/2 - r_1^3/8 + a_{12}r_2\sin(\theta_1 - \theta_2)/2,$$

$$(r_1/\varepsilon)d\theta_1/dt = \upsilon_1 r_1/2 + a_{12}r_2\cos(\theta_1 - \theta_2)/2,$$

$$(1/\varepsilon)dr_2/dt = \beta(r_2/2 - r_2^3/8) + a_{21}r_1\sin(\theta_2 - \theta_1)/2,$$

$$(r_2/\varepsilon)d\theta_2/dt = \upsilon_2 r_2/2 + a_{21}r_1\cos(\theta_2 - \theta_1)/2,$$

and by letting $\theta_{12} = \theta_1 - \theta_2$, we have

$$(1/\varepsilon)dr_1/dt = r_1/2 - r_1^3/8 + a_{12}r_2\sin(\theta_{12})/2,$$

$$(1/\varepsilon)dr_2/dt = \beta(r_2/2 - r_2^3/8) - a_{21}r_1\sin(\theta_{12})/2,$$

$$(2/\varepsilon)d\theta_{12}/dt = \upsilon_1 - \upsilon_2 + (a_{12}r_2/r_1 - a_{21}r_1/r_2)\cos(\theta_{12}).$$

Next, let $r_i = 2 + \delta_i$ for small δ_i, and $i = 1,2$ so

$r_i = 2 + \delta_i$ and $r_i^3 \approx 8 + 12\delta_i$

and

$(1/\varepsilon)d\delta_1/dt = -\delta_1 + a_{12}(1 + \delta_2/2)\sin(\theta_{12})$,

$(1/\varepsilon)d\delta_2/dt = -\beta\delta_2 - a_{21}(1 + \delta_1/2)\sin(\theta_{12})$,

$(2/\varepsilon)d\theta_{12}/dt = \upsilon_1 - \upsilon_2 + (a_{12}(1 + \delta_2/2 - \delta_1/2) - a_{21}(1 + \delta_1/2 - \delta_2/2))\cos(\theta_{12})$,

and we assume that the amplitude of each oscillator quickly reaches equilibrium, so that

$\delta_1 \approx a_{12}\sin(\theta_{12})$,

$\delta_2 \approx -a_{21}\sin(\theta_{12})/\beta$,

which leaves just one differential equation for θ_{12}:

$(2/\varepsilon)d\theta_{12}/dt = \upsilon_1 - \upsilon_2 + (a_{12} - a_{21})\cos(\theta_{12}) - (a_{12} + a_{21})(a_{12} + a_{21}/\beta)\sin(2\theta_{12})/4$.

Looking at the fixed points of this equation, one can see that many types of behaviors can occur. First, let $(a_{12} - a_{21})$ be large. If $\upsilon_1 - \upsilon_2$ is small, the oscillators will synchronize 90° out of phase.

Principle of Synchrony 15: *The relative coupling strengths can determine the phase difference between the oscillators.*

Another case is where $a_{12} \approx a_{21}$. Here, the term $(a_{12} - a_{21})\cos(\theta_{12})$ disappears, so that is why the last term has been kept. In this case, if $\upsilon_1 - \upsilon_2$ is small, the oscillators can stably entrain either in phase (0° phase difference) or out of phase (180° phase difference). Also note that, as β decreases, larger values of $\upsilon_1 - \upsilon_2$ can be accommodated.

Principle of Synchrony 16: *The smaller the "stiffness" of the oscillator, the larger the range of period differences that can be accommodated.*

We calculate the period of the coupled system from

$(r_1/\varepsilon)d\theta_1/dt = \upsilon_1 r_1/2 + a_{12}r_2\cos(\theta_{12})/2$.

With $r_1 \approx r_2 \approx 2$, we have

$(1/\varepsilon)d\theta_1/dt = \upsilon_1/2 + a_{12}\cos(\theta_{12})/2$.

So when they are in phase, the frequency is $1 + \varepsilon(\upsilon_1 + a_{12})/2$, and it is $1 + \varepsilon(\upsilon_1 - a_{12})/2$ when they are out of phase. *Thus, the phase configuration can determine the period of coupled oscillators.*

More generally, many problems of two coupled oscillators can be reduced to a differential equation for the phase difference between the oscillators (e.g., $d\theta_{12}/dt$), as well as

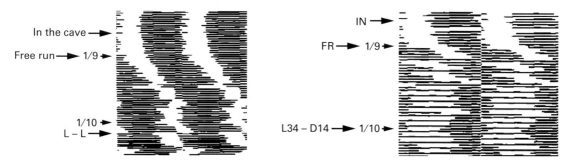

Figure 7.10
Human sleep in time isolation. Recorded daily temperature (left) and sleep (right) rhythms from a subject isolated from time cues in a cave. These rhythms are initially in phase, but then exhibit phase trapping (Kronauer et al. 1982). Taken from Chouvet et al. (1974).

differential equations for the amplitudes of the oscillators. Here, we assumed that the oscillators quickly reached their limit-cycle amplitudes. If we do not assume this, many other behaviors can be seen. One behavior occurs when a stably entrained state loses stability via a Hopf bifurcation (Jordan and Smith 1987). This is known as *phase trapping*, and it has even been seen in human experimental data (Gleit et al. 2013; Kronauer et al. 1982) (see figure 7.10 for a record of a man living in a cave). In this case, the phase difference between the oscillators (as well as their relative amplitude) oscillates over time with a period that can be significantly different than the period of the oscillators.

Patterns of synchrony can also be found by looking for future times where the system takes the same state, but the state of the oscillators is switched. This we will find in patterns of not only in-phase synchrony but also out-of-phase synchrony. An example of using these techniques to study locomotion is presented by Golubitsky et al. (1999).

7.8   Coupling Damped Oscillators

Here we illustrate a property of coupled limit-cycle oscillators that was recently pointed out by Wang and Slotine (2005).

Principle of Synchrony 17: *Damped oscillators easily synchronize.*

Consider n oscillators of the form

$$d\underline{x_i}/dt = f(\underline{x_i}) + u(\underline{x_i}, z(\underline{x}_1,\ldots,\underline{x_i},\ldots,\underline{x_n})).$$

These oscillators are identical in that they have the same coupling and the same coupled dynamics. Suppose we can rewrite

$$u(\underline{x_i}, z(\underline{x}_1,\ldots,\underline{x_i},\ldots,\underline{x_n}))=u_d(\underline{x_i}) + u_c(z(\underline{x}_1,\ldots,\underline{x_i},\ldots,\underline{x_n})).$$

We then have

$$dx_i/dt - f(x_i) - u_a(x_i) = u_c(z(x_1,\ldots,x_i,\ldots,x_n)).$$

This is an interesting form, which all n oscillators then take. While we do not know $u_c(z(x_1,\ldots,x_i,\ldots,x_n))$, we do know that it is some function of time. Let us define synchrony in the following way. Consider any initial conditions of

$$dx_i/dt - f(x_i) - u_a(x_i) = 0$$

and assume that the aforementioned system has a "contracting" property, in that, over time, all initial conditions go to the same value. The system is then no longer a sustained oscillator, perhaps because of the effects of the coupling term, but the system could be a damped oscillator. If this system has the contracting property, the related system

$$dx_i/dt - f(x_i) - u_a(x_i) = g(t) \equiv u_c(z(x_1,\ldots,x_i,\ldots,x_n))$$

also has this property (and not just for u_c but for any $g(t)$) (Wang and Slotine 2005). Thus, where we start an oscillator has no bearing on its final state. Since the only difference between the oscillators is their initial state, they must go to the same value and synchronize.

A few words of caution are needed. First, while the preceding argument shows that the different oscillators go to the same state over time, we do not know what this state is. Typically, $u_c(z(x_1,\ldots,x_i,\ldots,x_n))$ is an oscillation of some sort, but it could also go to a fixed point or show more complex behavior. Second, the argument reduces the coupled oscillator problem to one of different initial conditions of the same system. This may work, but only if the oscillators are homogeneous. Finally, this works typically for larger values of coupling, at least those that can overcome the individual oscillators' tendency to be sustained and instead contract.

7.9   Amplitude Death and Beyond

An interesting property of coupled limit-cycle oscillators is that when they are strongly coupled out of phase, they can drive each other's amplitude to zero (Mirollo and Strogatz 1990a). We illustrate this with two coupled van der Pol oscillators. When the oscillators start in phase, oscillations continue. When they start 180° out of phase, rhythms quickly dampen to zero amplitude. See figure 7.11 and code 7.6.

Principle of Synchrony 18: *Strongly coupled oscillators can annihilate each other's amplitude when coupled out of phase.*

7.10   Theory: Proof of Synchrony in Homogeneous Oscillators

Here, we present a simple proof of a celebrated result established by Mirollo and Strogatz (1990b) in their work on pulse-coupled oscillators, which answers a question posed by

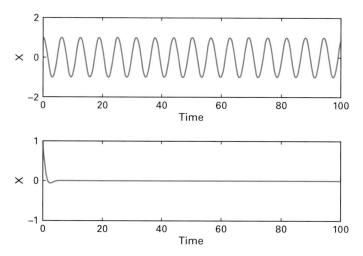

Figure 7.11
Amplitude death. Simulation of two van der Pol oscillators with in-phase coupling (top) and out-of-phase coupling (bottom). In the top, both oscillators start at phase point (1,0), and in the bottom, one oscillator starts at (1,0) and the other at (−1, 0). The latter case quickly leads to amplitude death. See code 7.6 for more details.

Peskin (1975). In order to preserve the homogeneity of the oscillators, we limit our consideration to cases that have an all-to-all coupling scheme.

We recall the basics of this system described in section 7.2: Consider oscillators with phase $0 < \phi_i \leq 1$. When $\phi_i = 1$, oscillator i "fires" and the phase of each oscillator j other than the firing oscillator is increased from ϕ_j to $\phi_j + p(\phi_j)$, which, if >1, causes the other oscillator to fire and synchronize. After firing, an oscillator's phase is reset to 0 even if other oscillators fire with it. In between firings, all oscillator phases are increased at the same rate until the next firing.

We require that $dp/d\phi > 0$ (and typically assume that $p(\phi) > 0$ although this is not required; see the following note), and use ϕ_i^m to denote the state of oscillator i just as the mth firing occurs. Consider the phase shift to oscillator i caused by the previous N firings, which we will call the "speed" of the oscillator:

$$s_r(\phi_i) = p(\phi_i^{r-1}) + p(\phi_i^{r-2}) + \cdots + p(\phi_i^{r-N}).$$

Claim: Unless $s_r(\phi_i) = s_r(\phi_j)$ for all i, j, a synchronization must occur in finite time.

Note: The requirement that $p > 0$ is not necessary for the proof but does require a different setup to account for the possibility that $\phi + p(\phi) < 0$. To fix this, we require that $\phi + p(\phi) \geq 0$. This allows for the possibility that $\phi + p(\phi) = 0$, which would mean that, at a firing (when the phase of the firing oscillator goes to zero), another oscillator goes to zero and the two oscillators synchronize in a way similar to the preceding case where one oscillator firing pushes another oscillator above 1.

Proof. Start the oscillators in any configuration, including those that are partially synchronized, and assume that no oscillators will further synchronize. We start our discussion just
before a firing occurs. We can find a "fastest" oscillator k so that $s_r(\phi_k) \geq s_r(\phi_i)$ for all i, and
a "slowest" oscillator l so that $s_r(\phi_k) \leq s_r(\phi_i)$ for all i, with $s_r(\phi_k) > s_r(\phi_l)$. Then, to mod 1, ϕ_k^{r-N}
$- \phi_l^{r-N} + s_r(\phi_k) - s_r(\phi_l) = \phi_k^r - \phi_l^r$ since the only time when the oscillators are advanced at different rates is during a firing. So, over the previous N firings, oscillator k phase-advances
with respect to oscillator l by an amount $\varepsilon = s_r(\phi_k) - s_r(\phi_l)$ mod 1, meaning, for example,
if $\phi_k^r > \phi_l^r$, then $\phi_k^r - \phi_l^r = \phi_k^{r-N} - \phi_l^{r-N} + \varepsilon$, or if $\phi_l^r > \phi_k^r$, then $\phi_l^r - \phi_k^r = \phi_l^{r-N} - \phi_k^{r-N} - \varepsilon$. Also,
by the definition of s, $s_{r+1}(\phi_k) - s_{r+1}(\phi_l) = s_r(\phi_k) - s_r(\phi_l) + p(\phi_k^r) - p(\phi_l^r) - p(\phi_k^{r-N}) + p(\phi_l^{r-N})$.
Since $dp/d\phi > 0$, there is a δ such that $s_{r+1}(\phi_k) - s_{r+1}(\phi_l) > s_r(\phi_k) - s_r(\phi_l) + \delta$. In other words,
if we consider the state at the next firing (say by oscillator j), oscillator k is phase-advanced
with respect to oscillator j when compared to the last time j fired, oscillator l is phase-
delayed when compared with the last time j fired, and the difference between the speeds of
oscillators k and l, as defined before, also increases.

We can now repeat the argument one firing later, by finding the fastest oscillator and the
slowest one. If they are oscillators k and l respectively, the difference between the speed
of the fastest and slowest oscillators is $s_{r+1}(\phi_k) - s_{r+1}(\phi_l)$. If the fastest or slowest oscillator has changed, the difference in speed is greater than this amount. At the next firing, the
speed difference must again increase by at least δ. Repeating the argument shows that after
each additional firing, the difference in speed between the fastest and slowest oscillators
increases again by δ. However, the difference in speed cannot grow arbitrarily large (this
would imply, for example, that there is an arbitrarily large phase difference between oscillator k and the other oscillators) without one oscillator overtaking another, an event that
would trigger synchronization. QED

This shows that eventually we will end in a state where all the oscillators proceed at the
same speed. Does this mean that all initial states will end up synchronized? The answer
to this is no. However, Mirollo and Strogatz show that if we pick an initial condition randomly, with probability 1, the end state will eventually be synchronized and not go to one
of these states. We can see this based on the following arguments.

Choose a random initial state and encode it as a set of phases $(\phi_1, \ldots, \phi_{N-1})$ of the $N-1$
oscillators when the Nth oscillator has fired. The chance of hitting a partially synchronized
state, where two oscillators have the exact same speed, is zero, so we can assume we do
not start at a partially synchronized state without any loss. Because of the monotonicity of
p, oscillators that have distinct phases before a firing will still have distinct phases after
the firing, unless they fire themselves, and this continues to hold for any finite number of
firings. Additionally, by the monotonicity of p, oscillators with distinct speeds before the
firing, will continue to have distinct speeds. In fact, even stronger properties hold. If we
multiply the strength of the firing signal from any oscillator by an integer multiple between
1 and N, they will still have distinct speeds with probability 1. Also, if we remove any

number of the oscillators, this would be equivalent to starting with fewer oscillators, and the oscillators will still have distinct speeds. But synchrony events can be thought of as removing oscillators and multiply the strength of the oscillator they synchronize to by an integer multiple. Thus, with probability 1, another synchrony even will happen, unless all oscillators are synchronized.

For the case of inhibitory firing, we consider the case where $dp/d\phi < 0$ and we also assume that either $d^2p/d\phi^2 > 0$ or $d^2p/d\phi^2 < 0$ (this concave-up or concave-down condition matches many models, including the original by Mirollo and Strogatz, although their example considers $dp/d\phi > 0$). With these assumptions, the assertion about the oscillators' behavior is:

As $t\to\infty$, $s_r(\phi_i)\to s_r(\phi_j)$ for all choices of i and j.

Proof: Similar arguments to the preceding proof hold up to noting that there exists δ such that now $s_{r+1}(\phi_k) - s_{r+1}(\phi_l) < s_r(\phi_k) - s_r(\phi_l) - \delta$, and thus the speeds of the oscillators will be brought closer together. The fastest and slowest oscillator speeds will converge until all oscillators have the same speed (note that δ can be bounded from below unless the oscillators have the same speed), which, by the definition of s, means that the oscillators are equidistant from each other. Two possible cases could stop this. First, two oscillators could synchronize, as described before. This will only happen a finite number of times, so each time this happens, we can restart our argument, and the oscillators will again approach the same period. The more difficult case is to determine what happens if after a firing the fastest or slowest oscillator changes.

Consider the fastest oscillator at time r (similar arguments hold for the slowest oscillator). Let it be oscillator k that is the fastest oscillator, a status that we assume switches to oscillator j at the next firing: $s_{r+1}(\phi_j) > s_{r+1}(\phi_k)$, and before the firing $s_r(\phi_j) < s_r(\phi_k)$. This means $s_{r+1}(\phi_j) - s_r(\phi_j) > s_{r+1}(\phi_k) - s_r(\phi_k)$ or $p(\phi_j^r) - p(\phi_j^{r-N}) > p(\phi_k^r) - p(\phi_k^{r-N})$. We also have $\phi_k^r - \phi_k^{r-N} > \phi_j^r - \phi_j^{r-N}$, since k was the fastest oscillator, which if $d^2p/d\phi^2 > 0$ means $\phi_k^r < \phi_j^r$ (if $d^2p/d\phi^2 < 0$, $\phi_k^r > \phi_j^r$). Note that oscillators cannot overtake each other (they can synchronize, which can happen only a finite number of times, we can just start over) so their ordering is fixed. Thus, k cannot later overtake j at a future time, which means that k cannot be faster than j. There are $_2C_N = N!/(2!(N-2)!)$ pairings, and only this many possibilities for oscillators to synchronize to each other. QED

Note: One amazing fact is that the preceding proofs also hold for oscillators with different coupling strengths. We can see this by replacing such a network, with a network of oscillators with the same strength, and just add more synchronized oscillators to each phase so that the same overall effect occurs. It also holds when oscillators have different periods. Thus, so long as there is a fastest and a slowest oscillator, there will be another synchrony event in finite time for excitatory coupling. However, the difficulty with period heterogeneity is that after a synchronization, the oscillators may not remain synchronized.

Code 7.1 Two Coupled Biochemical Feedback Loops

```
%This code shows the example of two interlocked feedback loops. The
%reader is encouraged to determine if they act as two independent loops.
%or one loop.
function Y = twoloopsexample(t, X)
eps = 100;
AmRNA = X(1); AmRNAc = X(2); Aprotein = X(3); Aproteinc = X(4);
BmRNA = X(5); BmRNAc = X(6);, Bprotein = X(7); Bproteinc = X(8);
complex = X(9);
Y(1) = 1/complex^3 - eps*AmRNA;
Y(2) = eps*AmRNA—eps*AmRNAc;
Y(3) = eps*AmRNAc—eps*Aprotein;
Y(4) = eps*Aprotein—eps*Aproteinc;
Y(5) = 1/complex^20 - BmRNA;
Y(6) = BmRNA—BmRNAc;
Y(7) = BmRNAc—Bprotein;
Y(8) = Bprotein—Bproteinc;
Y(9) = 20*Bproteinc*Aproteinc—eps*complex;
Y = Y';
%The following code is for the command line.
%[T, X] = ode15s(@twoloopsexample, 0:0.001:50, [1 1 1 1 1 1 1 1 1.1]);
%figure(14)
%subplot(1, 2, 1)
%plot(T(40000:40250)-40, X(40000:40250, 1), 'k')
%axis([0 0.25 0 0.02])
%title('AmRNA')
%xlabel('Time')
%subplot(1, 2, 2)
%plot(T(40000:50000)-40, X(40000:50000, 5),'k')
%axis([0 10 0 2000])
%title('BmRNA')
%xlabel('Time')
end
```

Code 7.2 Pulse-Coupled Oscillators

(see figure 7.2)

```
%This is a sample code for simulating pulse-coupled oscillators.
clear
numosc = 9;
numevents = 50;
% number of firings to consider
So = 1.5; gamma = 2; eps = .01;%this sets the parameters of the model
% Mirollo and Strogatz call I in the I&F model of 7.2 So and a gamma
frq = ones(1, numosc);% this is the speed of the oscillators
```

```
% This is an addition to the Mirollo and Strogatz model and is
%equivalent to a scaling of time of each oscillator.
pha = rand(1, numosc); % this sets the initial state of the oscillators
pharec(1,:) = pha; % This records the state of the oscillators
for ij = 1:numevents
 %advance the oscillator states to the time of the next firing
pha = frq*min((1-pha)./frq)+pha;
numfire = 0;
 % Determine which oscillators have just fired adding 1e-9 accounts
  %for possible numerical error
comp = (pha + 1e-9 >= ones(1, numosc));
oscfire = find(comp);
while (sum(comp) > numfire)
numfire = numfire + 1;%an oscillator fires
pha = log((exp(-gamma*pha) - eps*gamma/So))/(-gamma);%advances the
  %phases due to the firing See 7.2 for the derivation of this
  %formula
compa = comp;%keeps the previous value of comp
comp = (pha +1e-9 >= ones(1, numosc));% This determines which
  %oscillators have just fired and have values > 1
oscfire = [oscfire, find(comp—compa)];%adds to the list of firing
  %neurons those that just fired
pha = comp.*ones(1, numosc) + (1-comp).*pha;% This sets the
  %oscillators which have values > 1 to 1.
end
pharec(ij,:) = pha;%records the phases
numfiring(ij) = max(size(oscfire)); % records how many fired
pha = (1-(pha == ones(1, numosc))).*pha;%sets the phases of the
  %oscillators that have fired to 0
end
figure(2)
subplot(2, 1, 1)
plot(pharec, '*')
xlabel('Firing Number')
ylabel('Phases')
subplot(2, 1, 2)
plot(numfiring, '*k')
xlabel('Firing Number')
ylabel('Number Firing')
```

Code 7.3 Inhibitory Pulse-Coupled Oscillators

(see figure 7.3)

```
%This code is similar to code 7.2 but has the oscillators coupled through
%inhibition. The reader is encouraged to find the differences between
%these two codes.
```

```
clear
numosc = 25;%The number of oscillators
numevents = 50;% number of firings to consider
So = 1.5; gamma = 2; eps = -.1;%this sets the parameters of the model
frq = ones(1, numosc);% this is the speed of the oscillators
pha = rand(1, numosc); % this sets the initial state of the oscillators
pharec(1,:) = pha; % This records the state of the oscillators
for ij = 1:numevents
  %advance the oscillator states to the time of the next firing
pha = frq*min((1-pha)./frq)+pha;
numfire = 0;
  % Determine which oscillators have just fired adding 1e-9 accounts
  %for possible numerical error
comp = (pha + 1e-9 >= ones(1, numosc));% note inhibitory firing
  %does not change the number of firing oscillators
oscfire = find(comp);
for ijj = 1:sum(comp)
pha = log((exp(-gamma*pha) - eps*gamma/So))/(-gamma);%advances
  %the phases due to the firing
end
pha = comp.*ones(1, numosc) + (1-comp).*pha;
pha = max(pha, 0);% sets the oscillators with phases below 0 to 0
pharec(ij,:) = pha;%records the phases
numfiring(ij) = max(size(oscfire)); % records how many fired
pha = (1-(pha == ones(1, numosc))).*pha;%sets the phases of the
  %oscillators that have fired to 0
end
%The same code as in 7.2 can be used to plot the results.
```

Code 7.4 Noisy Coupled Oscillators

(see figure 7.7)
```
%This code is similar to code 7.2 but simulates the oscillators with
%noise. The reader is encouraged to find the differences between these
%two codes.
clear
xmat = zeros(100, 10000);
cnt = 1;
% This enumerates all possible consecutive combinations of 100 numbers
for ij = 1:100
for ik = 1:(101-ij)
for il = ik:(ik+ij-1)
xmat(il, cnt) = 1;
end
cnt = cnt+1;
end
```

```
end
ymat = xmat(:, 1:5050);
for ij = 1:5050
ymat(:, ij) = ymat(:, ij)/sum(ymat(:,ij));% This scales them so that we
% can use ymat to take the averages
end
% Here we generate 100 random numbers and sort them from the smallest
% to the largest. We then find the average of the two smallest, the
% next two smallest etc. then the first three smallest etc. all the way
% to the average of all.
for ij = 1:10000
recmat(ij,:) = sort(randn(1, 100))*ymat; %Use rand rather than randn
%for a constant rather than normal distribution.
end
%We next find the standard deviation of these numbers.
zmat = std(recmat);
xreshape(:, 1) = zmat(1:100);
cnt = 1;
for ij = 1:99;
cnt = cnt+ (101 - ij);
xreshape((ij+1):100, ij+1) = zmat(cnt:(cnt+99-ij));
end
figure(1)
surf(xreshape)
figure(2)
plot(zmat(1:100))
```

Code 7.5 Coupled Chain of Oscillators

(see figure 7.9)

```
%This code is similar to code 7.2 but causes the oscillators to have
%different periods and a nearest neighbor coupling. The reader is
%encouraged to find the differences between the two codes.
clear
numosc = 20;
numevents = 10000;% number of firings to consider
So = 1.5; gamma = 2; eps = .03;%parameters of the model
frq = ones(1, numosc) + -0.3*(1:numosc)/numosc;% oscillator speed
pha = zeros(1, numosc); % initial state of the oscillators
pharec(1,:) = pha; % This records the state of the oscillators
for ijj = 1:numosc
oscf(ijj, 1) = 1;
end
for ij = 1:numevents
pha = frq*min((1-pha)./frq)+pha;%advances to the next firing
numfire = 0;
```

```
comp = (pha + 1e-9>= ones(1, numosc));% This determines which
   %oscillators have just fired and have values >= 1
oscfire = find(comp);
   while (sum(comp) > numfire)
numfire = numfire + 1;%an oscillator fires
nextfire = min(oscfire(numfire)+1, numosc);%finds oscillator who
   %receives the signal and caps this if it is the last oscillator
pha(nextfire) = log((exp(-gamma*pha(nextfire)) - eps*gamma/So))/(-…
gamma);%advances the phase due to the firing note line continuation
compa = comp;%keeps the previous value of comp
comp = (pha + 1e-9>= ones(1, numosc));% This determines which
   %oscillators have just fired and have values > 1
oscfire = [oscfire, find(comp—compa)];%adds to the list of firing
   %neurons those that just fired
pha = comp.*ones(1, numosc) + (1-comp).*pha;% This sets the
   %oscillators which have values > 1 to 1.
end
pharec(ij,:) = pha;%records the phases
for ijj = 1:max(size(oscfire))
   oscf(oscfire(ijj), find(oscf(oscfire(ijj),:), 1, 'last')+1) = ij;
end
pha = (1-(pha == ones(1, numosc))).*pha;%sets the phases of the
   %oscillators that have fired to 0
end
%The same code as in 7.2 can be used to plot the results.
```

Code 7.6 Amplitude Death

(see figure 7.11)
```
%This gives an example of two oscillators which, when coupled, drop their
%amplitude to zero.
function Y = twovdpexample(t, X)
eps = .13;
c = 0.5;
xa = X(1); ya = X(2); xb = X(3); yb = X(4);
Y(1) = ya + eps*(xa - (4/3)*xa^3)+0.5*(xb—xa);
Y(2) = -xa+0.5*(yb—ya);
Y(3) = yb + eps*(xb - (4/3)*xb^3)+0.5*(xa—xb);
Y(4) = -xb+0.5*(ya—yb);
Y = Y';
%The following code is for the command line.
%subplot(2,1,1)
%[T, X] = ode45(@twovdpexample, 0:0.01:100, [1 0 1 0]);
%plot(T, X(:,1))
%[T, X] = ode45(@twovdpexample, 0:0.01:100, [1 0-1 0]);
%subplot(2, 1, 2)
%plot(T, X(:,1))
end
```

Exercises

General Problems
For these problems pick a biological model of interest, preferably one you study.

1. Couple multiple copies of your model and simulate them in a way that closely resembles what is found in the wild.

2. How many of the principles of synchrony from the chapter can you find in your model?

Specific Problems
1. Using code 7.1, determine how different the two feedback loops need to be to show different oscillations.

2. Simulate code 7.2 with different model parameters and numbers of oscillators. Determine how long synchrony takes.

3. The oscillator phases are advanced by the line in the code

pha = log((exp(-gamma*pha) - eps*gamma/So))/(-gamma);

Derive this line, based on the model proposed by Mirollo and Strogatz.

4. Simulate code 7.3 with noise. Show that noise can cause nonrhythmic oscillations to synchronize.

5. Study different patterns of intrinsic frequencies with code 7.5. What other patterns can occur?

6. Do pulse coupled van der Pol oscillators always synchronize? Explore numerically.

III

ANALYSIS AND COMPUTATION

8

Statistical and Computational Tools for Model Building: How to Extract Information from Timeseries Data

Mathematical models of biological clocks need to be fitted to data. Here, we begin by covering general techniques to find parameters from data, particularly from timeseries. We emphasize the benefits of two techniques in particular: first transforming the data to make finding parameters easier, and second developing a statistical model of the data and error, especially if the error at one time point depends on the error at other time points. Next, we consider the theoretical limits to determining information from timeseries and discuss how to determine the structure and dimensionality of a model from data. The remaining sections focus on the Kalman filter, the least squares method, and maximum likelihood estimates, and explore how these can be used together to analyze data where the error is correlated. These techniques are used together in an example determining a signal from noisy protein expression data.

8.1 How to Find Parameters of a Model

8.1.1 Two Types of Data

We consider how to develop and fit mathematical models to biological data. There are two basic reasons why one would want to do this: (1) to determine if our current understanding of the biology is sufficient to represent the known experimental data, and (2) to make predictions about future experiments. Both goals draw a distinction between two types of data: data about the underlying physiology and data about the overall behavior. Data about the underlying physiology characterize the "parts" of the system—for example, the fact that degradation rate of PER1 mRNA is about 1 hr^{-1} (Siepka et al. 2007). Data about the overall behavior tell us not about the parts, but about the system as a whole—for example, what rhythms appear and how they change under the influence of external stimuli. In addition to two types of data and two goals, our model parameters will come in two types: those that are set to values determined by underlying physiological data, and those that are adjusted in order to make the model predictions better match the overall system behavior.

8.1.2 Determining the Error of a Model Prediction

First, let us think about how to measure the goodness of fit of a model prediction to experimental data. Assume we have some data and a curve that represents the model predictions (see figure 8.1).

The key question is how we can measure the distance between the model predictions and the experimental data. This turns out to be a more difficult problem than one might initially expect.

The easiest way to approach this is to measure the vertical distances between the data and the curves and sum them. For example, let us fit a phase response curve. Assume the data (phase shifts) are represented by $x_1, x_2, \ldots, x_n$ and were taken at time points $t_1, t_2, \ldots, t_n$, and call the model curve $m(t)$. The error can be calculated either as

$$\sum_i |x_i - m(t_i)|$$

or

$$\sqrt{\sum_i (x_i - m(t_i))^2}.$$

Engineers often call this latter error the root-mean-square error. It has certain nice mathematical properties, including having a continuous derivative.

However, both of these error expressions weigh each data point equally in the fit. Often, certain data points are much better known than others. The easiest way to account for this is to scale any distance by the standard deviation σ of the data point. This yields the following error estimate:

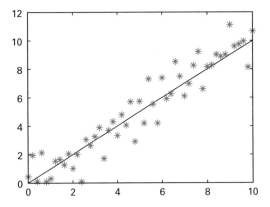

Figure 8.1
A linear regression to data.

$$\sqrt{\sum_i \left(\frac{x_i - m(t_i)}{\sigma_i} \right)^2}.$$

Even this formula is often not appropriate. A phase response curve measures both the phase at which a stimulus is applied and the response. Our actual measurements are an initial phase and a final phase. Both have error. For example, consider the model prediction and data point shown in figure 8.2.

In the vertical direction, the error of the prediction is ε. However, if our measurement of the phase is slightly too high, we could actually be making a perfect prediction of the phase. Somehow this needs to be accounted for.

Phase response curves calculate the difference between the experimentally measured initial and final phases. Calculating the error of the difference of two data points is some-what complicated. Instead, we can fit the phase transition curve, which plots phase against phase. If the initial and final phases are measured in the same way, they should have about the same error. In this way, vertical error bars become concentric circles around a data point, and we can find the minimum distance between the model prediction curve and the data (see figure 8.3).

8.1.3 Measuring Biochemical Rate Constants

Next, we consider how to determine rate constants for biophysical models. The first way is to simply measure them. For example, consider a protein that is produced and then degraded. A model for this is

$$dP/dt = s(t) - d_0 P,$$

where $s(t)$ is the production rate, P is the protein level, and d_0 is the degradation rate. Thus, the overall behavior of the system depends on both the transcription rate and the

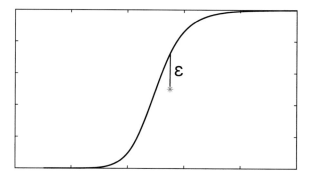

Figure 8.2
The vertical and horizontal error between data and a model may be different. Here, the vertical error, ε, is much larger than the horizontal error.

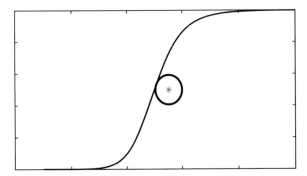

Figure 8.3
Equalizing the horizontal and vertical error.

degradation rate. The chemical cyclohexamine can block translations of transcripts. In this case our equation would be

$dP/dt = -d_0P,$

which has a solution

$P(t) = Ce^{-d_0t}.$

A better way of plotting this solution is to look at its logarithm. In this case we have

$\ln(P) = \ln(C) - d_0t.$

So if we plot $\ln(P)$ with respect to time, we should get a curve we can fit. The result of this transformation is shown in figure 8.4.

If such data are not available, we can measure these rates indirectly. For example, assume that transcription is rhythmic. We then have

$dP/dt = \sin(wt) - d_0P,$

where d_0 is the degradation rate and w is the frequency of the rhythm in transcription. We can directly calculate the rhythmic component of the solution of this equation,

$$\int_{-\infty}^{0} \sin(wt+t')e^{d_0t'}\,dt' = A(\alpha)\sin(wt + \theta(\alpha)),$$

where $A(\alpha)$ and $\theta(\alpha)$ are determined by the solution of the integral. The slower the degradation rate, d_0, the more θ changes, and so θ indirectly measures the degradation rate. We can calculate θ from the data as shown in figure 8.5.

The general principle of finding an explicit solution to a model, and using that solution to determine parameters is further illustrated in section 8.1.4.

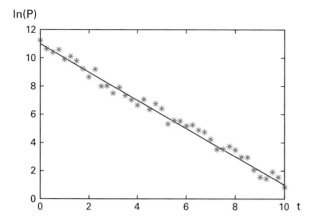

Figure 8.4
Transforming data can make easier fits. Here we take the logarithm (ln) of data on the clearance of a protein, which transforms it so it can easily be fit by a line.

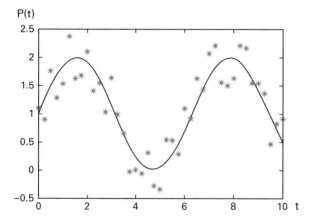

Figure 8.5
Fitting rhythmic data. Note that negative values are possible as the protein level here is calculated by subtracting the background fluorescence level.

But remember that all this prediction assumes a basic model. We could have easily postulated a different model. For example, consider a saturable degradation:

$dP/dt = s(t) - d_0 P/(K + P)$.

Here, a different fitting procedure would be needed, as there is no explicit solution. Such a procedure is considered in section 8.1.5.

8.1.4   Frontiers: Indirect Ways to Measure Rate Constants with Rhythmic Inputs

In section 8.1.3, we considered a simple case where a model's explicit solution was calculated and parameters determined from that solution. In this section, we present a more elaborate example of this method. A key feature of this example is that rhythmic inputs are used to probe the system. Much of this section is devoted to finding an explicit solution to a model and determining which rhythmic inputs would give the most information about parameters. Readers not interested in these details can skip this section.

Consider the problem of determining the parameters α and β, which are the rates of activation and deactivation of a photopigment due to light, as discussed in chapter 5.11. An interesting experiment was done to determine the values of α and β (Rimmer et al. 2000). First a continuous pulse of light, several hours in duration, was presented to human subjects, and the change in their circadian phase in response to this light pulse was determined. Next, the pulse was divided into several short pulses (intermittent light) and again presented to human subjects. These experiments enabled researchers to determine the difference in phase shifts between the continuous light pulse and the train of intermittent light pulses. Then it was possible to calculate how the difference in signaling between continuous light and intermittent light depends on α and β. Matching these results to data determines α and β (Kronauer et al. 1999).

When continuous light is applied, n approaches $\alpha/(\alpha + \beta)$. In the intermittent light case, after an initial transient, we can also assume an equilibrium is reached. We define τ_l as the length of time light is applied during one of the intermittent pulses. Let us also define τ_d as the length of dark time between intermittent pulses. Let t be the time just before an intermittent light pulse. The equilibrium that the system attains is the situation where $n(t + \tau_l + \tau_d) = n(t)$: in other words, after one light pulse and after the period of darkness following it, the system returns to the same state as it was in before the light pulse was applied.

We now calculate how much signaling occurs for the intermittent light stimulus, as compared with continuous light. For simplicity, we assume that n has reached the equilibrium $n(t + \tau_l + \tau_d) = n(t)$. Let us look at three different cases. First, assume that the time between pulses is long. In this case, all the photopigment goes to its inactive state just before the pulse and $n(t) \sim 0$. In a similar way, if the stimulus is long enough, the initial transients will not have a large overall effect when compared with the total stimulus. Finally, if the time in the light is very brief, $n(t) \sim 0$, since very few photoreceptive molecules were converted during the light episode. In all three cases, simple approximations can be made.

More generally, we assume that, throughout the protocol, the light intensity when the stimulus is applied is constant (and thus we replace $\alpha(I)$ with α). Our equation for n is

$$\frac{dn}{dt} = \alpha(1-n) - \beta n,$$

and we find the solution is

$$n(t) = \frac{\alpha}{\alpha + \beta} + Ce^{-(\alpha+\beta)t}.$$

When the light is turned off, we have

$$\frac{dn}{dt} = -\beta n.$$

So after the dark period, which lasts τ_d, we have

$$n(t + \tau_l + \tau_d) = \left(\frac{\alpha}{\alpha + \beta} + Ce^{-(\alpha+\beta)\tau_l} \right) e^{-\beta \tau_d}.$$

Equilibrium occurs if $n(\tau_l + \tau_d) = n(0)$. This means

$$\frac{\alpha}{\alpha + \beta} + C = \left(\frac{\alpha}{\alpha + \beta} + Ce^{-(\alpha+\beta)\tau_l} \right) e^{-\beta \tau_d}.$$

With some algebra, we have

$$\frac{\alpha}{\alpha + \beta} e^{\beta \tau_d} + Ce^{\beta \tau_d} = \frac{\alpha}{\alpha + \beta} + Ce^{-(\alpha+\beta)\tau_l},$$

$$Ce^{\beta \tau_d} - Ce^{-(\alpha+\beta)\tau_l} = \frac{\alpha}{\alpha + \beta} - \frac{\alpha}{\alpha + \beta} e^{\beta \tau_d},$$

$$C = \frac{\dfrac{\alpha}{\alpha + \beta}\left(1 - e^{\beta \tau_d}\right)}{e^{\beta \tau_d} - e^{-(\alpha+\beta)\tau_l}}.$$

The signaling that occurs during one light pulse, starting at this initial condition, is

$$\int_0^{\tau_l} \alpha(1-n)dt = \int_0^{\tau_l} \alpha \left(1 - \frac{\alpha}{\alpha + \beta} - Ce^{-(\alpha+\beta)t} \right) dt$$

$$= \frac{\alpha \beta}{\alpha + \beta} \tau_l - \frac{\alpha C}{\alpha + \beta}\left(1 - e^{-(\alpha+\beta)\tau_l}\right).$$

For the continuous light case, we find that, at equilibrium, $n = \alpha/(\alpha+\beta)$, and the light is on for $\tau_d + \tau_l$. So the cumulative signal is $(\tau_d + \tau_l)\alpha\beta/(\alpha+\beta)$. The ratio of the drive of the intermittent light to the continuous light is

$$\frac{\dfrac{\alpha\beta}{\alpha+\beta}\tau_l - \dfrac{\alpha C}{\alpha+\beta}\left(1-e^{-(\alpha+\beta)\tau_l}\right)}{\dfrac{\alpha\beta(\tau_l+\tau_d)}{\alpha+\beta}}.$$

Substituting for C and simplifying, we find

$$\frac{\tau_l + \dfrac{\alpha}{\beta(\alpha+\beta)}\left(\dfrac{\left(1-e^{-(\alpha+\beta)\tau_l}\right)\left(e^{\beta\tau_d}-1\right)}{e^{\beta\tau_d}-e^{-(\alpha+\beta)\tau_l}}\right)}{\tau_l+\tau_d}. \tag{8.1.1}$$

This ratio is a function of τ_d, τ_l, α, and β. It is always greater than $\tau_l/(\tau_d + \tau_l)$. The total signaling from intermittent light, as a fraction of that from continuous light, is determined by the choice of α and β. If we could experimentally measure this fraction, α and β could be determined. Unfortunately, this remains difficult in animals and impossible in humans. However, experiments could then be done to determine the difference in phase shifts between continuous and intermittent light. If we assume that the phase shift is proportional to the total signaling, α and β could be determined from these experiments. See Kronauer et al. (1999) for an example.

8.1.5   Finding Parameters to Minimize Error

Rate constants that cannot be directly measured must be measured from the system's behavior. Here is an algorithm to fit them:

1. Write down a mathematical model without specifying the exact value of the rate constants.

2. Determine the rate constants from experimental data or the literature when possible.

3. List the unknown parameters $(c_1, c_2, \ldots, c_n)$.

4. Develop a way to solve the system: $\underline{x}(t, c_1, c_2, \ldots, c_n)$.

5. Develop a way to measure goodness of fit $G(\underline{x}(t, c_1, c_2, \ldots, c_n))$.

6. Maximize $G(\underline{x}(t, c_1, c_2, \ldots, c_n))$ with respect to $(c_1, c_2, \ldots, c_n)$.

Steps 1–4 are model specific. Earlier sections in this chapter discuss step 5. We now discuss three types of methods that can be used to do the maximization in step 6. Only a brief overview of the methods is presented. Readers should consult the references for further details.

Gradient methods—These methods first determine how the error, which we define as $1/G$, rather than goodness of fit, changes with each parameter and then use this information to develop a better fit. The most celebrated example of this is Newton's method (Bradie 2006). We illustrate this for a goodness of fit to one parameter. Consider an initial guess of the parameter, c_1^0. Calculate $d(1/G)/dc_1 \equiv q_0$ at this initial parameter value. Then, update the parameter choice by calculating

$$c_1^1 = c_1^0 - 1/(G(\underline{x}(t, c_1^0))q_0).$$

Repeat this process with our new parameter value.

Search methods (Bradie 2006) — In these methods we start with a set of initial parameters $(c_1^0, c_2^0, \ldots, c_n^0)$ and a set of tolerances for each parameter $(\varepsilon_1^0, \varepsilon_2^0, \ldots, \varepsilon_n^0)$, which can be a guess at the standard error in our guess of the parameter. We then determine the goodness of fit for the parameter choice, as well as the goodness of fit if each of the parameters were individually increased or decreased by the tolerance. The choice of parameters with the best goodness of fit is then chosen. If it was the original set of parameters, the tolerance is decreased. This is shown in the following algorithm:

For $j = 1$ to n:

$a_1 = G(\underline{x}(t, c_1^0, c_2^0, \ldots, c_n^0))$

$a_2 = G(\underline{x}(t, c_1^0, c_2^0, \ldots, c_j^0 + \varepsilon_j^0, \ldots, c_n^0))$

$a_3 = G(\underline{x}(t, c_1^0, c_2^0, \ldots, c_j^0 - \varepsilon_j^0, \ldots, c_n^0))$

if $a_2 > a_1$

$\qquad c_j^0 = c_j^0 + \varepsilon_j^0$

elseif $a_3 > a_1$

$\qquad c_j^0 = c_j^0 - \varepsilon_j^0$

else

$\qquad \varepsilon_j^0 = \varepsilon_j^0/2$

end

end

This is called a coordinate search algorithm. There are better (i.e., faster) methods, e.g., Nelder Mead (Lagarias et al. 1998; Nelder and Mead 1965). The key feature of these other algorithms is that they vary more than one parameter at a time.

8.1.6 Simulated Annealing

A difficulty with the previous methods is that they may find a local maximum rather than the global maximum. Simulated annealing is a method designed to better find global minima (which of course can be applied to finding maxima by adding a minus sign to the target function). It proceeds in a similar way to gradient or search methods, except that each improvement in the parameters is accepted only with a certain probability. The idea here is that we would proceed to a minimum, but do so without necessarily accepting the next step each time a step is proposed. Occasionally, we would move from approaching one minimum to approaching a different minimum that could be better. In this way, a simulated annealing method could approach a global minimum, rather than a local minimum. More details can be found in (Gonzalez et al. 2007; Kirkpatrick et al. 1983).

Finally, it is important to note that it may be impossible to identify some parameters of a system from a given measurement, even if the error of the measurements is arbitrarily small. This can occur when one measures part of a system that does not influence the part of the system that we wish to study. Mathematical techniques now exist to determine whether the variables of a model one is interested in are identifiable. Further information on identifiability can be found in (Eisenberg and Hayashi 2014; Eisenberg et al. 2013). Section 8.2 explores these ideas in more depth, with a focus on rhythmic systems.

8.2   Frontiers: Theoretical Limits on Fitting Timecourse Data

8.2.1   How Much Information Is in Timecourse Data?

In this section, we assume that very accurate measurements can be made of both the input signal to a system, $s(t)$, and the state of a system itself, $p(t)$. If the equations of the system are of the form

$$df(p(t))/dt = s(t) - g(p(t)), \tag{8.2.1}$$

given any $s(t)$ and $p(t)$ (with the caveat that if $ds/dt = 0$, $d^2s/dt^2 \neq 0$ and if $dp/dt = 0$, $d^2p/dt^2 \neq 0$), f and g can be found uniquely (Forger 2011). However, few mathematical models are of the form (8.2.1). A much more common form for models (e.g., see chapter 2 or chapter 4) is

$$dp/dt = f(s(t)) - g(p). \tag{8.2.2}$$

In this form, f plays the role of the production rate and g plays the role of the degradation rate. Sometimes, the potentially nonlinear functions f and g can be determined uniquely simply from s and p.

How would one determine f and g? First, note that if we know p in a timecourse, we also know dp/dt. Thus, at each time point, we have s, p, and dp/dt. Moreover, let us assume that both p and s are rhythmic, and, for simplicity, that during one cycle dp/dt and ds/dt are only 0 twice, at the maximum and the minimum of their respective cycles. Then, during each cycle, p takes the same value twice (except at the maximum and minimum of the cycle), but at each of these points dp/dt is different. We start by guessing $g(p)$ at a particular time point t, and since dp/dt is known, we can find $f(s(t))$ at these two time points where p takes the same value by (8.2.2). Now that $f(s(t))$ is known at these two time points, we can find the parts of the cycle where s takes the same value and find $g(p)$ at those time points. Thus, we started at one time point t, and we found the value of $f(s(t))$ at that time point and another time point, t', where $g(p)$ has the same value. Knowing $g(p)$ at t' allowed us to find $f(s(t))$ at t' and a third time point, which we call $\varphi(t)$, where $f(s(t))$ takes the same value at t'.

The process can be continued until f and g are determined. Now, note that (8.2.2) holds as well if we subtract off the mean of $f(s(t))$ and $g(p)$. By doing this, we can remove the dependence on our initial guess of $g(p)$.

This process does not always work, particularly if during the iteration process we return to a point where the values of f and s have already been determined. In these cases, we need to choose f or g to be a particular value, but, unless it was previously chosen to be that exact value, the process cannot continue. We call this a conflicting point, for example, when p and s take the same values at two time points. More complex conflicting points can be seen, such as finding a conflict after proceeding through several steps of the algorithm.

Conflicting points are actually a good thing in that they can help us select models. Kim and Forger (2012) show that when this kind of conflict happens, a model of the form (8.2.2) cannot fit the timecourses s and p. Kim and Forger give simple conditions for when this occurs, and, from this, it often can be inferred that s does not cause the production of p, at least not in the form of (8.2.2). Such reasoning can be used to determine the interactions of the variables in a system. Moreover, we can use this method to determine the dynamics of a system, in particular, by sending a known signal $s(t)$ to a system and measuring the output. This is illustrated in figure 8.6. The general principle here is that *timecourses measured from biological systems contain much more information about a system than single time points.*

8.2.2 Completely Determinable Dynamics

We now extend the result presented in section 8.2.1 by considering the following question: Assuming we know the network diagram for a biochemical system, but do not know the reaction kinetics (neither their rate constants nor their form), can we determine the reaction kinetics from timecourse data over a small interval?

Consider a slightly generalized biochemical reaction network of the form

$$dp_i/dt = f_i(p_{i-1}) - g_i(p_i). \tag{8.2.3}$$

Each element of the system, p_i, where the elements could be protein concentrations, mRNA concentrations, etc., is controlled by a production rate $f_{i-1}(p_{i-1})$ that depends on the previous element of the system, and by a clearance rate $g_i(p_i)$. Here we do not require the

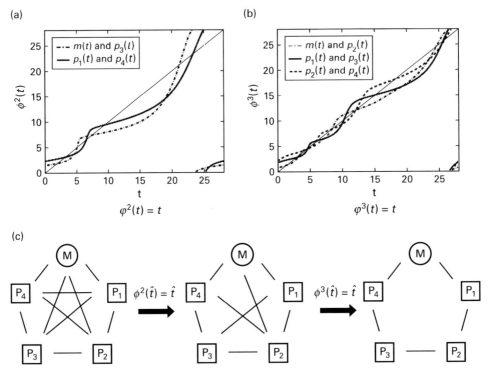

Figure 8.6
We first simulate an extended Goodwin model with 5 species, and record their timecourses. The model is then set aside, and any possible interaction among the 5 species is considered. From the recorded timecourses, the original interactions (shown on the bottom right) are reconstructed. For each interaction, we calculate the conflicting points, shown as crossings on the diagonal line representing $\varphi^2(t) = t$ in (a). Note that conflicting points occur between M and P_3 as well as between P_1 and P_4. This shows that these interactions are not possible. Next we consider an additional step of the algorithm, and this shows that any other potential interactions must be removed, except for those in the Goodwin model. Thus, consideration of conflicting points enables reconstruction of the interactions from the timecourses. Taken from Kim and Forger (2012). Copyright © 2012 Society for Industrial and Applied Mathematics. Reprinted with permission. All rights reserved.

structure of a simple feedback loop, nor do we require monotonicity. Section 8.2.1 shows that f_i and g_i can often be accurately determined from timecourse data (e.g., protein time-courses measured in a lab).

We repeat the method of section 8.2.1 with some generalizations, including the fact that we now do not require the solutions to be rhythmic. Consider the solution of $p_j(t)$ and $p_i(t)$ over some time interval $t_a \leq t \leq t_b$, where t_a and t_b are chosen so that p_j and p_i take each value twice (each having one minimum and one maximum value, p_{imin}, p_{imax}) during the time interval. Pick some initial time t, and guess any value of $g_i(p_i)$. This guess actually does not matter, since an additive constant can be added to $g_i(p_i)$ and $f_i(p_j)$ without changing the equations. Since we know $p_i(t)$, we know dp_i/dt as well, and this gives us the value of $f_i(p_j)$.

There is another time point, t', where $p_j(t) = p_j(t')$. We know dp_i/dt at t' as well, and so we then know $g_i(p_i)$ at t'. Now, there is another time point where p_i takes the same value as at t', and the process can be repeated. This gives the functions f_i and g_i.

We assumed at the outset that we know which interactions occur, and this reasoning may at first be counterintuitive: If we know the interactions, one might wonder why one would want to reconstruct them. However, note that we can use this reconstruction, made on the basis of the solution of the system from t_a to t_b, to determine the reaction rates and the behavior of the system at future times or the behavior of the system if it had been started in a different state. The method of Kim and Forger (2012) shows that, if f_i and g_i exist, as is the case here, the functions f_i and g_i constructed by that method will be unique, unless p_{i-1} and p_i take fixed values over a time interval, which would occur in our model only at a fixed point. Thus, we can reconstruct the f_i and g_i in the range between the minimum and maximum values of p_i in the time interval. The calculated f_i and g_i hold for all future time and for all p_i with $p_{i\min} < p_i < p_{i\max}$.

This is almost unbelievable. Normally, we would only expect a solution of our system to determine the behaviors of nearby solutions. Even this does not need to happen in chaotic systems. However, in the present case, by reconstructing the f_i and g_i, we start out knowing the trajectory of the system between t_a and t_b, and end up being able to determine exactly what will happen to the system for all initial states that fall within the minimum and maximum values of this trajectory. So, from a 1-D timecourse, we know exactly how the system will behave in an n-D region of phase space ($p_{i\min} < p_i(t) < p_{i\max}$).

In short, knowledge of the behavior of the biochemical network over a short time interval determines the behavior of the biochemical network at all other starting points and times, a property one can call *completely determinable dynamics*—a much stronger condition than completely observable systems considered elsewhere. This is the opposite of chaos, where even nearby trajectories show very different behavior. More generally, rather than thinking only about chaos, perhaps we should also think about the opposite: When can we completely determine a system's behavior based on one trajectory over a limited time interval?

8.2.3   Determining Model Dimension from Timecourse Data

One may be able to determine what the dimension of the underlying system is just from a single timecourse. We illustrate this with the Goodwin model. In figure 8.7, we plot the mRNA timecourse, along with the same mRNA timecourse shifted by 0.1 hours, as well as the same mRNA timecourse shifted by 0.2 hours. Surprisingly, one finds the dynamics lie on a two-manifold.

As we saw in section 8.2.1, a single timecourse can tell us much about the dynamics of a system. Note that knowing a timecourse and its value shifted a little earlier by Δt can give us a derivative, since $dM/dt \approx (M(t) - M(t - \Delta t))/\Delta t$. Similarly, we can estimate d^2M/dt^2 from the timecourse and its value shifted by two time lags. Now, consider a system that has

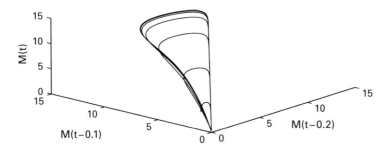

Figure 8.7
Simulations of the Goodwin model considering only the mRNA timecourse. The mRNA timecourse M is plotted along with its value 0.1 hours earlier and its value 0.2 hours earlier. The system contracts to a two-dimensional manifold. The code used was [T, X] = ode45(@good, 0:0.01:100, [1 1 1]); plot3(X(1:9980,1), X(11:9990,1), X(21:10000,1), 'k').

two variables, given by $dX/dt = \ldots$ and $dY/dt = \ldots$. We can often convert such systems to another system with dX/dt and d^2X/dt^2 allowing this kind of analysis. An example of this is given in the section 5.6 for the van der Pol oscillator. If we plot further time delays, this will not add new information, and the results will live on a two-manifold. Similar methods can be used to determine if chaotic behavior is present (Pravitha et al. 2001).

8.3   Discrete Models, Noise, and Correlated Error

8.3.1   An Introduction to ARMA Models for Error
We begin with a one-dimensional system of the form

$$x_{t+1} = a_1 x_t + \varepsilon_{t+1}. \tag{8.3.1}$$

Here, x_t is a dynamical system defined by parameter a_1, and this system is perturbed by noise, represented by ε_{t+1}, that is drawn from a normal distribution, independent of its values at other times, with mean c and standard deviation σ_ε.

While ε_t and ε_{t+1} are independent, x_t and x_{t+1} are not independent. Thus, one way to think about the system (8.3.1) is as a way to generate correlated noise. For readers who study stochastic systems, another way to think of (8.3.1) is as a discretization of a continuous stochastic system according to the Euler-Maruyama method (Kloeden and Platen 1992).

If $a_1 \neq 1$, in the limit of $t \to \infty$, the effect of the mean c of the noise on x_t will be to increase its mean by $c/(1-a_1)$. So consider a change of variable $x'_t + c/(1-a_1) = x_t$, as well as ε'_{t+1} drawn from a random distribution with mean zero and standard deviation σ_ε, which yields

$$x'_{t+1} = a_1 x'_t + \varepsilon'_{t+1}$$

Dropping the $'$, we arrive at a system the same as (8.3.1) but with ε_{t+1} having mean zero, so, up to a shift of x_t, we can assume that ε_{t+1} has mean zero.

Before further analyzing this system, let us first consider the behavior of (8.3.1) when no noise is present ($\varepsilon_{t+1} = 0$). First if $|a_1| < 1$, then $|x_{t+1}| < |x_t|$ and the system approaches 0 over time. If $|a_1| > 1$, the system shows unbounded growth. If $|a_1| = 1$ then the magnitude of x_t does not change over time. Of potential interest to us is the case where $a_1 < 0$, since the value of x_t then oscillates between negative and positive values. This oscillation is not of much use, since its period is fixed (i.e., it returns positive after two time units). However, if we allow x_t and a_1 to be complex, then the period can take more values. In particular, letting

$$z_{t+1} = a_r e^{ib} z_t + \varepsilon_{t+1},$$

one can get oscillations if $a_r = 1$. In this case, the phase of z_t in the complex plane will shift b radians at time $t+1$.

Equation (8.3.1) can be easily generalized to a higher-dimensional system. For example, consider

$$x_{t+1} = a_1 x_t + a_2 x_{t-1} + \varepsilon_{t+1}. \tag{8.3.2}$$

Here, it is helpful to put (8.3.2) in vector form:

$$\begin{bmatrix} x_{t+1} \\ x_t \end{bmatrix} = \begin{bmatrix} a_1 & a_2 \\ 1 & 0 \end{bmatrix} \begin{bmatrix} x_t \\ x_{t-1} \end{bmatrix} + \begin{bmatrix} \varepsilon_{t+1} \\ 0 \end{bmatrix},$$

or

$$\underline{x}_{t+1} = \underline{A}\,\underline{x}_t + \underline{\varepsilon}_{t+1}, \tag{8.3.3}$$

where $\underline{x}_t = (x_t, x_{t-1})$, $\underline{\varepsilon}_{t+1} = (\varepsilon_{t+1}, 0)$, and $\underline{A} = (a_1, a_2; 1, 0)$. Again, as in (8.3.1), it is best to think of ε_{t+1} as having mean zero. We could also allow $\underline{\varepsilon}_{t+1} = (\varepsilon_{t+1}, \varepsilon_t)$ or have x_{t+1} depend on a weighted average of ε_{t+1} and ε_t. Likewise, we could easily expand (8.3.2) such that $\underline{x}_t$ is m-dimensional, so that the next value of x will depend on the previous m values. In that case, $\underline{A}$ will be $m \times m$. The structure of $\underline{A}$ will have nonzero values in the top row, ones below the diagonal elements, and zeros everywhere else. In this expanded model, we could also have n nonzero values of $\underline{\varepsilon}_{t+1}$. Such models are called ARMA(m,n) models, where m refers to the dimension of $\underline{x}_t$ and n refers to the number of nonzero values in $\underline{\varepsilon}_{t+1}$ (perhaps with some additional weighting) (Jones 1980; Marmarelis 2004). Here, we choose $m = 2$ and $n = 1$ for simplicity.

As we saw before, we can think of (8.3.2) as a method of generating correlated noise, or, more likely, a discretization of a continuous dynamical system. We can also think of it as a linearization of a nonlinear discrete dynamical system. In such a case, we would draw a parallel to the linearization of continuous dynamical systems by finding fixed points and making a linear approximation to the system around these points to determine the long-term behavior of the system. As with continuous dynamical systems, one does have to be careful, since chaos or other complex behaviors can be seen in nonlinear discrete dynamical models of biological systems (Edelstein-Keshet 1988).

8.3.2 Correlated Errors and Masking

In addition to an underlying rhythmic signal, we often have noise generated by the biological process. Such noise is often correlated, meaning that its value at one time point is dependent to some extent on its value at a previous time point. This additional noise "masks" the underlying signal. We now generate a relatively simple model for the correlated error.

Again, consider the system from (8.3.3),

$$\underline{x}_t = \underline{A}\,\underline{x}_{t-1} + \underline{\varepsilon}_t, \tag{8.3.4}$$

where we assume $\underline{\varepsilon}_t$ has variance $\underline{\Sigma}$. Also consider measurement y_t, where

$$y_t = \underline{l}^{\mathrm{T}}\underline{x}_t + \xi_t + s_t \tag{8.3.5}$$

and ξ_t has standard deviation σ_ξ. Here we seek to best estimate a signal $s_t = s(t)$, from y_t, where the noise represented by $\underline{x}_t$ is correlated and the noise represented by ξ_t is not.

We could have also considered a system where the signal appears in the x_t equation:

$$\underline{\mathbf{X}}_t = \underline{A}\,\underline{\mathbf{X}}_{t-1} + \underline{\varepsilon}_t + \underline{s}_t. \tag{8.3.6}$$

Here the signal enters into a part of the system whose dynamics are represented by $\underline{A}$. This can be solved for in a similar way to the system described by (8.3.4)–(8.3.5) using the following arguments. We use the principle of superposition in linear systems to decompose (8.3.6) into two parts. Thus, we can consider

$$\underline{x}_t = \underline{A}\,\underline{x}_{t-1} + \underline{\varepsilon}_t, \tag{8.3.7}$$

$$\underline{x}'_t = \underline{A}\,\underline{x}'_{t-1} + \underline{s}_t, \tag{8.3.8}$$

with $\underline{\mathbf{X}}_t = \underline{x}_t + \underline{x}'_t$. Our observation equation would then be

$$y_t = \underline{l}^{\mathrm{T}}\underline{\mathbf{X}}_t + \xi_t = \underline{l}^{\mathrm{T}}\underline{x}_t + \xi_t + \underline{l}^{\mathrm{T}}\underline{x}'_t, \tag{8.3.9}$$

where ^{T} denotes the transpose. Solving for $\underline{s}_t$ in (8.3.4)–(8.3.5) is the same as solving for $\underline{l}^{\mathrm{T}}\underline{x}'_t$ in (8.3.7) and (8.3.9), and from (8.3.8) we could work out the signal that in (8.3.6)–(8.3.9) enters through the $\underline{x}$ equation.

The two previous formalisms are very useful when considering real-world signals. If the only noise that one had to consider were measurement noise, then a least-squares approach would work well (see section 8.6). However, most biological systems have internal noise. If this noise is correlated, we can characterize it using the formalism of (8.3.4)–(8.3.5). Another case is where we measure the output of a system that is driven by a signal we want to extract, as in the examples of inferring a core circadian signal from human temperature rhythms (Brown and Luithardt 1999; Brown et al. 2000) or the transcription rate of a gene from a GFP signal measuring protein. A similar case is shown in the example discussed in section 8.8. In these cases, the formalisms of equations (8.3.4)–(8.3.5) and (8.3.6)–(8.3.9) are equivalent, except that once a signal is extracted from (8.3.6)–(8.3.9), an additional step is required to extract $\underline{s}_t$ from $\underline{l}^T \underline{x}'_t$.

8.4 Maximum Likelihood and Least-Squares

An approach widely used in parameter estimation is maximum likelihood. Consider the measurement of a signal with Gaussian noise,

$$y_t = s_t + \varepsilon_t, \tag{8.4.1}$$

with ε_t again drawn from a Gaussian distribution with mean zero and standard deviation σ_ε. First, consider the measurement at time t separately from all other measurements.

In general, we wish to find parameters to minimize the square of the error ε_t. This is called least-squares. However, another way to approach this is from a probabilistic point of view. If s_t is fixed, the probability of measuring any particular value of y_t is determined by the probability of ε_t, and is given by

$$\frac{1}{\sigma_\varepsilon \sqrt{2\pi}} e^{-\frac{\varepsilon_t^2}{2\sigma_\varepsilon^2}}. \tag{8.4.2}$$

In terms of likelihood of the original system given y_t, we see that if we are given a measurement in y_t, we can find how likely any particular state of the system is by the preceding probability distribution. The most likely value of the measurement is a perfect measurement, where $\varepsilon_t = y_t - s_t = 0$. Moreover, if we know y_t and σ_ε, we can find the most likely value of s_t, in this case, simply y_t. Finding the maximum likelihood is the same as minimizing the error or least-squares in this case.

If all time points are independent from each other, then the preceding argument holds for the multidimensional case when all time points are considered. However, this is rare, and the problem is often more complex. For example, consider the common least-squares problem

$$\underline{y} = Q\,\underline{b} + \underline{\varepsilon}, \tag{8.4.3}$$

where $\underline{y}$ is a vector of observations and $\underline{\varepsilon}$ is a vector whose individual elements are drawn from the probability distribution (8.4.2) and our goal is to choose $\underline{b}$. Now assume that $\underline{\varepsilon}$ is drawn from the multivariate probability distribution

$$\frac{1}{\sqrt{(2\pi)^n \left| \underline{\underline{\Sigma}} \right|}} e^{-\frac{1}{2}\underline{\varepsilon}^T \underline{\underline{\Sigma}}^{-1} \underline{\varepsilon}}. \tag{8.4.4}$$

The key parameters here are in the covariance matrix $\underline{\underline{\Sigma}}$. If each element of $\underline{\varepsilon}$ is independent from each other element, then $\underline{\underline{\Sigma}}$ will be diagonal. Assuming that all elements of $\underline{\varepsilon}$ have the same statistics and that they are all independent, $\underline{\underline{\Sigma}}$ will then have the form of a constant times the identity matrix. In that case the probability distribution is

$$\frac{1}{\sqrt{(2\pi\sigma_e^2)^n}} e^{-\frac{1}{2\sigma_e^2}\underline{\varepsilon}^T \underline{\varepsilon}} = \frac{1}{\sqrt{(2\pi\sigma_e^2)^n}} e^{-\frac{1}{2\sigma_e^2}\|\underline{\varepsilon}\|^2}.$$

So finding $\underline{b}$ to minimize the value of $\|\underline{\varepsilon}\|^2$ will produce the maximum likelihood estimate. This is just what the method of least squares does, as it maximizes the amount of $\underline{y}$ that can be explained by $\underline{\underline{Q}}$.

8.5   The Kalman Filter

Here, we consider the problem of determining the state of the system at the current time based on the current and previous measurements of the system. This process is known as filtering. If, instead, we seek the value of the system in the future, similar techniques can be used, and the process is called prediction. Additionally, we may want to know the best estimate of the system in the past. This process is called smoothing. A filter is a process for taking data points $x_1, x_2, \ldots, x_n$ to $y_1, y_2, \ldots, y_m$.

We consider here one measurement of the system taken at many time points. It is possible that the measurements will not give information about all parts of the system, and this can be checked using the control-theory idea of observability. We assume here that the system studied is observable.

The Kalman filter is one of the most widely used tools in engineering (Bryson and Ho 1975; Kalman 1960). Many other references provide an in-depth and rigorous discussion of the Kalman filter. Here we will provide a brief overview. Our basic model is

$$\underline{x}_t = \underline{\underline{A}}\, \underline{x}_{t-1} + \underline{\varepsilon}_t,$$

where we assume $\underline{\varepsilon}_t$ has variance $\underline{\underline{\Sigma}}$ (this can actually vary with time, a complication we do not consider here). Thus, the system evolves to the next time point based on $\underline{\underline{A}}$. Now assume that we cannot directly measure $\underline{x}_t$. Instead, we measure y_t, where

$$y_t = \underline{l}^T \underline{x}_t + \xi_t,$$

where ξ_t, the error in measurement, has mean zero and standard deviation σ_ξ (and σ_ξ could vary with time). We typically do not directly measure all parts of the system (e.g., all parts of $\underline{x}_t$), and $\underline{l}$ tells us which parts can be measured. So, given a stream of measurements, $y_1, \ldots y_t$, how can we best estimate $\underline{x}_t$?

The Kalman filter assumes that the noise is Gaussian, and we have some estimate of $\underline{x}_{t-1}$ and a previous estimate $\underline{S}_{t-1|t-1}$ of the variance of $\underline{x}_{t-1}$. Since $\underline{\varepsilon}_t$ has mean zero, our best guess of $\underline{x}_t$ is $\underline{A}\,\underline{x}_{t-1}$, which we define as $\underline{x}_{t|t-1}$. Thus

$$\underline{x}_{t|t-1} = \underline{A}\underline{x}_{t-1}. \tag{8.5.1}$$

The variance of $\underline{x}_{t-1}$, $\underline{S}_{t-1|t-1}$, is the expected value of $\underline{x}_{t-1}\,(\underline{x}_{t-1})^{\mathrm{T}}$. The covariance of $\underline{x}_{t|t-1}$, $\underline{S}_{t|t-1}$, is affected by the noise (with variance matrix $\underline{\Sigma}$) and the expected value of $\underline{x}_{t|t-1}\,(\underline{x}_{t|t-1})^{\mathrm{T}}$, which is the expected value of $\underline{A}\underline{x}_{t-1}\,(\underline{x}_{t-1})^{\mathrm{T}}\underline{A}^{\mathrm{T}}$. Since $\underline{\varepsilon}_t$ and $\underline{x}_{t-1}$ are independent, their variances can be added, so

$$\underline{S}_{t|t-1} = \underline{A}\underline{S}_{t-1|t-1}\,\underline{A}^{\mathrm{T}} + \underline{\Sigma} \tag{8.5.2}$$

It is good to estimate the expected error in our estimate of the measurement. The expected variance of the difference between what we observe, y_t, and our estimate of this measurement, $\underline{l}^{\mathrm{T}}\underline{x}_{t|t-1}$, is

$$d_t = \underline{l}^{\mathrm{T}}\underline{S}_{t|t-1}\,\underline{l} + \sigma_\xi^2. \tag{8.5.3}$$

Again, the first term shows how the variance propagates, and the second term shows the variance from measurement error. It is also important to note that (8.5.2) and (8.5.3) do not require knowledge of y, and, thus, the reliability of the measurements can be calculated without knowledge of the actual measurements.

We now have two estimates of the system's state, our best estimate of the system's state, $\underline{x}_{t|t-1}$, and our measurement at time t (y_t). Given these two measurements, our best guess of the system's state would be a combination of the two, weighted by their variances. This is

$$\underline{x}_t = \underline{x}_{t|t-1} + \underline{S}_{t|t-1}\,\underline{l}\,(y_t - \underline{l}^{\mathrm{T}}\underline{x}_{t|t-1})/d_t. \tag{8.5.4}$$

The first term is the estimate without the measurement. We then correct any difference between expected measurements and actual measurement ($y_t - \underline{l}^{\mathrm{T}}\underline{x}_{t|t-1}$) by a factor $\underline{S}_{t|t-1}\underline{l}/d_t$, which is often referred to as the Kalman gain. If $\underline{x}_{t|t-1}$ has high variance (large $\underline{S}_{t|t-1}$), then we rely more on the new measurement y_t. However, if y_t has a large variance (large d_t due to large measurement error σ_ξ^2), then the new measurement is discounted. The variance in $\underline{x}_t$ is then given by

$$\underline{S}_{t|t} = \underline{S}_{t|t-1} - \underline{S}_{t|t-1}\,\underline{l}\,\underline{l}^{\mathrm{T}}\,\underline{S}_{t|t-1}/d_t. \tag{8.5.5}$$

Again, the idea is that the variance decreases with further observation, as the second term shows. This quantity does not depend on the actual observed y_t. Finally, it is often convenient to calculate the "log-likelihood" of the observation. This can be calculated by

$$\zeta_t = (y_t - \underline{l}^{\mathrm{T}}\underline{x}_{t|t-1})/d_t^{0.5}. \tag{8.5.6}$$

This is the difference between the expected observation and the actual observation scaled by the standard deviation we would expect in this measurement. In practice, it is the actual error found at a time point scaled so that the variance of this error is 1. This scaling is done by the $d_t^{0.5}$ term.

The Kalman filter also allows an easy way to study data sets with missing data. When a data point is missing, we can choose some dummy data and set the variance of that measurement to infinity. The filter will then ignore the dummy data and continue to estimate the state of the system.

Recapping, the Kalman filter gives us a way to calculate a state given measurements of an output of a system, as well as inherent and measurement noise. Another property of the Kalman filter is that ζ_t and ζ_j are independent for all $t > j$, which means that the error in the Kalman filter's estimate at time t cannot be decreased by considering the measurement at j. This provides an easy way to calculate the likelihood of a model given experimental measurements, as we will see later. Examples of this are shown in code 8.1 and figure 8.9 (in section 8.8).

8.6 Calculating Least-Squares

The least-squares method is efficient for determining the parameters of a model. We have mentioned the underlying idea of dividing a vector of observations into two components in section 8.4, $y = Q\,\underline{b} + \underline{\varepsilon}$. Here we give some computational details and an example of a basis matrix.

Assume we have a vector of observations, y, which are taken at intervals $t = \Delta t, 2\Delta t, \ldots,$ τ and would like to pick out particular features within these observations. For example, we may wish to determine how much of this signal can be explained by a sinusoidal signal with period τ. We can then construct a basis matrix Q whose columns are the library of signals we are looking for. For example, we can let Q be the following columns, where we have set $n\Delta t = \tau$:

1	$\sin((2\pi/\tau)\Delta t)$	$\cos((2\pi/\tau)\Delta t)$	$\sin((4\pi/\tau)\Delta t)$	$\cos((4\pi/\tau)\Delta t)$
1	$\sin((2\pi/\tau)2\Delta t)$	$\cos((2\pi/\tau)2\Delta t)$	$\sin((4\pi/\tau)2\Delta t)$	$\cos((4\pi/\tau)2\Delta t)$
…	…	…	…	…
…	$\sin((2\pi/\tau)n\Delta t)$	$\cos((2\pi/\tau)n\Delta t)$	$\sin((4\pi/\tau)n\Delta t)$	$\cos((4\pi/\tau)n\Delta t)$

The first column would seek a constant signal, the second and third columns would look for a sinusoidal signal with period τ, and the fourth and fifth columns would look for a signal with period $\tau/2$, in other words, the second harmonic. Again, what works can determine the choice of the columns.

The preceding example has an advantageous property in that the columns of Q are orthogonal as $\Delta t \to 0$, i.e.,

$Q_j^T Q_r = 0$ if $j \neq r$.

But this is not required in the following arguments, but can make our future calculations simpler.

Now consider our least-squares goal. Given Q and $\underline{y}$ we wish to choose $\underline{b}$ so that $\|\underline{\varepsilon}\|$ is minimized. This is satisfied if we can decompose $\underline{y}$ into two components, $\underline{y}_|$ and $\underline{y}_\perp$, where $\underline{y}_| = Q\,\underline{b}$ for some $\underline{b}$ and $\underline{0} = Q^T \underline{y}_\perp$. Another way of saying this is that $\underline{y}_|$ contains all parts of $\underline{y}$ that can be explained by Q and that $\underline{y}_\perp$ contains the remainder. From these desired properties of the two components, we reason that

$$\underline{y} = \underline{y}_| + \underline{y}_\perp \text{ and } \underline{0} = Q^T \underline{y}_\perp = Q^T (\underline{y} - Q\,\underline{b}),$$

so

$$Q^T \underline{y} = Q^T Q\,\underline{b},$$

and finally

$$(Q^T Q)^{-1} Q^T \underline{y} = \underline{b}.$$

The factor $(Q^T Q)^{-1} Q^T$ is sometimes called the pseudoinverse. It can exist even if the inverse of Q does not. If the columns of Q are orthogonal, then $Q^T Q$ is a diagonal matrix, and calculating its inverse is easy (just replace each of the diagonal entries with its inverse). If not, $\underline{b}$ can still be efficiently calculated, e.g., by using a QR factorization (see Trefethen and Bau 1997).

The key is that our estimate $\underline{b}$ is optimal in the sense that, if the noise is, e.g., Gaussian and independent from one time point to the next, no other choice of $\underline{b}$ could minimize the error.

8.7 Frontiers: Using the Kalman Filter for Problems with Correlated Errors

We now show some efficient techniques for timecourse analysis (Brown and Luithardt 1999; Brown et al. 2000). We have noted that often the elements of $\underline{\varepsilon}$ are not independent. If they are not, then in order to calculate the likelihood, we must calculate $\underline{\Sigma}^{-1}$ and the determinant of $\underline{\Sigma}$, both of which can be costly. In this case, the Kalman filter can be very helpful. Given a guess for the signal s_t, and measurements $\underline{y}_t$, we would like to evaluate the likelihood given by (8.4.4). Define the difference between the signal and the measurement as

$$\mathbf{Y}_t \equiv y_t - s_t = \underline{l}^T \underline{x}_t + \xi_t,$$

which is the correlated noise. If we use $\mathbf{Y}_t$ in the Kalman filter, this will represent correlated noise defined by (8.3.4)–(8.3.5). The Kalman filter (8.5.1)–(8.5.6) can be used to find the expected deviation of this variable,

$$\zeta_t = (\mathbf{Y}_t - \underline{l}^{\mathrm{T}} \underline{x}_{t|t-1})/d_t^{0.5},$$

and the likelihood can then be evaluated as

$$\frac{1}{\sqrt{(2\pi)^n \prod_t d_t}} e^{-\frac{1}{2}\sum_t \zeta_t^2}$$

while noting that $\sum_t \zeta_t^2$ is the sum of the elements ζ_t^2 and not to be confused with $\underline{\Sigma}$. However, our original problem was more complex. Suppose we wish to find a signal composed from a known library of signals, $\underline{s} = \underline{Q}\,\underline{b}$, from $\underline{y}$. We wish to find $\underline{b}$ that minimizes

$$(\underline{y} - \underline{Q}\,\underline{b})^{\mathrm{T}} \underline{\Sigma}^{-1} (\underline{y} - \underline{Q}\,\underline{b}).$$

This is difficult if $\underline{\Sigma}^{-1}$ is not diagonal. However, we could apply a Cholesky factorization (Trefethen and Bau 1997),

$$\underline{\Sigma}^{-1} = \underline{L}^{\mathrm{T}}\underline{L},$$

and then we would need to evaluate

$$(\underline{y} - \underline{Q}\,\underline{b})^{\mathrm{T}} \underline{\Sigma}^{-1} (\underline{y} - \underline{Q}\,\underline{b}) = (\underline{L}\,\underline{y} - \underline{L}\,\underline{Q}\,\underline{b})^{\mathrm{T}} (\underline{L}\,\underline{y} - \underline{L}\,\underline{Q}\,\underline{b}) = \| \underline{L}\,\underline{y} - \underline{L}\,\underline{s} \|^2.$$

In other words, we evaluate $(\underline{y} - \underline{Q}\,\underline{b})^{\mathrm{T}} \underline{\Sigma}^{-1} (\underline{y} - \underline{Q}\,\underline{b})$ by finding the vectors $\underline{L}\,\underline{y}$, $\underline{L}\,\underline{Q}\,\underline{b}$ and calculating the magnitude of their difference. The Kalman filter performs linear operations on a vector of numbers (measurements), so its action can be represented as a matrix. One can show these are exactly the operations required to multiply by the matrix $\underline{L}$.

One difficulty remains. We do not yet know the value of $\underline{b}$. Because operations are linear, rather than applying the Kalman filter (or multiplying by $\underline{L}$) to $(\underline{y} - \underline{s})$, we applied it to $\underline{y}$ and $\underline{s}$ separately and then subtracted. In fact, since $\underline{s} = \underline{Q}\,\underline{b}$, we could apply it to $\underline{y}$ and to the columns of $\underline{Q}$ to get $\underline{LQ}$, and then multiply by $\underline{b}$. This seems like more work (i.e., multiplying $\underline{L}$ by each column of $\underline{Q}$ rather than just multiplying $\underline{L}$ by $\underline{s}$), but we do not know $\underline{b}$. So the idea is to use the Kalman filter to calculate $\underline{L}\,\underline{y}$ and $\underline{LQ}$. This then becomes a least-squares problem. Given a vector $\underline{L}\,\underline{y}$ and a basis $\underline{LQ}$, how do we find $\underline{b}$ in such a way as to minimize the residual $\| \underline{L}\,\underline{y} - \underline{LQ}\,\underline{b} \|^2$? Thanks to the Kalman filter, calculating $\underline{L}\,\underline{y}$ and $\underline{LQ}$ takes only $O(n)$ operations (Jones 1980).

Sometimes, $\underline{Q}$ is not fully known. For instance, perhaps $\underline{Q}$ consists of columns whose form is sinusoidal, but whose period is not known. Similarly, the matrix $\underline{A}$ may not be known, or the standard deviations of ξ_t and ε_t. However, if we have a guess at these parameters, we can use the Kalman filter and the least-squares method to find a best fit and the likelihood. This process can be used in a standard minimizer, e.g., as described in section 8.1.5, to then find these parameters.

The example in the next section implements this method. We calculate Q and apply the Kalman filter to Q and to the data so that we can determine the likelihood of a model and estimate the state of the system.

8.8   Examples

A major problem in cell biology is to infer the activity of a gene from its products. For example, this kind of inference was used to describe how genes are often transcribed with burst-like kinetics (Suter et al. 2011). Here the authors measured protein levels, but were able to infer transcription rates using a simple model (see figure 8.8). We now present a more accurate methodology for this calculation using the preceding methods.

Let us consider a gene whereof the rate of transcription is a function of time. We choose the rate of transcription to be of the form $a_1\sin(t) + a_2\cos(t) + a_3\sin(2t) + a_4\cos(2t)$.

Following transcription, mRNA is translated into protein at rate b_1 and protein degraded at rate b_2. Protein affects the transcription of mRNA at rate b_3 (we will choose this to be zero) and mRNA is degraded at rate b_4. All processes are assumed to be linear. Gaussian noise is assumed in transcription, translation, and measurement, all with variance 1. This then has the form of (8.3.4)–(8.3.5) and (8.3.6)–(8.3.9).

We first simulate this model to get the measured protein timecourse Yz (see figure 8.9). Then we forget about the fact that these data came from the model. Our job is to determine

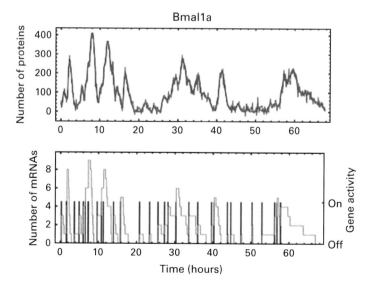

Figure 8.8
(Top) Measurement of the number of proteins produced by a gene. (Bottom) The inferred gene activity. Taken from Suter et al. (2011).

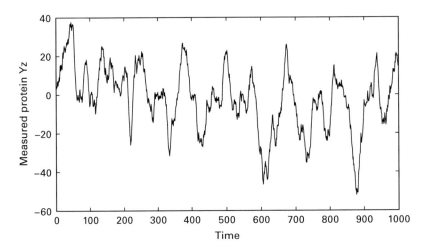

Figure 8.9
Noisy timecourse data from which we seek to extract model parameters and rates. Note that we have set the mean of the data to zero.

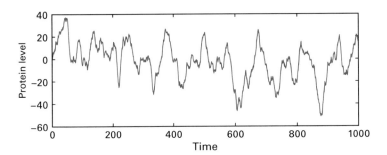

Figure 8.10
Estimated (red) and actual protein (blue) timecourses using the Kalman filter on data in figure 8.9. See code 8.1 and text for more details. Blue curve is underneath red curve.

the model parameters $b_1, \ldots, b_4$, the rate of transcription as a function of time (by finding $a_1, \ldots, a_4$), and the best estimates of the amount of mRNA and protein at each time point. We seek to extract all this information from a single noisy timecourse. Inspection of figure 8.9 shows how difficult this task is. The code to do this is provided in code 8.1.

Beginning with a guess of the model parameters, and using the Kalman filter, we are able to estimate the protein and mRNA levels. These are shown in figures 8.10 and 8.11. The Kalman filter also calculates the likelihood. Different parameters are then chosen until the likelihood is maximized. The Kalman filter also removes any correlation between the data points, giving us an estimate of the transcription signal. Using this estimated signal,

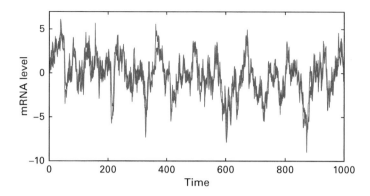

Figure 8.11
Estimated (red) and actual mRNA (blue) timecourses using the Kalman filter on data in figure 8.9. See code 8.1 and text for more details.

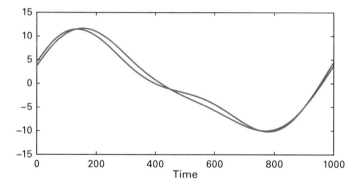

Figure 8.12
Estimated (red) and actual (blue) timecourses without noise using the Kalman filter on data in figure 8.8. These refer to the protein produced from the transcription signal. See code 8.1 and text for more details.

we can apply least squares, taking care to incorporate a basis transformed by the Kalman filter so that it will be applicable to the data transformed by the Kalman filter as described in section 8.7. Performing least squares in this manner gives us an estimate of the actual protein signal without noise, and the result is shown in figure 8.12.

Our original simulation had the values of b_1, ..., b_4 set to be 0.9, 1, 0, and 0.9. Our method estimates these parameters as 0.8535, 1.1789, 0.0000, and 0.9103. This is impressive, considering how noisy the data are (see figure 8.9) and how much information we extract from just one timecourse.

A similar methodology was used to analyze circadian variations of temperature rhythms in humans (see figure 8.13) (Indic et al. 2005).

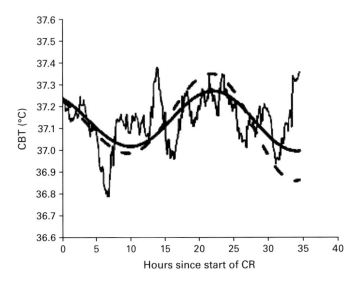

Figure 8.13
Measured core body temperature in a human, and its fit using similar methods to figures 8.9–8.12. Two models
(solid and dashed) are shown. Taken from Indic et al. (2005).

8.9 Theory: The Akaike Information Criterion

We now derive a method to choose between models due to Akaike. Our derivation is based
on Burnham and Anderson (1998) and Wasserman (2015). Note: some of the notation in
this section differs from the previous sections.

Two closely related concepts are information and likelihood. We have discussed
likelihood previously (sections 8.4, 8.5, and 8.7), so let us now consider information. We
assume that the truth can be represented as a probability density function, $f(\underline{x})$, which tells
us exactly what the probability that a particular event would occur is. We do not know $f(\underline{x})$.
However, we know $g(\underline{x}|\underline{c})$, which is a model of a particular event, with parameters indicated
by $\underline{c}$. The information in the model can be quantified by how far apart these probabilities
are, weighted for the likelihood that each state would occur. We indicate this by

$$\int f(\underline{x}) \log\left(\frac{f(\underline{x})}{g(\underline{x}\mid\underline{c})}\right) d\underline{x} \tag{8.9.1}$$

using $\log(f(\underline{x})/g(\underline{x}|\underline{c}))$ to measure the difference between the model and the truth. This
measure goes to zero (since $\log(1) = 0$) as the model becomes a better approximation of
reality. However, in practice it is never actually zero since $\underline{c}$ must be estimated from the
data. Calling $\underline{y}$ the data, the expected value of this is

$$\int f(\underline{y}) \int f(\underline{x}) \log\left(\frac{f(\underline{x})}{g(\underline{x}\mid\underline{c}(\underline{y}))}\right) d\underline{x}\, d\underline{y}$$

where, with a possible abuse of notation, we denote by $f(\underline{x})$ the probability of $\underline{x}$ occurring, and by $f(\underline{y})$ the probability of getting a certain data value. This is > 0 since, even if $f(\underline{x}) = g(\underline{x})$, $\underline{c}(\underline{y})$ would still need to be estimated.

Working with the information expression (8.9.1), we have

$$\int f(\underline{x}) \log\left(\frac{f(\underline{x})}{g(\underline{x}\,|\,\underline{c})}\right) d\underline{x} = \int [f(\underline{x})\log(f(\underline{x})) - f(\underline{x})\log(g(\underline{x}\,|\,\underline{c})]d\underline{x}.$$

The first term is unknowable, but luckily constant. We focus on the second term. The model that captures the most information is the model that maximizes

$$K = \int f(\underline{x}) \log(g(\underline{x}\,|\,\underline{c}))d\underline{x}.$$

In practice one has actual data, whose log-likelihood can be measured as

$$L = \frac{1}{n}\sum_{i=1}^{n} \log(g(\underline{y}_i\,|\,\underline{c})).$$

How are these two quantities related? For the K term, we can first assume that there is some set of parameters, $\underline{c}_0$, that maximizes the likelihood. Then, by taking a Taylor series of the parameters around $\underline{c}_0$, we have

$$\log(g(\underline{x}\,|\,\underline{c})) \approx \log(g(\underline{x}\,|\,\underline{c}_0)) + (\underline{c} - \underline{c}_0)^{\mathrm{T}}\,\underline{s}(\underline{x},\underline{c}_0) + \frac{1}{2}(\underline{c}-\underline{c}_0)^{\mathrm{T}}\,\underline{\underline{H}}(\underline{x},\underline{c}_0)(\underline{c}-\underline{c}_0),$$

where $\underline{s}(\underline{x},\underline{c}_0) = \partial \log(g(\underline{x}|\underline{c}))/\partial\underline{c}\big|_{\underline{c}=\underline{c}_0}$ and $\underline{\underline{H}}(\underline{x},\underline{c}_0)=\partial^2 \log(g(\underline{x}|\underline{c}))/\partial\underline{c}^2\big|_{\underline{c}=\underline{c}_0}$. Now since $\underline{c}_0$ is a maximum, $\underline{s}(\underline{x},\underline{c}_0) = 0$.

We can also approximate L, given a fixed estimate of the parameters $\underline{c}$, by using a Taylor series:

$$L \approx \frac{1}{n}\sum_{i=1}^{n} \log\left(g\left(\underline{y}_i\,|\,\underline{c}_0\right)\right) + (\underline{c}-\underline{c}_0)^{\mathrm{T}}\,\underline{s}\left(\underline{y}_i,\underline{c}_0\right) + \frac{1}{2}(\underline{c}-\underline{c}_0)^{\mathrm{T}}\,\underline{\underline{H}}\left(\underline{y}_i,\underline{c}_0\right)(\underline{c}-\underline{c}_0).$$

Here $\underline{s}$ is not zero, because we are considering actual data points $\underline{y}_i$ and estimating $\underline{c}$ from them.

We next consider the difference between K and L term by term. As $n \to \infty$ we have

$$\frac{1}{n}\sum_{i=1}^{n} \log\left(g\left(\underline{y}_i\,|\,\underline{c}\right)\right) \approx \int f(\underline{x})\log(g(\underline{x}\,|\,\underline{c}_0))d\underline{x}$$

and

$$\frac{1}{n}\sum_{i=1}^{n}\frac{1}{2}(\underline{c}-\underline{c}_0)^{\mathrm{T}}\,\underline{\underline{H}}\left(\underline{y}_i,\underline{c}_0\right)(\underline{c}-\underline{c}_0) \approx \int f(\underline{x})(\underline{c}-\underline{c}_0)^{\mathrm{T}}\,\underline{\underline{H}}(\underline{x},\underline{c}_0)(\underline{c}-\underline{c}_0)d\underline{x}.$$

Thus, the difference between the information and the likelihood is

$$\frac{1}{n}\sum_{i=1}^{n}(\underline{c}-\underline{c}_0)^{\mathrm{T}}\underline{s}(\underline{y}_i,\underline{c}_0)=(\underline{c}-\underline{c}_0)^{\mathrm{T}}\underline{S}_n$$

where

$$\underline{S}_n \equiv \frac{1}{n}\sum_{i=1}^{n}\underline{s}(\underline{y}_i,\underline{c}_0).$$

Moreover, $(\underline{c}-\underline{c}_0)$ and $\underline{S}_n$ are related. Another way of saying this is that we cannot use the same data to estimate $(\underline{c}-\underline{c}_0)$ and $\underline{S}_n$ and expect them to be independent estimates. To illustrate this, again let $\underline{c}_0$ be the maximum likelihood estimator where $\underline{S}_n(\underline{c}_0) = 0$ and, by Taylor series around $\underline{c}_0$,

$$\underline{S}_n(\underline{c})^{\mathrm{T}} \approx (\underline{c}-\underline{c}_0)\underline{H}(\underline{y}_i,\underline{c}_0).$$

The matrix $\underline{H}(\underline{y}_i,\underline{c}_0)$ is sometimes called the Fisher information matrix, and is equal to $\partial \underline{S}_n/\partial \underline{c}$. This argues that the difference between the information and the likelihood is

$$(\underline{c}-\underline{c}_0)^{\mathrm{T}}\underline{S}_n \approx (\underline{c}-\underline{c}_0)^{\mathrm{T}}\underline{H}(\underline{y}_i,\underline{c}_0)(\underline{c}-\underline{c}_0).$$

The expected value of this expression, over all possible data, which also means over $\underline{c}$ since this is estimated from the data, can be worked out to be the rank of $\underline{H}$ (Akaike 1974).

So, rather than the likelihood, it is better to consider the information in mathematical models. If the assumptions of the Akaike information criterion (AIC) hold, then we can calculate the log likelihood as before, and simply add to this the rank of $\underline{H}$, which is also the number of parameters, to calculate the information. The model with the lowest AIC is then often chosen. If two models have a similar AIC (e.g., a difference of 2 or less; see Burnham and Anderson 1998), then it may be better to choose both models, or at least test the predictions of both.

It is unfortunate that many studies use the AIC without consideration of its important assumptions. This can lead to erroneous results. In particular, the AIC assumes that each parameter has a unique influence on the likelihood. It also assumes that models are good in that the models are near the maximum likelihood estimator, and it assumes that we have a large number of data points. These assumptions are indeed restrictive, and need to be carefully considered.

8.10 A Final Word of Caution about Stationarity

Throughout this chapter, we have made the assumption that the basic properties of the timeseries, as well as the parameters of a model describing the timeseries, do not change over time. This is called the stationarity assumption and may not be correct. For example,

(a)

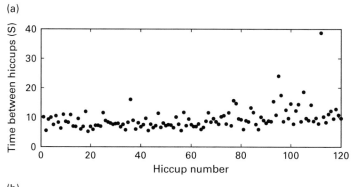

(b)

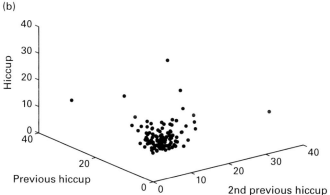

Figure 8.14
(Top) The time between hiccups during a late night spell. (Bottom) A plot looking for correlations between the time between hiccups and the previous times between hiccups.

in figure 8.14, the author recorded the time between his hiccups during a particularly long late night spell.

One aspect of this data set that makes its analysis difficult is that the properties of the system seem to change over time. For example, the mean (8.2) and standard deviation (2.1) of the first 1/3 of the data set (points 1 through 40) are quite different from the mean (11.5) and standard deviation (5.66) of the last 1/3 of the data set (points 81 through 120).

Code 8.1 Fitting Protein Data

(See figures 8.8–8.11)

```
%This code shows how to efficiently extract a signal from noisy data.
%This code first sets up the problem. The function Kalmanminimizea
%performs the Kalman Filter and least squares to calculate the
%likelihood. Since many random numbers are generated in this program,
```

```
%running it different times will result in different answers
clearvars
clearvars -global
global Q Yz xrecx brec
A = [0.9 1
0 0.9];% a matrix with all our parameters. It simulates transcription
%and translation in a biochemical network. The reader is encouraged to
%work out the model from the code.
x(1,:) = [0 0]'; %initial conditions
xx(1,:) = [3.8275 0.4718]';
numpts = 1000;% sets the number of datapoints
%the following code creates the matrix Q
Q(1:(numpts + 1), 1) = sin((1:(numpts + 1))*(2*pi/(numpts + 1)));
Q(1:(numpts + 1), 2) = cos((1:(numpts + 1))*(2*pi/(numpts + 1)));
Q(1:(numpts + 1), 3) = sin(2*(1:(numpts + 1))*(2*pi/(numpts + 1)));
Q(1:(numpts + 1), 4) = cos(2*(1:(numpts + 1))*(2*pi/(numpts + 1)));
Q(1:(numpts + 1), 1) = Q(1:(numpts + 1), 1)/norm(Q(1:(numpts + 1), 1));
Q(1:(numpts + 1), 2) = Q(1:(numpts + 1), 2)/norm(Q(1:(numpts + 1), 2));
Q(1:(numpts + 1), 3) = Q(1:(numpts + 1), 3)/norm(Q(1:(numpts + 1), 3));
Q(1:(numpts + 1), 4) = Q(1:(numpts + 1), 4)/norm(Q(1:(numpts + 1), 4));
bact = [2 1 0.5 0.25]';% these are the a parameters in the text they will
%be predicted later by brec
transsig = bact'*Q';% This is the transcription signal.
for ij = 1:numpts%This calculates the timecourses
x(ij+1,:) = A*x(ij,:)' + randn(2, 1);%adds noise to the timecourses
xx(ij+1,:) = A*xx(ij,:)' + [0, transsig(ij)]';%adds the transcription
%signal
end
Yz = x(:,1) + xx(:, 1) + randn(numpts+1, 1);% adds the processed
%transcription signal, internal noise and measurement noise. We now use
%this signal forgetting how it was calculated and determine parameters
a = fminsearch(@kalmanminimizea, [1 1 1 1])%find the minimum error
a %show the predicted values
kalmanminimizea(a); % here we run this once just to make sure that brec
%and xrecx are set
figure(1)
plot(Yz)%plot the timecourse
figure(2)
hold off
plot(x(:, 1)+xx(:,1))%plots the actual protein timecourse
hold on
plot(xrecx(:, 1), 'r')%plots the predicted protein timecourse
figure(3)
hold off
plot(x(:, 2)+xx(:,2))%plots the actual mRNA timecourse
hold on
plot(xrecx(:, 2), 'r')%plots the predicted mRNA timecourse
```

```
figure(4)
hold off
plot(xx(:, 1))%plots the actual protein signal
hold on
plot(brec'*Q', 'r')%plots the predicted protein signal
function like = kalmanminimizea(a)
global Q Yz xrecx brec
A = [abs(a(1)) abs(a(2))
abs(a(3)) abs(a(4))];%sets up the matrix with the best guess of the
%model parameters passed by our minimizer
gamma = [1 0
0 1];%determine the noise level.
%the following lines initialize the code
zrec(1) = 0;
xrec(1,:) = [0
0];
l = [1
0];
xyx = size(Q);
for ik = 1:(xyx(2)+1)
xnewprev(ik, 1) = 0;
xnewprev(ik, 2) = 0;
end
S = [0 0
0 0];
dtot = 0;
xyy = size(Yz);
lenY = xyy(1) - 1;
for ij = 1:lenY% These next three lines implement the Kalman Filter
Spred = A*S*A' + gamma;
d = l'*Spred*l + 1;
S = Spred-Spred*l*l'*Spred/d;
for ik = 1:xyx(2)
Y(ij+1) = Q(ij+1,ik);% Takes each column of Q to implement the KF
xnew = [xnewprev(ik, 1)
xnewprev(ik, 2)];
xpred = A*xnew;
xnew = xpred + Spred*l*(Y(ij+1) - l'*xpred)/d;
xnewprev(ik, 1) = xnew(1);
xnewprev(ik, 2) = xnew(2);
Qkalman(ij+1, ik) = (Y(ij+1) - l'*xpred)/(d^(1/2));% A new column of Q
end
Y(ij+1) = Yz(ij+1);%Takes the data and below implements the KF
ik = xyx(2) + 1;
xnew = [xnewprev(ik, 1)
xnewprev(ik, 2)];
xpred = A*xnew;
```

```
xnew = xpred + Spred*l*(Y(ij+1) - l'*xpred)/d;
xnewprev(ik, 1) = xnew(1);
xncwprcv(ik, 2) - xnew(2);
Ykalman(ij+1) = (Y(ij+1) - l'*xpred)/(d^(1/2));%The revised data
xrecx(ij+1,:) = xnew;
dtot = dtot + log(d);%calculates the variance used in the likelihood
end
b = ((Qkalman'*Qkalman)^(-1))*Qkalman'*Ykalman';%Performs least squares
brec = b;%Our best estimate of the parameters after the transformation
SN = Ykalman-b'*Qkalman';
like = dtot + lenY*log(SN(2:(lenY+1))*SN(2:(lenY+1))');%calculates the
%likelihood
end
```

Exercises

General Problems
For these problems pick a biological model of interest, preferably one you study.

1. Paying attention to the concerns listed in section 8.2, design practical experiments to test the parameters of your model, and/or discuss the experiments that were used to find the parameters of the model.

2. For question 1, how many samples are needed? How long will this take?

3. Following up on question 1, how will the experimental data be used to determine model parameters? How will you account for experimental error? How will you choose parameters of the model that are not informed by the experimental protocol?

4. To fit data from your system with least squares, what library (i.e., what columns of Q from section 8.4) would you choose?

5. Describe the noise that data on the timecourses from your system might have. Are they correlated?

6. Choose a 2×2 matrix $\underline{A}$ and generate correlated noise based on the framework of (8.3.4)–(8.3.5). If your system has correlated noise, match $\underline{A}$ to this correlated noise. If not, you can use the $\underline{A}$ given in the sample code. Add this to a simulated timecourse of the model. Now perform least squares with your choice of Q on these data.

Specific Problems
1. Demonstrate that the expression 8.1.1 is always < 1.

2. Demonstrate that the expression 8.1.1 is maximized for large α and small β.

3. Demonstrate the claim in 8.7 that multiplying by the matrix $\underline{L}$ is equivalent to performing the Kalman filter.

4. Add additional comments to code 8.1 explaining how the Kalman filter and least squares are used.

5. Code 8.1 uses three methods for analyzing data: (a) The Kalman filter, (b) least squares, and (c) parameter fitting using MATLAB's fminsearch function.

    a. Perform least squares on the signal Yz. You may assume the columns of Q are orthogonally and properly scaled. Note how your answer changes from the original answer we found when fitting the code.

    b. In the Kalmanminimizea function, time how long it takes to do the least squares calculation using MATLAB's tic and toc commands (i.e., tic; COMMAND; toc). Repeat this with MATLAB's fminsearch. Which is quicker? Which is more accurate?

    c. Consider what happens when you use the Kalman filter on Yz. Verify that the deviations (where Y = Yz)

        z(ij) = (Y(ij+1) – l'*xpred)/(d^(1/2))

        are normally distributed. Explain why having them not randomly distributed would mean that the Kalman filter was not optimal. Calculate z(1:1000)'*z(1:1000) and z(1:1000)'*z(2:1001), which should be two scalars (if not, type in z = z'; and recalculate). The first quantity should be much larger than the second. Explain why this indicates that the elements of z may be independent.

9

How to Shift an Oscillator Optimally

In many applications, one wishes to move a biological clock, or a model thereof, from one state to another in an optimal way. We first present an experimental protocol for determining many kinds of optimal stimuli. We then review techniques from optimal control theory applied to biological oscillator models. Several approaches are presented to illuminate theoretical concepts, including influence functions and the control Hamiltonian. We then present examples ranging from squid giant axon electrophysiology to the human circadian clock.

9.1   Asking the Right Biological Questions

Chapter 6 discussed how to classify the effects of perturbations on biological oscillators. We now ask a related, but just as important, question: How can we bring a system from one state to another in some optimal way? The assumption here is that this is done by a time-varying signal, whose values need to be optimally chosen at each time point. The focus of this chapter is time-varying stimuli. This is what is most difficult to determine since an infinite number of choices (the values at each time) need to be determined. Simple questions of determining what is best among a finite, particularly small, number of choices (e.g., a red, green, or blue stimulus) can be tested in a direct way, and are not considered here. Some possible optimal stimuli we consider here include how to

1. phase-shift an oscillator in minimum time;

2. phase-shift a clock with minimal side effects;

3. entrain to a new time zone in minimum time (the problem of jet lag);

4. maximize the effect of a stimulus;

5. minimize the amplitude of an oscillator;

6. start an oscillator with minimum effort.

The techniques described in this section can work for any biological system and are described in more detail in Forger and Paydarfar (2004) and Gelfand and Fomin (1963).

However, they are not guaranteed to give us an optimal solution. In the rest of this chapter, we describe mathematical methods that can guarantee some sort of optimal solution. The downside of these mathematical techniques is that they tend to be difficult to understand, difficult to apply computationally, and workable only for oscillators with relatively few variables.

Optimality is usually impossible to study experimentally because it would be impossible to study every possible signal, but that should not discourage us from trying. To determine an optimal signal experimentally, we would need to test all possible signals to a system, which could require an infinite number of experiments. However, statements about optimality can be made about models with a finite amount of simulation, as we shall see. We can test some signals and infer from them properties of optimal stimuli. We state this as an experimental algorithm:

1. Choose biologically plausible signals. Think carefully about the signals the system might encounter in the wild. Testing these plausible signals can be most informative about the system's real-world behavior.

2. Perturb the system with a random combination of these biologically plausible signals.

3. Choose the duration and intensity of the signals so that they can be reliably tested in an unbiased way. A very efficient approach is to present a long continuous signal to the system and continuously record the response. In this way, the response at each time point can be viewed as measuring the response to the previous signal, and in a weak way, a separate experiment from the next time point.

4. Choose relatively weak stimuli and a rare-event criterion for selecting a relatively small number of effective stimuli. For example, consider a neuron to which a continuous electrical signal is presented. Presenting low-amplitude signals will mean the neuron fires only rarely. When it does fire, it resets its response and forgets its prior history. Thus, if we choose too strong a stimulus, first, we will have many signals to consider, and second, a firing event may be affected by a previous firing event if it happens too frequently. A more general way to determine criteria for selecting stimuli is to plot the signal desirability vs. the outcome desirability and then set a strict criterion for what signals to choose.

5. To determine properties of optimal signals, carefully consider which features the effective stimuli have in common. Perhaps all signals contain a similar average signal? In this case, averaging the signal can be useful to determine the optimal candidate. Perhaps the signals can be separated into multiple groups of similar effective stimuli. If so, more than one signal might be optimal or desirable.

An example of this algorithm is presented in Forger et al. (2011). See figure 9.1.

Several concepts are worth mentioning here. First, while we may want the best possible signal when compared with any other signal (mathematically this is called a global

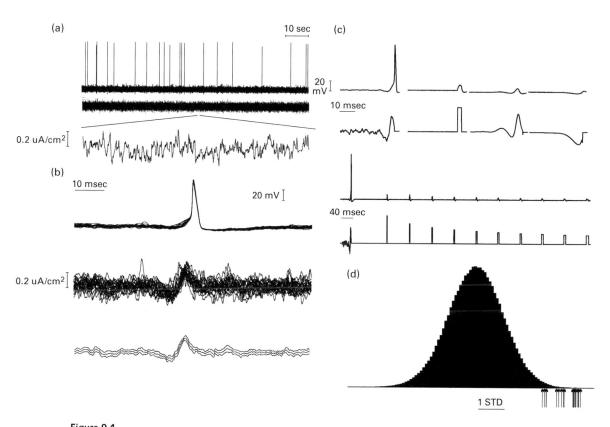

Figure 9.1
Experimentally finding optimal signals to excite a neuron. (a) A signal generated from randomly timed postsynaptic currents was fed into a squid giant axon at low magnitude, so action potentials occurred only rarely, and the resulting voltage was recorded. (b) At each action potential, the current that caused the action potential was saved and superimposed. Averaging these signals gave us a candidate for an optimal stimulus shape. Shown are the superimposed voltage traces (top) superimposed input signals (middle) and mean signal ± the standard error of the mean (bottom). (c) The optimal stimulus shape was applied to the neuron. For comparison, other current pulses of the same magnitude were tried as well. Only the optimal signal caused an action potential. (d) Here we plot a histogram for each timepoint of how similar the preceding signal was to the optimal signal. Times when the neuron fired are shown by arrows. Every time the neuron fired, the current had a shape very similar to the optimal signal. See Forger et al. (2011), from which this image was taken.

optimum), sometimes it is sufficient to consider a signal that is better than any other signal similar to it (mathematically this is called a local optimum). In the case of a local optimum, any small changes to the signal will result in a less desirable outcome, but there may be other signals that are significantly different and also perform better. Many biological systems have multiple local optima, which may be biologically important, as we will see (Forger et al. 2011).

As we mentioned in our algorithm, only certain types of stimuli may be available. For example, neurons receive signals that are composed of postsynaptic currents of preset shapes. Light levels chosen to phase-shift circadian clocks are never less than 0, and must be kept below blinding light levels. Thus, *the* optimal signal may not be achievable in biological practice if it is beyond these practical limits. Instead the system may, more often, respond to a local, but not global, optimum, one that can easily be achieved. An illustration can be seen in Forger et al. (2011), where a neuron with one shape of postsynaptic current (PSC) inputs responds differently to different signals composed of other shapes of PSCs. The mean value of the signal can affect the optimal signal by influencing the dynamics of the system. Taken together, these findings point to a general rule that *optimal signals in biology depend on the available stimuli.* Thus, to consider only the global optimal signal, as a mathematician might, might miss the most important signal in the real world.

Many problems in optimality can be proposed in two ways. For example, consider the problem of finding the stimulus of a certain magnitude that will phase-shift an oscillator more than all other signals of that magnitude. Assume that a signal phase-shifts the oscillator by a set amount. Another problem we could have solved would have been finding, out of all possible signals that phase-shift the oscillator by that amount, which of them has the lowest magnitude. Both problems will yield the same answer, which illustrates another key feature of optimality: *optimal problems often have a "dual" or alternate problem that will likely yield the same answer* (Winston 1994).

Considering the dual version of a problem can be very useful in that it may be much easier to solve than the original problem. In the preceding example, it is easy to screen signals to determine their magnitudes and then apply only those signals of a fixed magnitude. However, it may be impossible to determine which signals will phase-shift by a fixed amount.

It is important to realize that *some problems may be impossible to solve.* For example, we could ask for a stimulus of a certain magnitude that shifts a system from state *a* to state *b* in minimal time. But it is possible that no stimulus of that magnitude can shift the oscillator from *a* to *b*. This needs to be carefully considered when formulating the problem.

Having discussed methodologies for determining optimal signals, we now return to experimental systems and discuss how these optimal signals can be used in actual biological processes. Finding optimal stimuli can be useful in practical man-made applications, such as delivering a drug with maximal effectiveness or minimizing side effects. However, do natural systems use optimal stimuli similar to those we calculate in our models? For example, do neuronal networks allow for firing with the least energy cost? Future work will answer these questions.

We now formulate optimal problems mathematically and develop tools for their study. Readers who would like to first seek an example of the kinds of problems that can be studied may wish to first read section 9.10 for motivation.

9.2 Asking the Right Mathematical Questions

We now reformulate some of the problems in section 9.1 mathematically. We also use a convention that $x(t)$ refers to x at time t whereas x refers to x at all times. Consider the following system:

$$d\underline{x}/dt = \underline{f}(\underline{x}(t), u(t)),$$

where $\underline{x} \equiv (x_1, x_2, \ldots, x_n)$ is the system state, $\underline{f} \equiv (f_1, f_2, \ldots, f_n)$ represents the system's dynamics, and u is a control. We assume u is a scalar for simplicity, even though our arguments can be extended to multidimensional $\underline{u}$. We also consider autonomous systems even though nonautonomous systems are an easy extension of our arguments.

Typically, one would like the model to go from one state to another at a fixed time t_f:

$$\underline{x}(t_0) = \underline{x}_0 \text{ and } \underline{x}(t_f) = \underline{x}_f.$$

However, this can be relaxed somewhat. For example, we could have a goal that

$$g(\underline{x}(t_0)) = C, \text{ or } g(\underline{x}(t_f)) = C,$$

where g is a function defining the phase of the oscillator. Thus, we would want to go to a particular isochron (region of constant phase) at a fixed time t_f. Such problems have variable end conditions.

Another type of problem is a pursuit problem. Here we would want our system to match the state of another system at t_0 or t_f. This is the case of jet-lag problems, where we want a system entrained to another time zone at t_0 to match the state of a system entrained to the new time zone at a future time t_f. Mathematically, this is written as

$$|\underline{x}^*(t_0) - \underline{x}(t_0)| = 0 \quad \text{or} \quad |\underline{x}^*(t_f) - \underline{x}(t_f)| = 0,$$

where $\underline{x}^*$ is the target system and t_f can vary.

Sometimes, one encounters minimal time problems. For example, we may have a pursuit problem, as stated earlier, but would also like to find the minimum value of t_f for which the previous equation is satisfied.

Associated with the goal is a cost function, which we call $L(\underline{x}, u, t)$. It varies with t, and the total cost is the integral of this cost function:

$$\int_{t_0}^{t_f} L(\underline{x}, u, t') dt'. \tag{9.2.1}$$

Let us mention a few examples of the cost function. First, L could simply be constant:

$L(\underline{x},u,t) - 1$.

In this case the total cost is t_f, so seeking to minimize the total cost means we are solving a minimum-time problem.

A second common example of a cost function is

$L(\underline{x},u,t) = \|u\|$.

In this case we aim to minimize the root-mean-square (L^2 norm) of the input.

We could also add a cost to the final state:

$\varphi(\underline{x}(t_f))$.

The line between cost and constraint can be blurred. For example, we may want the control bounded, but we can accomplish this with a change to the cost rather than with a constraint. For example, if u is a light level, then $u < 0$ would not make sense. Moreover, we may impose a maximum possible light level (e.g., to avoid blindness or because of constraints imposed by a lighting system). One way to do this is to add a term to the cost function that approaches ∞ at these limits, but is zero elsewhere. For the example of light levels, we could add to $L(\underline{x},u,t)$ the two terms

$\varepsilon/u + \varepsilon/(\gamma - u)$,

where γ is the maximum light level. As the light level approaches γ, the cost goes to infinity, preventing u from actually reaching this value.

Amending the cost is not always the easiest or most accurate way to bound a control. For example, suppose we want to bound the control value to be not less than zero. We could replace u with u^2 in the equations, as this is, by definition, nonnegative.

A final way to impose limits on the possible values of u is to cause the signal not to have an effect on the equations outside of a certain region of parameter space (e.g., by setting the influence functions, as defined later, to zero).

One always needs to be very careful about the definition of the problem. A slight change in the problem definition could mean the difference between an impossible problem and a very easy problem. In fact, some problems may be impossible to solve. For example, consider the following equation, which we have seen several times before:

$dn/dt = \alpha(1 - n) - \beta n$,

with α, $\beta > 0$ and we would like to choose α such that $n(0) = 0.5$ and $n(t_f) = 5$. This is impossible since n can never be > 1 if n starts at 0.5 (verify this for yourself).

Finally, sometimes it helps to consider a dual problem, which is an alternate problem with the same answer (Winston 1994). Being more precise about the example of a dual problem given in the last section, we may wish to find which $u(t)$, among all functions u for which $\|u\|=b$, causes the greatest phase shift. We could also ask which u causes this phase shift with the minimum $\|u\|$.

9.3   Frontiers: A Geometric Interpretation of Optimality

Recapping the mathematical formulation of the preceding problem, we have

$$d\underline{x}/dt = \underline{f}(\underline{x}(t),u(t))$$

over a time interval $t_0 \leq t \leq t_f$, with t_f possibly variable.

The constraints on the control u could be that it is within some bounded set, $u_{min} \leq u(t) \leq u_{max}$, that it have a maximum amplitude,

$$\|u(t)\| \leq b,$$

and/or some other constraint on u. The system state $\underline{x}$ could have an initial or final constraint, such as $g(\underline{x}(t_f)) = b$. While we consider a final constraint here, an initial constraint can be considered in a similar fashion.

We minimize $\varphi(\underline{x}(t_f)) + \int_{t_0}^{tf} L(\underline{x},u,t')dt'$, where $L(\underline{x},u,t')$ is the cost at any time, and also consider a possible final cost $\varphi(\underline{x}(t_f))$.

We now determine how to best minimize the cost, and later make sure we meet the constraints. We cannot directly change $\underline{x}$, but we can change u and possibly t_f, both of which will affect $\underline{x}$.

We first consider how u can be changed to decrease the cost. When, possibly after making some such changes, we find that u cannot be changed to further minimize the cost, we then say we are at an optimal u.

First, if L depends just on u, we may write it as $L(u)$ (an example would be $L = u^2$), and then we can minimize L directly by changing u. By the method of steepest descent, choosing

$$\Delta u = -b_{\Delta u_a}\partial L/\partial u, \tag{9.3.1}$$

where $b_{\Delta u_a}$ is chosen so that $\|\Delta u\| = \varepsilon_{\Delta u_a}$, will do the best at lowering L for all signals of magnitude $\varepsilon_{\Delta u_a}$.

We can also minimize the final cost $\varphi(\underline{x}(t_f))$, by moving $\underline{x}(t_f)$ in the $-\partial\varphi/\partial\underline{x}$ direction. In fact, any small change in $\underline{x}(t_f)$ will decrease the cost, $\varphi(\underline{x}(t_f))$, by

$$\underline{\Delta x}(t_f) \cdot \partial\varphi/\partial\underline{x}.$$

But what about changes in $\underline{x}$ at previous times? Ideally, we would like some function, $\underline{\lambda}(t)$, where $\underline{\Delta x}(t) \cdot \underline{\lambda}(t)$ tells us how much a change in $\underline{x}(t)$ will affect the final cost $\varphi(\underline{x}(t_f))$.

A related question is: How can we best change u over $t_0 \leq t \leq t_f$ to make this change in the final state of $\underline{x}$?

This can best be answered with influence functions, which are described in more detail in the next section. Some readers may find it best to skip to that section now, since some details of the influence functions will appear later. Influence functions, $\underline{\lambda}$, tell us what small changes to $\underline{x}$ at any time t ($\equiv \underline{\Delta x}(t)$) have the same effect on the final state of $\underline{x}$ ($\equiv \underline{\Delta x}(t_f)$). They do this based on the following relation:

$$\underline{\Delta x}(t) \cdot \underline{\lambda}(t) = \underline{\Delta x}(t_f) \cdot \underline{\lambda}(t_f),$$

which provides a scaling of changes to $\underline{x}$ at different times. So if we choose $\underline{\lambda}(t_f) = \partial\varphi/\partial\underline{x}|_{tf}$, the influence function will tell us how much the final cost decreases for a small change in $\underline{x}$ at any previous time. With a fixed-magnitude change $\|\underline{\Delta x}\| = \varepsilon_\lambda$, the most effective way to decrease the cost would be to choose $\underline{\Delta x}(t) = b_\lambda \underline{\lambda}(t)$, where b_λ is chosen so that $\|\underline{\Delta x}\| = \varepsilon_\lambda$.

Thus, to minimize the final cost, we ask how $u(t)$ can be changed in order to change $\underline{x}$ in such a way that $\underline{\Delta x}(t) = b_\lambda \underline{\lambda}(t)$. A small change to $\underline{\Delta x}(t)$ will come from a change to $\underline{f}(\underline{x},u)$, i.e.,

$$\underline{\Delta x}(t) \sim \underline{\Delta f}(\underline{x},u).$$

Hence, we seek to change $\underline{f}(\underline{x},u)$ so that it is more in line with $\underline{\lambda}(t)$. This would be a step toward minimizing the cost by making the final state end up more in the $\partial\varphi/\partial\underline{x}|_{tf}$ direction. This is not possible if

$$\underline{\lambda}(t) \cdot \underline{f}(\underline{x},u)$$

does not change for any small change to u. We will come back to this quantity, but for the moment let us denote it as

$$H(t) = \underline{\lambda}(t) \cdot \underline{f}(\underline{x},u) \tag{9.3.2}$$

and postpone further discussion to section 9.5, observing for the present only that if $H(t)$ also does not change with changing u, i.e.,

$$\partial H/\partial u = 0$$

at all times, then we are at a minimum. Otherwise, if we consider all possible changes to u of the same magnitude, $\|\Delta u\| = \varepsilon_{\Delta u_b}$, the change that will best minimize the cost will be

$$\Delta u = -b_{\Delta u_b}\partial H/\partial u, \tag{9.3.3}$$

with $b_{\Delta u_b}$ chosen so that the magnitude $\|\Delta u\| = \varepsilon_{\Delta u_b}$.

So we have two ways to minimize the cost by changing u, given by (9.3.1) and (9.3.3). We can add them together (perhaps weighting one more than another), to best minimize the cost:

$$\Delta u = -b_{\Delta u_a}\partial L/\partial u - b_{\Delta u_b}\partial H/\partial u.$$

If the cost depends on t, we can find Δu to move the final state in the f direction rather than, or in addition to, the $\partial \varphi / \partial \underline{x}|_{tf}$ direction. This will "stretch" the trajectory, meaning we will get to the final state earlier in time, and go past it at t_f. We can then reduce t_f to get to the final state we wish.

Finally, if L depends on $\underline{x}$ at times other than the final time, we can also find ways to optimally change the state of $\underline{x}$ at times other than the final time in a similar way. The details of this are beyond our scope, and can be found elsewhere (e.g., in Bryson and Ho 1975). This concludes our discussion of how to optimally decrease the cost.

Next, we find out how to satisfy the constraints. One possibility is that we reach the boundary $u(t) = u_{min}$ or $u(t) = u_{max}$. In this case, we can just stay at the boundary. There are some problems for which the optimal solutions are always at the boundary of u (although the optimal stimuli can switch from u_{min} to u_{max} at certain time points). Such problems are called bang-bang problems, and for them special methods are available (Meier and Bryson 1990; Serkh and Forger 2014). They typically occur when L does not depend on u, and f is monotonic with respect to u, so $\underline{\lambda}(t) \bullet \underline{f}(\underline{x},u)$ will be maximized at the boundary.

Another constraint could be that $g(\underline{x}(t_f)) = b$ or even $\underline{x}(t_f) = \underline{x}_f$. We consider a final constraint, recognizing that an initial constraint would follow similar arguments. So, if the final state has been moved by $\Delta \underline{x}(t_f)$, we can find $\Delta \underline{x}_{cor}$, which will bring the final state back within the constraints. If we need $\underline{x}(t_f) = \underline{x}_f$, then $\|\Delta \underline{x}_{cor}\| = \|\Delta \underline{x}(t_f)\|$, but otherwise we can find $\|\Delta \underline{x}_{cor}\| < \|\Delta \underline{x}(t_f)\|$ so that $g(\underline{x}(t_f) + \Delta \underline{x}(t_f) + \Delta \underline{x}_{cor}) = b$. We now ask how we can optimally enact the change $\Delta \underline{x}_{cor}$ at the final time. This can also be done with influence functions. We let $\underline{\lambda}(t_f) = \Delta \underline{x}_{cor}$, and the change $\underline{\lambda} \bullet \partial \underline{f}(\underline{x},u)/\partial u = \partial H/\partial u$ would be the optimal way to correct for the change in endpoint.

When we can no longer decrease the cost, we are at an optimal solution. We note four possibilities where changing u cannot decrease the cost. First, as described, it is possible that changing u just does not affect the cost (e.g., the condition $\partial H/\partial u = 0$). Second, we may already be at a boundary where any improvement to the cost would put us outside the permissible range of u. Third, consider the constraint $\|u(t)\| < b$ and assume that we are at the boundary $\|u(t)\| = b$. If the avenue for decreasing the cost is by changing Δu in the direction of $u(t)$, the only way to improve u is to lower the magnitude of u first, so that a change can then be added. In this case, what is subtracted must be added back, resulting in no change in the cost. Fourth, we can decrease the cost of the endpoint by changing u by $\underline{\lambda} \bullet \partial \underline{f}(\underline{x},u)/\partial u$, which is $\partial H/\partial u$. But if $b_{\Delta u_a} \partial L/\partial u + \partial H/\partial u = 0$, then any improvement is undone by the fixing of the final state. This case will be discussed in justifying (9.5.4).

Summarizing, we would pick a starting $u(t)$ and simulate the system $\underline{x}$ as well as the influence functions $\underline{\lambda}$. Based on this, we minimize the cost, and apply corrections if this causes us to violate the constraints, using the influence functions to guide optimal perturbations. The new $u(t)$ is then simulated and the process continued until we reach an optimum.

Working through some examples allows us to build intuition about $H(t)$. We justify (9.3.2) and $\partial H/\partial u = 0$. Consider how to minimize t_f. First, consider a perturbation of the system along the $\underline{f}(\underline{x},u)$, direction, since a change

$$\Delta \underline{x} = \underline{x}(t+\Delta t) - \underline{x}(t) = \Delta t \underline{f}(\underline{x},u)$$

would move the system forward in time by an amount Δt. We could have enacted the same change by moving the system by Δt at any time point. Since all of the times when we could apply this change would yield the same effect, we would have

$$\Delta t \underline{\lambda}(t) \bullet \underline{f}(\underline{x},u) = b,$$

assuming $\underline{\lambda}(t)$ points in the direction of $\underline{f}(\underline{x},u)$. This shows why $H(t)$ is constant in this special case. Moreover, this equation could be used to define (9.3.2) assuming $\underline{\lambda}(t)$ points in the direction of $\underline{f}(\underline{x},u)$. If we choose $\underline{\lambda}(t_f)$ proportional to $\underline{f}(\underline{x}(t_f),u)$, then the preceding equation defines $\underline{\lambda}(t)$ at other times.

Now consider the case where $\underline{\lambda}(t)$ does not point in the direction of $\underline{f}(\underline{x},u)$. This means that we can move in a direction orthogonal to $\underline{f}(\underline{x},u)$ and still affect the final time. How is this possible? This can occur by "hitching a ride" on another trajectory. By this we mean that we can perturb the system at some time to another nearby trajectory. If this trajectory travels quicker than the current trajectory, then at a later time, we can again move the system in a direction orthogonal to $\underline{f}(\underline{x},u)$, and return to the original trajectory with a more advanced or delayed phase than if we had stayed on the original trajectory.

To track this, consider the set of all points in a region of $\underline{x}(t_0)$ that can be reached by moving orthogonally to $\underline{f}(\underline{x}(t_0))$. We map these points forward in time, and they act in a way similar to isochrons, except that, instead of all mapping to a single point as time approaches infinity, they are defined by their state at time zero. We call these generalized isochrons.

One possible outcome would be that, as time proceeds forward, the points always remain orthogonal to $\underline{f}(\underline{x}(t))$. In this case, if we changed u at one time to move the system orthogonal to $\underline{f}(\underline{x}(t))$, and then apply another change in u at a later time to move us back to the original trajectory, we will not have advanced forward or backward in time. However, if the points do not always remain orthogonal to $\underline{f}(\underline{x}(t))$, then we can introduce a small perturbation to u moving us away from the original trajectory in a direction orthogonal to $\underline{f}(\underline{x}(t))$, and then, at a later time, move back to the original trajectory, again in a direction orthogonal to $\underline{f}(\underline{x}(t))$ and find that we have arrived at a different time point than if we had stayed on the original trajectory. In such a case, u would not be optimal. For this nonoptimality *not* to be the case, we must have $\Delta t \underline{\lambda}(t) \bullet \underline{f}(\underline{x},u) = b$, which explains the importance of the quantity in (9.3.2). This quantity, called a Hamiltonian, is discussed further in section 9.5. In terms of phase-space geometry, this reasoning also explains why a time-optimal trajectory must always intersect generalized isochrons at the same angle.

9.4   Influence Functions

Here we present one way of deriving the influence functions. Two alternate derivations are provided in the next section, and practical examples of influence functions are given in the examples at the end of the chapter.

Calculating optimal perturbations can be greatly helped by calculating influence functions. An influence function can measure many things. For example, if we perturb the system by a small amount at time t in a particular direction, how does our final state or final cost decrease? By putting the question this way, we can consider costs throughout an interval simultaneously and thus quickly find the solutions to optimization problems. We first build intuition about influence functions in the context of oscillators.

Consider the following system:

$$d\underline{x}/dt = \underline{f}(\underline{x}(t),u(t)).$$

Now consider a solution $\underline{x}(t)$ and perturbations around it, $\underline{x}^* = \underline{x} + \Delta\underline{x}$. We know that, to first order,

$$d\underline{x}^*/dt = d\underline{x}/dt + d\Delta\underline{x}/dt = \underline{f}(\underline{x}^*,u,t) \approx \underline{f}(\underline{x},u,t) + \partial\underline{f}/\partial\underline{x}|_x \Delta\underline{x},$$

so the way in which this perturbation would change the system at future times is given by

$$d\Delta\underline{x}/dt \approx \partial\underline{f}/\partial\underline{x}|_x \Delta\underline{x}$$

where $\partial\underline{f}/\partial\underline{x}|_x$ is a matrix function of time, which we can evaluate with the unperturbed system (indicating this with a double underline would make our notation overly complex). The idea is that the system is evolving, and we are considering small changes to the system state that are small enough for the changes in the system's behavior to be linear with the change in $\underline{x}$. From this we can track how small changes in $\underline{x}$ at time t_0 lead to changes in $\underline{x}$ at time t_f.

A slightly more complicated case is to consider continuous small perturbations to $\underline{x}$, $\Delta\underline{x}$, and u, δu, which will induce small changes $\Delta\underline{x}$. So consider the following two systems:

$$d\underline{x}/dt = \underline{f}(\underline{x}(t)),$$

$$d\underline{x}/dt + d\Delta\underline{x}/dt = d\underline{x}^*/dt = \underline{f}(\underline{x}^*(t),u(t)) \approx \underline{f}(\underline{x}(t),u(t)) + \partial\underline{f}/\partial\underline{x}|_x \Delta\underline{x} + \partial\underline{f}/\partial u|_x \delta u + \cdots.$$

The first-order terms yield the following equations:

$$d\Delta\underline{x}/dt \approx \partial\underline{f}/\partial\underline{x}|_x \Delta\underline{x} + \partial\underline{f}/\partial u|_x \delta u. \tag{9.4.1}$$

This has the form of a driven linear differential equation. Define $\underline{\lambda}(t)$ by

$$d\underline{\lambda}/dt \equiv -\partial\underline{f}/\partial\underline{x}|_x \underline{\lambda} \tag{9.4.2}$$

and choose $\underline{\lambda}(t_f)$ to be the unit vector in which direction we would like the system to move at the final time. We can then measure how much perturbations to the system at previous times affect the system in this final direction at t_f:

$$\int_{t_0}^{t} \underline{\lambda}(t') \frac{\partial f}{\partial u}\bigg|_x \delta u \, dt'.$$

This provides a definition for the influence function, i.e., how much a change at some t' affects the system at a later t. Two alternate definitions will be provided in the next section.

Before the discussion continues, some remarks can help clarify influence functions. Based on the preceding, we can express the solution Δx as

$$\underline{\Delta x}(t) = \underline{\Delta x}(t_0) + \int_{t_0}^{t_f} \underline{\lambda}(t') \frac{\partial f}{\partial u}\bigg|_{t'} \delta u(t') dt'.$$

Thus, the interpretation of $\underline{\lambda}(t')$ is that it tells us how $\underline{\Delta x}(t)$ is affected by $\partial \underline{f}/\partial u|_x \, \delta u(t')$ and ultimately by $\delta u(t')$. It translates a signal in the past (time t') to an effect on time t. To illustrate this, consider the following example, where we assume both $\underline{x}$ and u have just one dimension. We then have

$$\Delta x(t) = \Delta x(t_0) + \int_{t_0}^{t} \lambda(t') \frac{\partial f}{\partial u}\bigg|_{t'} \delta u(t') dt'.$$

We then ask that given a signal with a fixed amplitude, $\|\delta u\|$, what is the maximum change in $\Delta x(t)$ that can be achieved? The answer is when

$$\underline{\lambda}(t') = b \frac{\partial f}{\partial u}\bigg|_{t'} \delta u(t')$$

for some constant b. For example, consider

$dx/dt = x + u.$

Consider all signals with $\|u\| = 1$. We then have

$d\lambda/dt = -\lambda,$

from which we obtain $\lambda = be^{-t}$ and likewise $u = be^{-t}$.

For this simple one-dimensional problem, the initial condition of λ is not that important; it can simply be chosen so that $\|u\| = 1$. However, choosing the initial conditions for a multidimensional problem is more complex. Often the goal is to choose them so that a particular $\Delta x(t)$ is chosen. In this sense, an n-variable system has $2n$ conditions that must be satisfied for an optimum: the n initial conditions of $\underline{x}$, and the n initial conditions of $\underline{\lambda}$.

However, often the initial conditions of $\underline{\lambda}$ are not given. For example, we could be given a set of initial $\underline{x}$ and a set of final $\underline{x}$, and need to find the initial $\underline{\lambda}$ that will yield the correct final $\underline{x}$.

9.5   Frontiers: Two Additional Derivations of the Influence Functions

Here we derive the influence functions in two other ways.

In our first approach, we wish to construct a function of the state variables, $\underline{x}$, of an n-variable system, in such a way that this function's value does not change on the optimal path. The function will turn out to be the Hamiltonian mentioned in section 9.3. Again, assume $\underline{x}$ changes according to the equations

$$d\underline{x}/dt = \underline{f}(\underline{x}(t), u(t)).$$

We now seek out a new set of variables $\underline{\lambda}$ whose changes balance any changes in $\underline{x}$ so that $\underline{\lambda} \cdot \underline{f}(\underline{x}(t), u(t))$ is constant. To do this, let

$$H \equiv \underline{\lambda} \cdot \underline{f}(\underline{x}(t), u(t)).$$

Therefore,

$$dH/dt = d\underline{\lambda}/dt \cdot \underline{f}(\underline{x}(t), u(t)) + \underline{\lambda} \cdot (\partial \underline{f}/\partial \underline{x} \cdot d\underline{x}/dt).$$

Here we wish to optimize this function so that $\partial H/\partial u = 0$, which is achieved when $\underline{\lambda}$ is orthogonal to $\partial \underline{f}/\partial u$, so a term with $\partial H/\partial u (du/dt)$ is not needed.

Assuming f is autonomous, expanding this out, and collecting terms that contain f_i, we have

$$dH/dt = f_1 d\lambda_1/dt + f_1 \Sigma_j (\partial f_j/\partial x_1)\lambda_j + \cdots + f_i d\lambda_i/dt + f_i \Sigma_j (\partial f_j/\partial x_i)\lambda_j + \cdots.$$

This expression will always be zero if we allow

$$d\lambda_i/dt = -\Sigma_j (\partial f_j/\partial x_i)\lambda_j, \tag{9.5.1}$$

which derives the influence function in a second way, as may be seen by comparing (9.5.1) and (9.4.2).

If we solve for both $\underline{x}$ and $\underline{\lambda}$, then, as before, we can construct an H whose value does not change. Now let us consider a one-dimensional system to gain intuition. With x and λ one-dimensional, we have

$$dx/dt = f(x),$$

$$d\lambda/dt = -(\partial f/\partial x)\lambda,$$

and we consider $\lambda f(x) = H$. Then $dH/dt = d\underline{\lambda}/dt \cdot \underline{f}(\underline{x}(t), u(t)) + \underline{\lambda} \cdot (\partial \underline{f}/\partial \underline{x} \cdot d\underline{x}/dt) = -f(x)(\partial f/\partial x)\lambda + \lambda(\partial f/\partial x)dx/dt = 0$.

Another way to generate these equations is by using Lagrange multipliers. We again wish to minimize a cost function as in (9.2.1),

$$J = \int_{t_0}^{t_f} L(\underline{x}, u)\, dt$$

where we denote by J the total cost. For a fixed u, we require the system to also obey the differential equations

$$d\underline{x}/dt = \underline{f}(\underline{x}(t), u(t))$$

(one equation for each component of the vector $\underline{x}$). We can add the differential equations to the cost function through Lagrange multipliers $\underline{\lambda}$, so that we will then minimize

$$J = \int_{t_0}^{t_f} L(\underline{x}, u) + \underline{\lambda} \cdot (\underline{f} - d\underline{x}/dt)\, dt \qquad (9.5.2)$$

where we note that, when minimized, the term multiplying $\underline{\lambda}$ is zero. The interpretation here is to consider all possible $\underline{x}$ and to choose those where $\underline{f} - d\underline{x}/dt = 0$. Integrate the last term by parts, and we have

$$J = \int_{t_0}^{t_f} (L(\underline{x}, u) + \underline{\lambda} \cdot \underline{f} + d\underline{\lambda}/dt \cdot \underline{x})\, dt - \underline{\lambda} \cdot \underline{x}\big|_{t_f} + \underline{\lambda} \cdot \underline{x}\big|_{t_0}$$

We wish the preceding expression to be minimized for changes in $\underline{x}$ in that we cannot decrease the cost by *changing* $\underline{x}$, or in other words, what we require is that the expression's partial derivative with respect to $\underline{x}$ be zero. This yields

$$dJ = 0 = \int_{t_0}^{t_f} ((\partial L / \partial \underline{x}) \Delta \underline{x} + \underline{\lambda} \cdot (\partial \underline{f} / \partial \underline{x}) \Delta \underline{x} + d\underline{\lambda}/dt \cdot \Delta \underline{x})\, dt - \underline{\lambda} \cdot \Delta \underline{x}\big|_{t_f} + \underline{\lambda} \cdot \Delta \underline{x}\big|_{t_0}$$

(again this, for the moment, assumes that $du = 0$). This yields the same equations for $d\underline{\lambda}/dt$ as in (9.4.2) and (9.5.1),

$$\frac{d\lambda_i}{dt} = -\frac{\partial L}{\partial x_i} - \sum_j \frac{\partial f_j}{\partial x_i} \lambda_j, \qquad (9.5.3)$$

and so we have our third derivation of the influence function. We can also generate another equation that helps to interpret $\underline{\lambda}$, even though it is often not used. Note that

$$\underline{\lambda} \cdot \Delta \underline{x}\big|_{t_f} = \underline{\lambda} \cdot \Delta \underline{x}\big|_{t_0},$$

and this holds for all possible $\Delta \underline{x}$. Thus, let us choose $\underline{\lambda}(t_f)$ (we will explain how this is done later) and, from the differential equations, $\underline{\lambda}(t_0)$ can then be found by integrating the equations backward in time. This then tells us how changes in $\underline{x}$ propagate forward in time along optimal solutions (and the same u).

One might introduce a cost to the terminal points $\varphi(\underline{x}(t_f))$, which needs to be minimized. In this case we have

$$J = \int_{t_0}^{t_f} \left(L(\underline{x},u) + \underline{\lambda} \cdot \underline{f} + d\underline{\lambda} / dt \cdot \underline{x} \right) dt - \underline{\lambda} \cdot \underline{x}\big|_{t_f} + \underline{\lambda} \cdot \underline{x}\big|_{t_0} + \varphi(\underline{x}(t_f)).$$

One could then require that

$$\underline{\lambda}\big|_{t_f} = \partial \varphi / \partial \underline{x}\big|_{t_f}$$

Changing $\underline{x}$ at other times would change the cost in the following way:

$$dJ = \underline{\lambda} \cdot \underline{\Delta x_f}.$$

Likewise, suppose we do not want a change in $\underline{x}$ to change the final state of $\underline{x}$ in a particular direction (e.g., we do not want to change the final voltage to a neuron). We then choose this direction in which not to perturb $\underline{x}$ as $\underline{\lambda}(t_f)$. We then require

$$0 = \int_{t_0}^{t_f} \underline{\lambda} \cdot \underline{\Delta x} dt$$

Proceeding in this way, we can also find conditions where the changes in $\underline{x}$ would not affect the final state of the system.

Now, most importantly, we wish for J to be minimized among all choices of u. This would require that: $dJ/du = 0$. Since u appears only in $L(\underline{x},u)$ and $\underline{f}(\underline{x},u)$, we then have $dH/du = 0$ (where the Hamiltonian is defined with L). Since any du is possible, this yields an algebraic equation that must be satisfied at all times:

$$\frac{\partial \underline{f}}{\partial u} \cdot \underline{\lambda} + \frac{\partial L}{\partial u} = 0 \tag{9.5.4}$$

Thus, our system now has the form of a set of differential equations for $\underline{x}$, a set of differential equations for $\underline{\lambda}$, and an algebraic equation for u.

9.6   Adding the Cost to the Hamiltonian

It is also possible to add the cost directly to the Hamiltonian, creating a simple notation for the problem. Consider a control, u, and some cost, $L(\underline{x},u)$, associated with the control. We wish to minimize the integral of $L(\underline{x},u)$ over the time interval and also have

$$d\underline{x}/dt = \underline{f}(\underline{x},u).$$

One way to calculate this is to consider an additional variable, y, such that

$$y(t_0) = 0 \quad \text{and} \quad dy/dt = L(\underline{x},u).$$

It is easy to see that the value of y is the integral of the cost, $L(\underline{x}, u)$, for a choice of u. Now if we consider this as one of the differential equations, we have $\underline{x} \equiv (y, x_1, x_2, \ldots)$ and the associated $\underline{\lambda} \equiv (\lambda_y, \lambda_1, \lambda_2, \ldots)$. Construct the Hamiltonian as before, $H \equiv \underline{\lambda} \cdot \underline{f}(\underline{x}, t)$, and choose $\underline{\lambda}$ so that H is constant along model trajectories. This gives the following additional differential equation:

$$d\lambda_y/dt = -\Sigma_j (\partial f_j/\partial y)\lambda_j.$$

However, we note that y was a dummy variable that did not originally appear in the model's equations. Thus, $\partial f_j/\partial y = 0$, and therefore

$$d\lambda_y/dt = 0.$$

Without loss of generality, let us then choose $\lambda_y = 1$. Likewise, we can separate out the f_y term and have

$$d\lambda_i/dt = -\partial L/\partial x_i - \Sigma_j (\partial f_j/\partial x_i)\lambda_j.$$

The interpretation of these sensitivity functions is that they tell us how the cost changes for a perturbation in the x_i variable at a previous time point.

9.7 Numerical Methods for Finding Optimal Stimuli

The optimal control problem we have discussed often results in a boundary value problem. We may know $\underline{x}(t_0)$ and $\underline{x}(t_f)$, but not know $\underline{\lambda}(t_0)$. Here, we outline two schemes for solving this problem (Bryson and Ho 1975).

Scheme 1: The shooting method. We can guess at $\underline{\lambda}(t_0)$ and find the corresponding $\underline{x}(t_f)$. If that is not the $\underline{x}(t_f)$ we wish, we can try another $\underline{\lambda}(t_0)$, and continue until we find the correct one. One would never do this in such a naïve way, but instead use Newton's method (Bradie 2006) or some other method to find the appropriate value of $\underline{\lambda}(t_0)$.

The shooting method has a fatal flaw: $\underline{\lambda}(t)$ tends to be very unstable and grows very quickly. This means that extremely small changes in $\underline{\lambda}(t_0)$ lead to very large changes in $\underline{x}(t_f)$, so large that they may ruin the computation. The stability of $\underline{\lambda}(t)$ can be greatly improved by simulating backward from t_f to t_0. The difficulty with this is that $\underline{x}(t)$ must be simulated forward in time. This leads to scheme 2.

Scheme 2: A variation on Picard iteration. Picard iteration starts with a guess of a solution, and then uses it in simulations to determine better and better numerical solutions. We can do this in the following way. First, choose any reasonable u. Next, from the u simulate $\underline{x}$ forward in time (and note $d\underline{x}/dt$ does not originally depend on $\underline{\lambda}$). Now that $\underline{x}$ and u are known, we can solve $\underline{\lambda}$ backward in time. From this solution of $\underline{\lambda}$, u can be updated, $\underline{x}$ simulated forward in time, and $\underline{\lambda}$ simulated backward in time. This is repeated until the solution is found.

One additional potential concern is that the preceding scheme may not satisfy a constraint, for example, that the final state be at a particular point. We can, however, add to u a small perturbation that brings us toward satisfying the constraint. The best way to do this is with an influence function that, at t_f, points the system in the direction closest to satisfying the constraint. Thus, when u is updated, we can also add a small change to bring it closer to the constraint. Further details of this method are presented in (Bryson and Ho 1975) and (Serkh and Forger 2014). Such methods are best seen in actual examples, as shown in the next section.

9.8 Frontiers: Optimal Stimuli for the Hodgkin–Huxley Equation

We next consider the optimal stimuli to cause an action potential in the Hodgkin–Huxley equations, describing the computations as done in code 9.1. This is a very important problem, as neuronal systems have large energy constraints (Clay et al. 2012) and need to cause action potentials with the least amount of energy. We illustrate this with two methods that can be widely applied. Section 9.9 shows another example using these methods.

First we consider the applied current of length 20 ms, u, of least magnitude,

$$\int_0^{20} u(t)^2 \, dt,$$

to take the Hodgkin–Huxley model to a voltage of 10 mV. Following methods described by Chang and Paydarfar (2014) that build on methods described by Bryson and Ho (1975) and by Serkh and Forger (2014), we first choose an initial guess of u, here where $u(t) = 1$, and simulate the Hodgkin–Huxley model (the hhop function). The influence functions are then simulated (the Roptm function). In particular, representing the Hodgkin–Huxley equations as

$$d\underline{x}/dt = \underline{f}(\underline{x})$$

for this problem, the influence equations obey

$$d\underline{\lambda}/dt = -\partial \underline{f}/\partial \underline{x} \, \underline{\lambda}.$$

We wish to track how changes in u affect the final state of the voltage, so we choose the final state of this system to be [1 0 0 0]. This equation is solved backward in time.

To minimize u optimally, we should decrease it by $-2u$ as described by the following line from code 9.1:

deltau = -2*u*0.001;

where 0.001 is a scaling factor determining how quickly we converge. We next see how much this change in u would change the final state of the voltage. While we could solve this by direct simulation, it is more efficient to use the influence functions in the following way:

q = (tottim/numpts)*yaa(:,1)'*yaa(:,1);

sva = deltau*yaa(:,1)*tottim/(numpts*q);

where q is a scaling factor. This code integrates the change in u, which corresponds to a change in V since the u term appears in the dV/dt equation in this model, along with the influence function, determining how a change in u, at time t, would change the final state of V.

We now know that the best way to change u would be by adding deltau to u. This would change the final voltage by sva(1). To correct for this change, we add to u the additional term

-(sva +sv(1)-10)*yaa(:,1)'/q

which is the optimal way to change the final value of the voltage to make up for the difference between the final voltage without the change (sv(1)) and the value we seek (10).

This process is repeated 500 times until the proposed change in u has a magnitude of less than 0.001 the magnitude of u. The optimal current and corresponding voltage trace predicted by the Hodgkin–Huxley model are shown in figure 9.2.

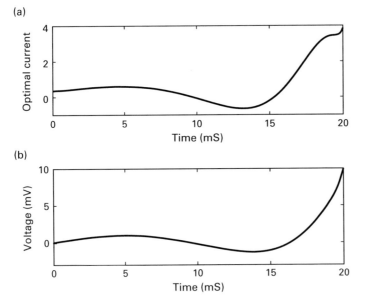

Figure 9.2
An optimal current (a) bringing the Hodgkin–Huxley model from rest to a voltage of 10 (b) with the least amount of current u. See code 9.1 for more details.

Another way to solve the problem is to solve the equations directly rather than in an iterative manner. We use this direct method now to solve a slightly different problem, using code 9.2, similar to that described in Forger et al. (2011).

We first simulate the Hodgkin–Huxley model starting from a state of rest with a 6-ms pulse of current. The amplitude of this current pulse is adjusted so that it has the minimal current possible to cause an action potential. We now consider the final state of this system at the end of the pulse and ask whether we can reach that state with another input current u, with lower magnitude, as defined earlier.

Consider equation (9.5.2) derived in section 9.5:

$$J = \int_{t_0}^{t_f} L(\underline{x}, u) + \underline{\lambda} \cdot (\underline{f} - d\underline{x}/dt)dt$$

We have the same influence equations and the same Hodgkin–Huxley equations. But we also have the algebraic equation (9.5.4),

$$\frac{\partial \underline{f}}{\partial u} \cdot \underline{\lambda} + \frac{\partial L}{\partial u} = 0$$

In this case, this equation yields

$$u = \lambda_1/2,$$

linking the influence equations and the Hodgkin–Huxley equations. We now have a boundary-value problem taking the Hodgkin–Huxley equation, combined with the influence equations, from the rest state to the state found at the end of the optimal 6-ms pulse. We do not have initial or final conditions for the influence equations, and we have both initial and final conditions for the Hodgkin–Huxley model. This is solved in code 9.2 and shown in figure 9.3.

9.9   Examples: Analysis of Minimal Time Problems

Finally, we present two examples of models that allow minimal time problems to be addressed analytically. Our first example comes from Nabi and Moehlis (2012). It has a form similar to some of the models studied in section 5.3:

$$d\phi/dt = f(\phi) + Z(\phi)u,$$

where $f(\phi)$ represents the dynamics of the oscillator and $Z(\phi)$ the sensitivity of the oscillator to a stimulus u. Here, we enforce a constraint that $|u| < a$ and require that $d\phi/dt' > 0$. We wish to determine u over the interval $0 \le t < 2\pi$ so that $\phi(0) = 0$ and $\phi(t_f) = 2\pi$ for minimum t_f.

(a)

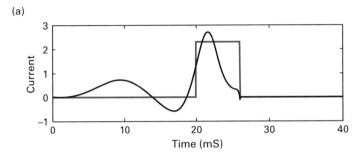

(b)

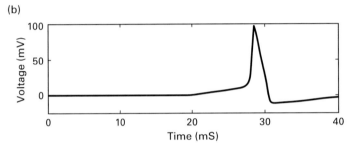

Figure 9.3
(a) We first find the 6-ms pulse of current with the minimal amplitude to cause an action potential (blue). An optimal stimulus of any shape is then found which brings the system to the same state as the optimal pulse with minimal magnitude (black). The corresponding voltage is shown (b).

We can define the cost function as $L = 1$ since this is a minimum time problem. The Hamiltonian is then

$$H = 1 + \lambda(f(\phi) + Z(\phi)u).$$

To minimize this expression we chose $u = -a$ if $\lambda Z(\phi)$ is positive and $u = a$ if $\lambda Z(\phi)$ is negative. If you think about it, this solution is intuitive. The function λ determines how changes at any t affect the final cost, and this is always positive for $f(\phi) > 0$. So we decide whether to apply a positive or negative stimulus based on the sign of $Z(\phi)$, which determines whether the stimulus increases or decreases the phase of the oscillator.

Nabi and Moehlis also consider a slightly more complicated problem of also including the constraint that u has mean zero. To account for this, we add another state variable x, where $dx/dt = u$. This then gives us a new Hamiltonian:

$$H = 1 + \lambda_1(f(\phi) + Z(\phi)u) + \lambda_2 u.$$

This problem is simpler than one might think since $d\lambda_2/dt = 0$. To minimize H, we look for the transition points between when u takes the values a and $-a$, where the problem is not constrained by the bounds on u. This occurs when $\partial H/\partial u = 0$ or

$$\lambda_1 Z(\phi) = -\lambda_2.$$

We switch between the maximum and minimum values of u, or vice versa, when this equation holds.

Another oscillator we considered in section 5.4 was the harmonic oscillator,

$$\frac{dx}{dt} = x_c + u,$$

$$\frac{dx_c}{dt} = -x.$$

We note that if u is constant, we can change variables so that $x'_c = x_c + u$, and are left with the harmonic oscillator whose solutions are $x'_c = a\sin(t+b)$. So for fixed u, the system oscillates with constant amplitude around the point $(0, -u)$.

We now consider possible optimal solutions where $-1 \leq u(t) \leq 1$ following section 2.6.4 of Schättler and Ledzewicz (2012). We wish to find the control $u(t)$ that takes the model from an initial state $\underline{x}(t_0) = \underline{x}_0$ to the origin in minimal time.

There are two influence functions and their equations are

$$\frac{d\lambda_1}{dt} = -\lambda_2,$$

$$\frac{d\lambda_2}{dt} = \lambda_1,$$

or

$$\frac{d^2\lambda_2}{dt^2} = -\lambda_2,$$

and the associated Hamiltonian is

$$H = 1 - \lambda_2 x + \lambda_1(x_c+u).$$

We find $\partial H/\partial u = 0$ when $\lambda_2 = 0$. Since λ_1 and λ_2 obey the equations for a harmonic oscillator (i.e., it has solutions $a\sin(t+b)$), the time-optimal u switches between -1 and 1 every π time units. When $u = 1$, as discussed, the system oscillates around $(0, -1)$. When $u = -1$, the system oscillates around $(0, 1)$. We can start this signal at any point (e.g., with $u = 1$ or $u = -1$ and at a time point of our choosing before the next switch), and we choose this so that the final state is at the origin. This is illustrated in figure 9.4.

9.10   Example: Shifting the Human Circadian Clock

In section 9.9, we saw an analytical example of time optimal controls for the harmonic oscillator. Our model of the human circadian clock presented in section 5.10 is based on

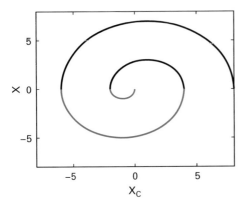

Figure 9.4
Time-optimal solutions to bring the harmonic oscillator to the origin. Two solutions are shown. The black region indicates where $u = -1$. The blue region indicates where $u = 1$. Modified from Schättler and Ledzewicz (2012).

a model similar to the harmonic oscillator. One then wonders if the problem of jet-lag, or getting the human circadian clock from one state to another in minimal time, has similar solutions.

To solve this jet-lag problem, Serkh and Forger (2014) determined the optimal schedules to shift (reentrain) the human circadian clock for all possible time zone shifts, and also for many possible maximal lighting levels, using the methods described in this chapter. This expands on previous work of Dean et al. (2009) and uses the methods of Bryson and Ho (1975) described earlier. It was also shown that this problem is bang-bang, so that individuals need only consider the maximal (e.g., corresponding to daytime) or minimal (corresponding to nighttime) light levels. In particular, the goal was to get to the appropriate isochron in minimal time. The amplitude was treated as a cost, where final amplitudes were required to be within 10 percent of the entrained value. Thus, by adjusting when one begins or ends a day, one can get over jet lag in minimal time. Sample schedules are shown in figure 9.5. This was also incorporated into a widely used iPhone app, ENTRAIN (http://www.entrain.org), which can simulate an individual's circadian clock to determine which schedules to follow; see figure 9.6.

Code 9.1 Optimal Stimulus for the Hodgkin–Huxley Equations

(See figure 9.2)
```
%This code gives an example of codes to calculate optimal stimuli. The
%reader should figure out the code and add more comments.
clear
global yy tt u tottim
tottim = 20;%length of the stimulus
```

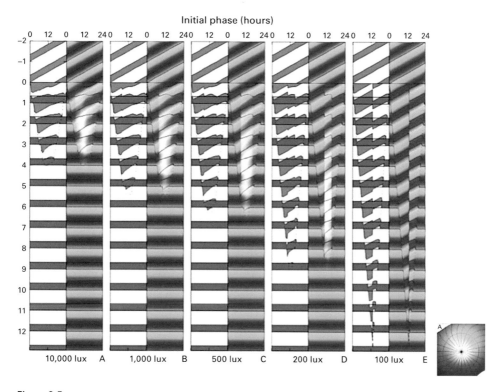

Figure 9.5
Optimal schedules to overcome jet lag. Schedules are based on the maximal lighting available, as shown on the bottom of the plots. Individuals are assumed to come from all possible time zones and enter a common time zone at time 0. (Left) Schedules can be read from the top down, based on the initial phase (time zone from which they arrive). Before time zero, individuals follow a schedule of 16 hours of light and 8 hours of darkness per day from their original timezone. White regions refer to times when it is optimal to seek light, and dark regions are times when light should be avoided. At the dotted lines, an individual is predicted to be entrained. (Right, with colors) Simulations of the model of the human circadian clock to these schedules. Phase is encoded by color, with red being when the clock thinks it is night. The amplitude of the clock is encoded by the strength of the color, with white being zero amplitude. Time is given in hours horizontally and days vertically. See legend on bottom right and Serkh and Forger (2014) for more details from which this image was taken.

(a)   (b)

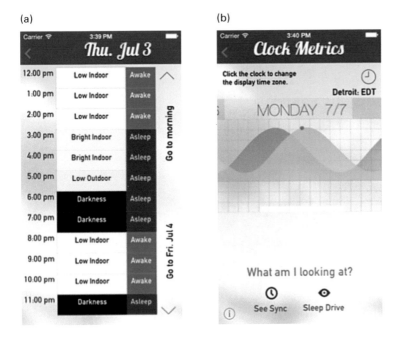

Figure 9.6
Screen shots of the ENTRAIN iPhone app. Images provided by Olivia Walch.

```
numpts = 2001;%number of timepoints in the stimulus
tt = 0:(tottim/(numpts - 1)):
tottim;u = ones(1, numpts);
%we choose an initial guess as 1 for all values of %u
for ij = 1:500
options = odeset('MaxStep', 0.1);
[tt, yy] = ode45(@hhodeop, tt, [0.0036, 0.053, 0.3177, 0.5960], options);
%simulates the HH equations forward in time
sv = yy(numpts,:);%takes the final timepoint
deltau = -2*u*0.001;%optimal decrease without regard to constraint
[ta, ya] = ode45(@Roptm, tt, [1, 0, 0, 0], options);%simulates
%influence equations
yaa = fliplr(ya')';%flips it because we integrated backwards in time
q = (tottim/numpts)*yaa(:,1)'*yaa(:,1);%scaling factor for influence eqn
sva = deltau*yaa(:,1)*tottim/(numpts*q);% how much the final point will
%change just by changing u
u = u+deltau -(sva +sv(1)-10)*yaa(:,1)'/q;%corrects for u
end
function dydx = hhodeop(t, y)
global u tt
uu = interp1(tt, u, t, 'cubic');%finds the value of u
```

```
gm = 0.3; gna = 120.0; gk = 36.0; ek = -12.0; ena = 115.0; vm = 10.613;
aaba = gna*(y(2)^3)*y(4)*(ena-y(1)) + gk*(y(3)^4)*(ek-y(1)) + gm*(vm -…
y(1)) +uu;
aabb = am(y(1))*(1.0 - y(2)) - bmm(y(1))*y(2);
aabc = an(y(1))*(1.0 - y(3)) - bn(y(1))*y(3);
aabd = ah(y(1))*(1.0 - y(4)) - bh(y(1))*y(4);
dydx = [aaba; aabb; aabc; aabd];
end
function dxdt = Roptm(t, z) %equations for the influence functions
%these rely on many functions defined below that are derived from the
%Hodgkin-Huxley equations
global yy tt tottim
y(1) = interp1(tt, yy(:,1), tottim-t, 'cubic');
%finds the values of v, m, n and h
y(2) = interp1(tt, yy(:,2), tottim-t, 'cubic');
y(3) = interp1(tt, yy(:,3), tottim-t, 'cubic');
y(4) = interp1(tt, yy(:,4), tottim-t, 'cubic');
gm = 0.3; gna = 120.0; gk = 36.0; ek = -12.0; ena = 115.0; vm = 10.613;
aza = (-(gna*(y(2)^3)*y(4) + gk*(y(3)^4) + gm));
azb = (3*gna*(y(2)^2)*y(4)*(ena-y(1)));
azc = (4*gk*(y(3)^3)*(ek-y(1)));
azd = (gna*(y(2)^3)*(ena-y(1)));
A = [aza azb azc azd
(adm(y(1))*(1.0 - y(2)) - bdm(y(1))*y(2)) -(am(y(1)) + bmm(y(1))) 0 0
(adn(y(1))*(1.0 - y(3)) - bdn(y(1))*y(3)) 0 -(an(y(1)) + bn(y(1))) 0
(adh(y(1))*(1.0 - y(4)) - bdh(y(1))*y(4)) 0 0 -(ah(y(1)) + bh(y(1)))];
dxdt = A*z;%simulate backwards in time
end
function y = ah(v)
y = 0.07*exp(-v/20.0);
end
function y = am(v)
if v == 25
y = 1.0;
else
y = (25.0 - v)/(10.0*((exp((25.0 - v)/10.0)) - 1.0));
end
end
function y = an(v)
if v == 10.0
y = 0.1;
else
y = (10.0 - v)/(100.0*(exp((10.0 - v)/10.0) - 1.0));
end
end
function y = bh(v)
y = 1.0/(1.0 + exp((30.0 - v)/10.0));
```

```
end
function y = bmm(v)
y = 4.0*(exp(-v/18.0));
end
function y = bn(v)
y = 0.125*exp(-v/80.0);
end
function y = adh(v)
y = -0.0035*exp(-v/20.0);
end
function y = adm(v)
if v == 25
y = 0.5;
else
y = exp(v/10.0)*(10.0*exp(v/10.0) - exp(5.0/2.0)*(-15.0+
v))/(100.0*(exp(5.0/2.0)-exp(v/10.0))^2);
end
end
function y = adn(v)
if v == 10
y = 0.5;
else
y = exp(v/10.0)*(10.0*exp(v/10.0) - exp(1.0)*v)/(1000.0*(exp(1.0) -
exp(v/10.0))^2);
end
end
function y = bdh(v)
y = exp(3.0 + v/10.0)/(10*(exp(3.0)+exp(v/10.0))^2);
end
function
y = bdm(v)
y = -(2.0/9.0)*exp(-v/18.0);
end
function
y = bdn(v)
y = -0.0015625*exp(-v/80.0);
end
```

Code 9.2 An Alternate Method to Calculate Optimal Stimuli for the Hodgkin–Huxley Model

(See figure 9.3)

```
% This presents an alternate method to that in code 9.1. The reader is
%encouraged to compare and contrast these codes.
global sigstr hhinit hhfin
hhinit = [0.0036, 0.053, 0.3177, 0.5960];
bvpinitcond = [.0036 0.0530 0.3177 0.5960 0.0240 0.4098-28.2935-19.7606];
```

```
sigstr = 2.01; % We first the 6 mSec pulse with minimum amplitude
%that causes an AP
options = odeset('MaxStep', 0.1);
[t, y] = ode45(@hhode, 0:0.25:50, hhinit, options);
if max(y(:,1)) < 60
while max(y(:,1)) < 60
sigstr = sigstr + 0.01;
[t, y] = ode45(@hhode, 0:0.25:50, hhinit, options);
end
end
hhfin = y(27,:);
%the stimulus ends at point 25, but we wait two
%timepoints to allow the ode to see the end of the stimulus
solinit = bvpinit(linspace(0,40,100),bvpinitcond);% This sets up the
%initial mesh and the initial conditions
sigstrrec = sigstr;
options = bvpset('NMax', 4000);
sol = bvp4c(@hhcalvar, @hhbc, solinit, options);
xx = linspace(0, 40, 200);
yy = deval(sol, xx);
ya = yy(5,:)/2; % This is the optimal signal
%The following function sets up a multivariate function that will be
%zero when the problem is solved
function dydx = hhbc(ya, yb)
global hhinit hhfin
aaba = ya(1) - hhinit(1); aabb = ya(2) - hhinit(2);
aabc = ya(3) - hhinit(3); aabd = ya(4) - hhinit(4);
aabe = yb(1) - hhfin(1); aabf = yb(2) - hhfin(2);
aabg = yb(3) - hhfin(3); aabh = yb(4) - hhfin(4);
dydx = [aaba; aabb; aabc; aabd; aabe; aabf; aabg; aabh];
end
% The following functions, hhode, and hhcalvar solve the Hodgkin-
%Huxley equations and the influence functions. They rely on functions
%(e.g. am, adm) that are given in Code 9.1
function dydx = hhode(t, y)
global sigstr
gm = 0.3; gna = 120.0; gk = 36.0; ek = -12.0; ena = 115.0; vm = 10.613;
if t < 6;
ysr = sigstr;
else
ysr = 0;
end
aaba = gna*(y(2)^3)*y(4)*(ena—y(1)) + gk*(y(3)^4)*(ek—y(1));
aaba = aaba + gm*(vm -y(1)) + ysr;
aabb = am(y(1))*(1.0 - y(2)) - bmm(y(1))*y(2);
aabc = an(y(1))*(1.0 - y(3)) - bn(y(1))*y(3);
aabd = ah(y(1))*(1.0 - y(4)) - bh(y(1))*y(4);
```

```
dydx = [aaba; aabb; aabc; aabd];
end
function dydx = hhcalvar(t, y) % The following functions are the
%influence functions and differential equations
gm = 0.3; gna = 120.0; gk = 36.0; ek = -12.0; ena = 115.0; vm = 10.613;
aaba = gna*(y(2)^3)*y(4)*(ena-y(1)) + gk*(y(3)^4)*(ek-y(1));
aaba = aaba + gm*(vm -y(1)) + y(5)/2.0;
aabb = am(y(1))*(1.0 - y(2)) - bmm(y(1))*y(2);
aabc = an(y(1))*(1.0 - y(3)) - bn(y(1))*y(3);
aabd = ah(y(1))*(1.0 - y(4)) - bh(y(1))*y(4);
aabe = y(5)*(gna*(y(2)^3)*y(4) + gk*(y(3)^4) + gm);
aabe = aabe-y(6)*(adm(y(1))*(1.0 - y(2)) - bdm(y(1))*y(2));
aabe = aabe-y(7)*(adn(y(1))*(1.0 - y(3)) - bdn(y(1))*y(3));
aabe = aabe-y(8)*(adh(y(1))*(1.0 - y(4)) - bdh(y(1))*y(4));
aabf = -y(5)*3*gna*(y(2)^2)*y(4)*(ena-y(1)) + y(6)*(am(y(1));
aabf = aabf + bmm(y(1)));
aabg = -y(5)*4*gk*(y(3)^3)*(ek-y(1)) + y(7)*(an(y(1)) + bn(y(1)));
aabh = -y(5)*gna*(y(2)^3)*(ena-y(1)) + y(8)*(ah(y(1)) + bh(y(1)));
dydx = [aaba; aabb; aabc; aabd; aabe; aabf; aabg; aabh];
end
```

Exercises

General Problems
For these problems pick a biological model of interest, preferably one you study.

1. Design an experiment based on the algorithm presented in section 9.1 to experimentally approximate optimal signals.

2. Find some goal of the system (e.g., achieve a phase shift in minimal time) and code up one of the schemes described in section 9.7 to find an optimal signal.

Specific Problems
1. Consider the following model:

$$dx/dt = (2\pi/24) x_c + \varepsilon(x - (4/3)x^3) + B,$$

$$dx_c/dt = -(2\pi/24)x.$$

This represents a simple mathematical model of the human circadian pacemaker, with B representing the level of light. Write down a series of differential (and possibly algebraic) equations that, when solved, cause the system to move from $x = 1$, $x_c = 0$ to $x = 0$, $x_c = 1$ in 72 hours while minimizing B^2.

2. Determine the stimulus that moves the radial isochon clock as described in chapter 5 to increase its amplitude by a maximal amount for a fixed stimulus strength.

3. In section 9.4, we derived $\Delta x(t) = \Delta x(t_0) + \int_{t_0}^{t} \lambda(t') \frac{\partial f}{\partial u}\bigg|_{t'} \delta u(t')dt'$. Verify which stimulus of a set amplitude, $\|\delta u\|$, causes the maximal $\Delta x(t)$.

4. Modify code 9.1 so that it causes an action potential in the least time for currents of a fixed magnitude.

5. Modify code 9.1 and code 9.2 so that they solve the same problem as each other. Show that they find the same answer.

6. Modify code 9.1 so that it uses the Morris-Lecar model. Solve the problem with parameters for the model to be type 1 and type 2. How are these answers different?

7. Modify code 9.1 to solve for some of the schedules shown in figure 9.5.

10

Mathematical and Computational Techniques for Multiscale Problems

In this chapter, we extend classical asymptotic, numerical, and analytical methods previously used on smaller models to large, multiscale models of biological clocks. New methods are presented for averaging models with many variables by using the geometry of the phase space. A numerical method is presented for rapidly simulating multiscale models. After showing how Poincaré maps can be used to reduce models, we end with a discussion of the Poincaré–Bendixson theorem that can apply to some high-dimensional models of biological clocks.

10.1   Simplifying Multiscale Systems

Multiscale problems are very difficult. Chapter 4 considered how individual biophysical elements in a feedback loop work together to generate oscillations. Chapter 7 considered how individual oscillations can couple together to form complex behaviors. Here we study general methods for simplifying, simulating, or analyzing a complex multiscale model of a biological oscillator.

We explore this in depth in three ways. In section 10.1 we use the method of averaging, which is based on a separation of fast and slow scales. A large portion of this chapter is devoted to this method of averaging, which, historically, has played an important role in understanding models that show oscillations (Andronov et al. 1949; Hale 1963; Hayashi 1964; Stoker 1950). We also note some of the pitfalls of averaging, and correct them with second-order averaging.

It is very tempting to use solutions of linear systems to understand these multiscale problems. As an example, consider the Goodwin model. In figure 10.1, we plot the solution of this model, along with the solution of the system linearized around a fixed point. These two solutions have quite a bit of similarity. The linear solution exists on a plane, and the actual solution exists on what is called mathematically a 2-manifold (a two-dimensional manifold similar to a surface), but it is actually quite close to being a plane.

Many complicated things can happen to a two-dimensional manifold. It can develop folds, where solutions start on one branch, and then leave the manifold to fall to another

(a)

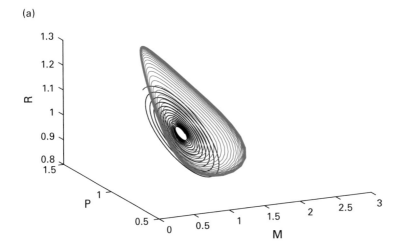

(b)

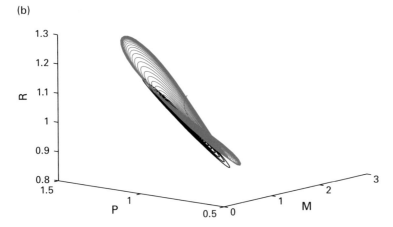

Figure 10.1
Simulations of the Goodwin model. Plots in (a) and (b) are views, from two different directions in state space, of parametric plots of solutions of the Goodwin model. The actual solution is shown in blue, while the solution linearized around the fixed point is shown in black. These two solutions have much in common.

branch (Cosnard et al. 1983; Strogatz 2000). However, many biological problems can be solved without considering these special cases. Much of this chapter will assume a simple geometry based on the idea that oscillations in high-dimensional systems quickly contract to a two-dimensional manifold. We will say more in section 10.2 about how these simple geometries arise and how they can be approximated. In these systems, we show that, in general, as the amplitude of oscillations increases, the period of the oscillations also increases.

Section 10.3 uses a piecewise linear approximation to the solution of differential equations, to solve multiscale problems both analytically and numerically. This gives a surprisingly accurate numerical method, which is then compared with standard methods.

A third approach for simplifying multiscale systems, which is based on Poincaré maps, is presented to simplify oscillators in section 10.4. We end in section 10.5 with two techniques to rule out oscillations that apply to large-dimensional systems. All of these sections use techniques that are used typically for two-dimensional systems, and they show how these techniques can be applied to much larger systems.

10.1.1   The Method of Averaging

The method of averaging gives an approximate solution to a model whose nonlinear terms have a much smaller magnitude than the linear terms, or a model whose dynamics can be approximated by a linear system. We call this the quasilinear limit, as much of the dynamics come from linear terms. The solution can then be approximated by the solution of the linear system (where the magnitude of the nonlinear terms is set to zero) combined with a small correction whose behavior is much slower than the linear solution since its magnitude is much less. Other texts on averaging include (Sanders et al. 2007; Strogatz 2000). Readers familiar with the method of averaging can skip to the next section.

To start, we consider the van der Pol equations to illustrate the method:

$$\frac{dx}{dt} = \mu\left(x_c + \varepsilon\left(x - \frac{4}{3}x^3\right)\right),$$

$$\frac{dx_c}{dt} = -\mu x.$$

The quasilinear limit occurs when $\varepsilon \to 0$. We then have

$$\frac{dx}{dt} = \mu x_c,$$

$$\frac{dx_c}{dt} = -\mu x.$$

or

$$\frac{d^2x}{dt^2} + \mu^2 x = 0,$$

where the solutions are

$x(t) = c_1 \sin(\mu t),$

$x_c(t) = c_2 \cos(\mu t).$

Another way of representing this is in polar coordinates. Here we find

$x(t) = r \sin(\mu(t + \theta)),$

$x_c(t) = r \cos(\mu(t + \theta)).$

This is called the standard form of the problem, and we require $r \geq 0$. The standard form is an accurate solution when $\varepsilon \rightarrow 0$ over a correspondingly small time interval. (Over a long time interval the effects of a term with magnitude ε can accumulate.)

The method of averaging assumes the form of the linear solution, but allows the parameters to slowly vary. In this case, r and θ vary with time:

$x(t) = r(t)\sin(\mu(t + \theta(t))),$

$x_c(t) = r(t)\cos(\mu(t + \theta(t))).$

This is sometimes called the method of variation of parameters (Sanders et al. 2007). It is a change of coordinates of the system, and thus far involves no approximations. The new variables are r, the amplitude of the oscillation, and θ, the change in phase of the oscillation. We then find that

$$dx/dt = (dr/dt)\sin(\mu(t + \theta(t))) + r(t)\mu\cos(\mu(t + \theta(t)))\,((d\theta/dt) + 1), \qquad (10.1.1a)$$

$$dx_c/dt = (dr/dt)\cos(\mu(t + \theta(t))) - r(t)\mu\,\sin(\mu(t + \theta(t)))\,((d\theta/dt) + 1). \qquad (10.1.1b)$$

We can then multiply (10.1.1a) by $\sin(\mu(t+\theta(t)))$ and the second equation by $\cos(\mu(t + \theta(t)))$. Adding the two equations would then yield

$dr/dt = \sin(\mu(t + \theta(t)))\,dx/dt + \cos(\mu(t + \theta(t)))\,dx_c/dt.$

Likewise, we could multiply (10.1.1a) by $\cos(\mu(t + \theta(t)))$ and the second equation by $-\sin(\mu(t + \theta(t)))$. We would then find

$r\mu(d\theta/dt + 1) = \cos(\mu(t + \theta(t)))\,dx/dt - \sin(\mu(t + \theta(t)))\,dx_c/dt.$

We also have

$$\frac{dx}{dt} = \mu\left(r\cos(\mu(t+\theta)) + \varepsilon\left(r\sin(\mu(t+\theta)) - \frac{4}{3}r^3\sin^3(\mu(t+\theta)) \right) \right),$$

$$\frac{dx_c}{dt} = -\mu r \sin(\mu(t+\theta))$$

so we find that

$$\frac{dr}{dt} = \mu r \cos(\mu(t+\theta))\sin(\mu(t+\theta)) + \mu\varepsilon\left(r\sin^2(\mu(t+\theta)) - \frac{4}{3}r^3\sin^4(\mu(t+\theta))\right)$$
$$- \mu r \sin(\mu(t+\theta))\cos(\mu(t+\theta)),$$

or

$$\frac{dr}{dt} = \mu\varepsilon\left(r\sin^2(\mu(t+\theta)) - \frac{4}{3}r^3\sin^4(\mu(t+\theta))\right).$$

The key to this equation is that it goes to zero as ε approaches zero (it is $O(\varepsilon)$). So the 0th-order approximation in ε to dr/dt is $dr/dt = 0$. Likewise, for the phase equation

$$1 + \frac{d\theta}{dt} = \left(\frac{1}{\mu r}\right)\left(\begin{array}{c} \mu r \cos^2(\mu(t+\theta)) + \mu\varepsilon(r\sin(\mu(t+\theta))\cos(\mu(t+\theta)) \\ -\cos(\mu(t+\theta))\frac{4}{3}r^3\sin^3(\mu(t+\theta))) - \mu r \sin(\mu(t+\theta))(-\sin(\mu(t+\theta))) \end{array}\right),$$

we find that

$$1 + \frac{d\theta}{dt} = \left(\cos^2(\mu(t+\theta)) + \sin^2(\mu(t+\theta)) + \varepsilon\left(\sin(\mu(t+\theta))\cos(\mu(t+\theta)) - \cos(\mu(t+\theta))\frac{4}{3}r^2\sin^3(\mu(t+\theta))\right)\right)$$

or

$$\frac{d\theta}{dt} = \varepsilon\left(\sin(\mu(t+\theta))\cos(\mu(t+\theta)) - \cos(\mu(t+\theta))\frac{4}{3}r^2\sin^3(\mu(t+\theta))\right).$$

We also find that this equation is $O(\varepsilon)$. So the 0th-order approximation in ε to $d\theta/dt$ is $d\theta/dt = 0$. We can also think about this from the perspective of time scales. Both r and θ are approximately constant on the time scale of the 0th approximation.

We can then look at these two time scales separately. First, let r and θ be constant as we look at the fast time scale of the standard form. This is given by linear equations. On the slow time scale of r and θ, we are not interested in the fast dynamics of the linear system. They "average" out, or the r and θ system only feels the average effect of the standard form. We can average over any fast time scale w, for example, by replacing the effect of the standard form by its average effect over a cycle. For instance, take the first term in the $d\theta/dt$ equation,

$$\varepsilon\sin(\mu(t+\theta))\cos(\mu(t+\theta)),$$

averaged over a period, and we find

$$\varepsilon \frac{\mu}{2\pi} \int_0^{\frac{2\pi}{\mu}} \sin(\mu(t'+\theta))\cos(\mu(t'+\theta))dt' = 0,$$

where we have assumed, over this fast time scale, that θ is constant since it is not on the fast time scale. Thus, this term has no first-order effect on the oscillation when this averaging assumption is made. Let us also look at the second term in the $d\theta/dt$ equation:

$$\varepsilon \cos(\mu(t+\theta))\frac{4}{3}r^2 \sin^3(\mu(t+\theta)).$$

Note that

$$\varepsilon \frac{\mu}{2\pi} \int_0^{\frac{2\pi}{\mu}} \cos(\mu(t'+\theta))\frac{4}{3}r^2 \sin^3(\mu(t'+\theta))dt' = 0.$$

So we find that, with the averaging assumption,

$d\theta/dt \approx 0.$

Again, this is an approximation, since we replaced the actual system's behavior with the averaged system's behavior. It is useful because it tells us that any phase deviations measured by θ are small.

However, the dr/dt equation is a different story. The first term in the dr/dt equation is

$$\mu \varepsilon r \sin^2(\mu(t+\theta))$$

Again, assuming that r is constant, we find that

$$\mu \varepsilon r \frac{\mu}{2\pi} \int_0^{\frac{2\pi}{\mu}} \sin^2(\mu(t'+\theta))dt' = \frac{\mu \varepsilon r}{2}.$$

Likewise, for the second term,

$$-\frac{4}{3}\mu \varepsilon r^3 \sin^4(\mu(t+\theta)),$$

we find that

$$-\frac{4}{3}\mu \varepsilon r^3 \frac{\mu}{2\pi} \int_0^{\frac{2\pi}{\mu}} \sin^4(\mu(t'+\theta))dt' = -\frac{\mu \varepsilon r^3}{2}.$$

So the averaged dr/dt equation is

$$\frac{dr}{dt} = \frac{\mu\varepsilon}{2}(r - r^3).$$

This has two steady-state solutions. One is at $r = 0$. The other solution is at $r = 1$. Now we see why the van der Pol model has a factor of 4/3 in the equations. This allows the average amplitude at the limit cycle to be 1. One can verify this by changing this factor and re-deriving the averaged equations. So start with

$$\frac{dx}{dt} = \mu(x_c + \varepsilon(x - a_1 x^3)),$$

$$\frac{dx_c}{dt} = -\mu x.$$

where a_1 replaces the 4/3 factor. Our averaged equation would then be

$$\frac{dr}{dt} = \frac{\mu\varepsilon}{2}\left(r - \frac{3a_1}{4}r^3\right),$$

which tells how to change the amplitude of the limit cycle (i.e., $dr/dt = 0$, $r \neq 0$) of the van der Pol model. The terms in this equation come from the original terms $\mu\varepsilon(x - a_1 x^3)$, and thus we call these terms amplitude recovery terms. The example in the next section shows how these can be modified.

We now see the first-order effect of all parameters of the van der Pol oscillator. The outermost coefficient, μ, determines the period of the oscillator. The coefficient of the amplitude recovery terms, ε, determines how quickly the system approaches its limit cycle. The coefficient of the cubic nonlinear term, a_1, determines the amplitude of the limit cycle.

Additional terms can be added into the van der Pol oscillator to change its behavior. These will appear in the dx/dt or dx_c/dt equations, which will then appear in

$$dr/dt = \sin(\mu(t + \theta(t)))(dx/dt) + \cos(\mu(t + \theta(t)))\, dx_c/dt,$$

$$r\mu(d\theta/dt + 1) = \cos(\mu(t+\theta(t)))(dx/dt) - \sin(\mu(t + \theta(t)))\, dx_c/dt.$$

Substituting for dx/dt and dx_c/dt, we can proceed with averaging. The following integral formula helps us evaluate the averaged equations:

$$\frac{\mu}{2\pi}\int_0^{2\pi/\mu} \cos^p(\mu(t'+\theta))\sin^q(\mu(t'+\theta))\, dt' = \prod_{i=1}^{\frac{p\ or\ q}{2}} \frac{2i-1}{2i}, \textit{ if } p \textit{ or } q \textit{ is even and the other zero},$$

$$\frac{\mu}{2\pi}\int_0^{\frac{2\pi}{\mu}} \cos^p(\mu(t'+\theta))\sin^q(\mu(t'+\theta))\, dt' = 0 \text{ if p or q is odd,}$$

$$\frac{\mu}{2\pi}\int_0^{\frac{2\pi}{\mu}}\cos^p(\mu(t'+0))\sin^q(\mu(t'+\theta))dt' = \left(\prod_{i=1}^{\frac{p}{2}}\frac{2i-1}{q+2i}\right)\left(\prod_{j=1}^{\frac{q}{2}}\frac{2j-1}{2j}\right) \text{ if otherwise.}$$

Consider an a_1x^{2q} term in the dx/dt equation or an $a_1x_c^{2q}$ in the dx_c/dt equation. This would appear respectively as

$$a_1r^{2q}\sin^{2q+1}(\mu(t+\theta)) \quad \text{or} \quad a_1r^{2q}\cos^{2q+1}(\mu(t+\theta))$$

in the dr/dt equation. Looking at the preceding formula, we see that both terms are zero when averaged. In the $d\theta/dt$ equation they appear as

$$a_1r^{2q-1}\sin^{2q}(\mu(t+\theta))\cos(\mu(t+\theta)) \quad \text{or} \quad -a_1r^{2q-1}\cos^{2q}(\mu(t+\theta))\sin(\mu(t+\theta)).$$

Both terms average to zero as well. Terms in the dx/dt equation that are of the form a_1x^{2q}, or terms in the dx_c/dt equation of the form $a_1x_c^{2q}$, have no first-order effect. Finally, it can be shown that terms in the dx/dt equation that are of the form $a_1x_c^{2q+1}$, or terms in the dx_c/dt equation that are of the form a_1x^{2q+1}, have no first-order effect on the dr/dt equation, but they do affect the $d\theta/dt$ equation.

From the method of averaging, we can see how the van der Pol model is the simplest quasilinear oscillator. The $-\mu x$ term in the dx_c/dt equation and the μx_c term in the dx/dt equation give us the standard form. Two terms in one equation are required to have a limit cycle. Since a term of the form a_1x^2 in the dx/dt equation would have no first-order effect, a linear term and a cubic term are used.

One should also note that the averaged van der Pol equations can be solved for instead of the original equations. One solution is given by Sanders et al. (2007):

$$x(t) = \frac{r_0e^{\frac{\varepsilon\mu t}{2}}}{\left(1+r_0^2\left(e^{\varepsilon\mu t}-1\right)\right)^{1/2}}\sin(\mu(t+\theta)).$$

where r_0 is the amplitude of the limit cycle of the oscillator. The first term of this equation is an approximate solution to

$$\frac{dr}{dt} = \frac{\mu\varepsilon}{2}\left(r-\frac{3a_1}{4}r^3\right),$$

which again implies the averaging assumption (that the system can be described on a fast time scale by a linear system) is valid.

10.1.2 Example: Applying the Method of Averaging to Human Circadian Rhythms

Jewett and Kronauer (1998) proposed an alternate model of the human circadian pacemaker. It has the following form:

$$\frac{dx}{dt} = \mu\left(x_c + \varepsilon\left(\frac{1}{3}x + \frac{4}{3}x^3 - \frac{256}{105}x^7 \right) \right),$$

$$\frac{dx_c}{dt} = -\mu x.$$

When averaged, this equation becomes

$$\frac{dr}{dt} \varepsilon\mu\left(\frac{r}{6} + \frac{r^3}{2} - \frac{2r^7}{3} \right).$$

In section 5.11, we derived another model of the human circadian clock. A comparison of the dr/dt equation of this model with the original model is shown in figure 10.2.

Originally, it was proposed that the difference in the amplitude recovery of these models significantly affected their behavior. As one can see in figure 10.2, they are not too different (Indic et al. 2005; Serkh and Forger 2014).

10.1.3 What Is Lost in the Method of Averaging?

Consider the original van der Pol model:

$$dx/dt = \mu(x_c + \varepsilon(x - a_1x^3)),$$

$$dx_c/dt = -\mu x.$$

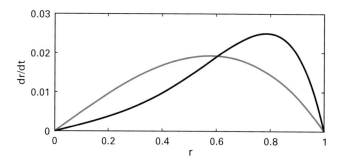

Figure 10.2
Comparisons of the averaged amplitude dynamics of two models of the human circadian clock. The curve with the higher amplitude is for the model presented here. The other curve is for the model presented in 5.11.

One can verify that this model has the same averaged equations as

$$dx/dt = \mu x_c,$$

$$dx_c/dt = -\mu(x + \varepsilon(x_c - a_1 x_c^3)),$$

which also have an equivalent averaged equation as the radial isochron clock,

$$\frac{dr}{dt} = \frac{\mu\varepsilon}{2}(r - r^3),$$

$$d\phi/dt = \mu,$$

when we let $x = r\cos(\phi)$ and $x_c = r\sin(\phi)$.

Differences can be seen in the behavior of these models, which depend on the size of ε. For the radial isochron clock, the isochrons are radial lines outward from the origin. For the other two models, they are different. In the first, the amplitude recovery terms appear only in the dx/dt equation, whereas in the second, they appear in the dx_c/dt equation. So consider a stimulus that pushes the system in the positive x direction starting at $x = -1$ and $x_c = 0$. In the first model, the amplitude recovery terms will push the system back in the negative x direction. In the second model, the system will be pushed in the x_c direction. The final state will thus be different. This difference has been shown to be important for modeling human circadian rhythms (Forger et al. 1999).

The bottom line is that modelers should be aware of the differences between the averaged model and the original model. In some scenarios, the averaged model is suitable for the original model, and in others it is not.

We can illustrate this with the Goodwin model in figure 10.3. We simulate this model, as well as the linear solution, which is used to average the equations. Note that the linear solutions appear as an ellipse with growing amplitude. Oscillations occur around a fixed point that is at the center of the oscillation. The actual solution has more complex dynamics. One will note that the solution is farther from the central point when M and P are large, and closer when M and P are small. This can be accounted for with second-order averaging.

10.1.4 Frontiers: Second-Order Averaging

Motivation: When we averaged the van der Pol equation, we found that $d\theta/dt \approx 0$. This means that the averaged system has the same period as the system in standard form ($2\pi/\mu$). If we simulate the van der Pol equation, we find that the period is slightly different from this. In models of the human circadian clock, we aim to have very accurate predictions of the period. Second-order averaging allows us to do better. This accounts for two effects: first, how actual solutions need not occur around a central point as in the linearized system (and discussed in section 10.1.3); second, we account for the changing amplitude within a cycle. In the end, we will arrive at equations for dr/dt and $d\theta/dt$ that are more accurate, but harder to derive.

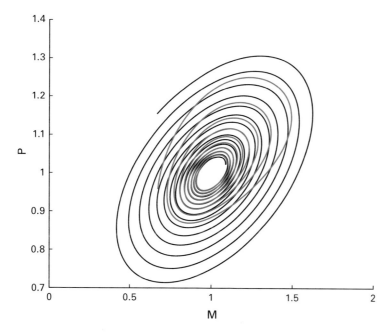

Figure 10.3
Solutions of the Goodwin model showing the linearized and actual solutions. Note that oscillations occur around a center point in the linearized system (black), whereas the actual solution (blue) does not have this property.

The easiest way to understand second-order averaging is to consider the two key time scales. The first time scale is the fast time scale, which consists of the linear terms. These are considered large when compared with the slow time scale, and the slow time scale uses the motion on the fast time scale to calculate an effect of the nonlinear terms. This yields the averaged equations, through two steps:

1. Solve first the fast time scale, ignoring the nonlinear terms.
2. Using the fast time scale, solve for the slow time scale, averaging the equations as before.

Two corrections can now be made. First, we can correct the assumption that the slow time scale only feels the effect of the average of the fast time scale. At any time point, take the difference between the solution of the averaged equation and the solution we averaged. Averaged over one cycle, the difference between these two solutions is zero, but at any time point, it is likely nonzero. A linear correction would be to take, at each time point, the difference between these solutions, and multiply it by the derivative of how the differential equation changes with respect to a change in the parameters.

Writing this down mathematically, before averaging we had

$$d\underline{x}/dt = \underline{f}(\underline{x},t),$$

which we averaged to be

$$d\underline{x}_{av}/dt = \frac{1}{\tau}\int_{t}^{t+\tau} \underline{f}(\underline{x}_{av},t')dt'$$

We can also consider an error at time t_a,

$$\underline{x}_{err}(t_a) = \int_{t}^{t_a} \underline{f}(\underline{x}_{av},t')dt' - \underline{x}_{av}(t_a),$$

where we have made the approximation that $\underline{x}_{av}$ is a good enough solution to the model. A first-order correction for the averaging would be

$$\frac{1}{\tau}\int_{t}^{t+\tau} \underline{x}_{err}\partial\underline{f}/\partial\underline{x}dt'$$

The second correction accounts for the different time scales. Over the fast time scale, we assumed that the slow variables were constant. A better approximation would be to use the averaged equations rather than assume the variables are constant. Mathematically, this would be

$$\frac{1}{\tau}\int_{t}^{t+\tau} \partial\underline{f}/\partial\underline{x}\,\underline{f}_{av}dt'$$

with

$$\underline{f}_{av} = \frac{1}{\tau}\int_{t}^{t+\tau} \underline{f}(\underline{x},t')dt'.$$

By doing this, one is left with a new equation, similar to the averaged equation, but with extra terms. Going through this calculation, the averaged van der Pol equation, to second order, is

$$\frac{dr}{dt} \approx \frac{\mu\varepsilon}{2}(r-r^3),$$

$$\frac{d\theta}{dt} = (\mu\varepsilon)^2\left(-1+\frac{3}{2}r^2-\frac{11}{32}r^4\right).$$

Thus, the dr/dt equation does not change, but the $d\theta/dt$ equation does change.

10.2 Frontiers: Averaging in Systems with More Than Two Variables

The preceding arguments assume that the model has two dimensions. Most systems we study will have more than two dimensions. Here we describe how the method of averaging can still be used following the arguments of Forger and Kronauer (2002), a method called averaging on approximate manifolds.

While biological oscillators typically have more than two dimensions, many of the dimensions might not matter. In fact, after brief initial transients, models of biological clocks often oscillate on a two-dimensional surface in the phase space for reasons we will soon see. We can see this in figure 10.4.

We then can restrict the system to the two-dimensional surface (manifold), or an approximation of it, and then use the method of averaging. The following method does this by approximating the manifold, using the system linearized around a fixed point.

Before actually using the method, it is helpful to review the behavior of a system

$$\frac{d\underline{x}}{dt} = \underline{f}(\underline{x})$$

linearized around a fixed point. We first look for the fixed points, $\underline{\bar{x}}$, of this system, which are those points where

$$\underline{f}(\underline{\bar{x}}) = 0.$$

We also note that f can be represented as a Taylor series where the first-order terms dominate. We then have

$$\underline{f}(\underline{x}) = \underline{\underline{A}}(\underline{x} - \underline{\bar{x}}) + \varepsilon \underline{B}(\underline{x}),$$

where all the nonlinear terms are in $\underline{B}$ where $\underline{B}(\underline{x}) \to 0$ as $\underline{x} \to \underline{\bar{x}}$. Sufficiently close to the fixed point (or for a sufficiently linear f),

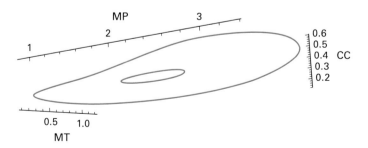

Figure 10.4
Two limit-cycle solutions from a 10-variable model by Leloup and Goldbeter (1998) of the *Drosophila* circadian clock. One can see they both lie on a two-dimensional manifold.

$$\underline{f}(\underline{x}) \approx \underline{\underline{A}}(\underline{x} - \underline{\bar{x}}).$$

We can then solve this system for the eigenvalues of $\underline{\underline{A}}$, which we call $(\lambda_1, \ldots, \lambda_n)$. Note that trajectories that start in one eigenspace stay there for all future times. For instance, assume we have a one-dimensional eigenvector $\underline{q}_j$ for a given eigenvalue λ_j. If we start in the space, we have

$$\underline{x} - \underline{\bar{x}} = b_j e^{\lambda_j t} \underline{q}_j.$$

Moreover, if we have oscillations near the fixed point with eigenvalues $\pm i\omega$ we find

$$\underline{x} - \underline{\bar{x}} = b_j e^{\omega t} \underline{q}_j + b_j e^{-\omega t} \underline{q}_j$$

$$= 2 b_j (\text{Re}(\underline{q}_j) \cos(\omega t) - \text{Im}(\underline{q}_j) \sin(\omega t)),$$

where $\text{Re}(\underline{q}_j)$ and $\text{Im}(\underline{q}_j)$ are the real and imaginary parts of the eigenvector $\underline{q}_j$. Now we can order all the eigenvalues and typically find, for oscillating systems, that all have negative real part except for $\pm i\omega$. In this case, the nonlinear system typically moves toward the eigenspace corresponding to $\text{Re}(\underline{q}_i)$ and $\text{Im}(\underline{q}_i)$ (see Guckenheimer and Holmes 1983) for details on dividing systems in this way).

Near the fixed point, this space is a plane, but as we go away from the fixed point, the plane may pick up curvature (2-manifold), a second-order effect not considered here. We assume that the contraction to the 2-manifold is quick, so it can be ignored. Thus, we can simply consider the system on this 2-manifold. While we do not know the manifold exactly, we can approximate it by the plane defined by $\text{Re}(\underline{q}_i)$ and $\text{Im}(\underline{q}_i)$ (i.e., the eigenspace defined near the fixed point). Formally we have

$$\underline{x} = \begin{bmatrix} Eigen \\ basis \end{bmatrix} \underline{z} + \underline{\bar{x}} \equiv \underline{\underline{E}} \, \underline{z} + \underline{\bar{x}}.$$

We can define the last two variables of $\underline{z}$ to be the plane to which we restrict the dynamics. Taking the derivative, we find that

$$\frac{d\underline{z}}{dt} = \underline{\underline{E}}^{-1} \frac{d\underline{x}}{dt} = \underline{\underline{E}}^{-1} \underline{f}\left(\underline{\underline{E}} \, \underline{z} + \underline{\bar{x}}\right).$$

Focusing just on the last two variables of $\underline{z}$, i.e., the dynamics in the eigenspace corresponding to the two largest eigenvalues, we let

$$\underline{z} \approx \begin{bmatrix} 0 \\ \cdots \\ 0 \\ r e^{i(\omega t + \theta)} \\ r e^{-i(\omega t + \theta)} \end{bmatrix}.$$

Also let the corresponding two components of $\underline{E}^{-1}\underline{f}\left(\underline{\underline{E}}\ \underline{z}+\overline{\underline{x}}\right)$ be $g_{n-1}(r,\theta,t)$ and $g_n(r,\theta,t)$. So we have

$$\frac{d}{dt}\left(re^{i(\omega t+\theta)}\right)\approx g_4\left(r,\theta,t\right),$$

$$\frac{d}{dt}\left(re^{-i(\omega t+\theta)}\right)\approx g_5\left(r,\theta,t\right).$$

As we have done before, we can solve for dr/dt and $d\theta/dt$. We then can average and find the following expressions:

$$\frac{dr}{dt}\approx\frac{\omega}{2\pi}\int_0^{2\pi/\omega}\frac{1}{2}\left(g_4\left(r,\theta,t\right)e^{-i(\omega t+\theta)}+g_5\left(r,\theta,t\right)e^{i(\omega t+\theta)}\right)dt,$$

$$\frac{d\theta}{dt}\approx\frac{\omega}{2\pi}\int_0^{2\pi/\omega}\frac{1}{2ir}\left(g_4\left(r,\theta,t\right)e^{-i(\omega t+\theta)}-g_5\left(r,\theta,t\right)e^{i(\omega t+\theta)}\right)dt.$$

10.2.1 Increasing the Amplitude of Oscillations in Biochemical Feedback Loops Increases the Oscillation Period

In this section, we apply the method of averaging on approximate manifolds to derive a very general result: as the amplitude of oscillations in a biochemical feedback loop increases, the period of these oscillations increases. This was discussed in chapter 3.

We start with our basic equation for a feedback loop (see chapter 4 for details):

$$\frac{dP_j}{dt}=f_{j-1}\left(P_{j-1}\right)-g_j\left(P_j\right).$$

Taking the derivative with respect to time, we have

$$\frac{d^2P_j}{dt^2}=\frac{df_{j-1}}{dt}-\frac{dg_j}{dP_j}\frac{dP_j}{dt},$$

$$\frac{d^2P_j}{dt^2}\frac{dt}{dP_j}\frac{dP_j}{dg_j}+1=\frac{df_{j-1}}{dg_j},$$

which is valid at all times except at the points where $dP_j/dt=0$. The term df_{j-1}/dg_j has a nonstandard interpretation, as f_{j-1} depends on P_{j-1} and g_j depends on P_j. We solve this by multiplying the equations for all j together and rearranging terms to get df_j/dg_j. This gives us

$$\prod_j\frac{d^2P_j}{dt^2}\frac{dt}{dP_j}\frac{dP_j}{dg_j}+1=\prod_j\frac{df_j}{dg_j}.$$

At each time point, we are given $\prod_j (df_j/dg_j)$ and dg_j/dP_j. We wish to determine the speed at which we proceed through the limit cycle. To do this, let $P_j = a_j + c_j e^{v(t)+i\phi(t)}$, since we are not using the linearized system. Letting a_j and c_j be fixed is an approximation good to first order by the method of averaging on approximate manifolds (Forger and Kronauer 2002), i.e., assuming that solutions occur on some plane. To first order, for example if $v(t) \approx at$ and $\phi(t) \approx \mu t$, we also have $(d^2 P_j/dt^2)(dt/dP_j) = (dv/dt) + i(d\phi/dt)$, which can be thought of as just separating out the real and imaginary parts of this expression, or

$$\prod_j \left(\frac{dv}{dt} + i\frac{d\phi}{dt} \right) \frac{dP_j}{dg_j} + 1 = \prod_j \frac{df_j}{dg_j}.$$

While dP_j/dg_j and df_j/dg_j depend on the value of P_j, the right-hand side of this equation is a negative number because of the structure of a negative feedback loop, and we assume $dP_j/dg_j > 0$. To first order by Taylor series, we can assume that dv/dt and $d\phi/dt$ are linearly related. In fact, they are always linearly related if $dP_j/dg_j \gg 1$ or if $dP_j/dg_j = dP_k/dg_k$. What this means is that, to first order, regardless of the oscillations in dP_j/dg_j and df_j/dg_j, the mean of dv/dt and $d\phi/dt$ will differ by a fixed amount. For oscillations to have a fixed amplitude and not grow or decay, the mean of dv/dt over one period is fixed, so the mean of $d\phi/dt$ must also be fixed. Then, calculating the period, we have

$$\int_0^{2\pi} \frac{1}{\frac{d\phi}{dt}} d\phi.$$

Since $1/x$ is a concave-up function, the period increases as the amplitude of oscillations in $d\phi/dt$ increases, due to oscillations in dP_j/dg_j and df_j/dg_j or oscillations of P_j. Thus, as the amplitude of P_j increases, dP_j/dg_j and df_j/dg_j vary more from their mean, unless they have the specific form P_j^m for some value of m.

10.2.2   A Numerical Method for Simulating Coupled High-Dimensional Clock Models
Consider a model for the noisy dynamics of a biological oscillator within a cell:

$$d\underline{x} = \underline{f}(\underline{x},t)dt + \underline{\underline{\sigma}}d\underline{\xi}.$$

$\underline{x}$ can have hundreds of variables, the equations take the form of stochastic differential equations as described in section 5.8, and we consider N coupled cells $N > 1,000$.

The idea here is to take a population density approach. Instead of simulating each oscillator, we simulate a population density function tracking the probability that an oscillator is in any particular state. This is especially helpful considering the fact that the oscillators are noisy. The population density function obeys the following equation:

$$\frac{\partial \rho}{\partial t} + \nabla \cdot \left(\underline{f} \, \rho \right) + \nabla \cdot \left(\underline{C} \, \rho \right) + \nabla \cdot \left(-\underline{\underline{D}} \nabla \rho \right) = 0$$

where $\underline{C}$ is the coupling function, for example $\int_{R^D} c(\underline{x}, \underline{y}, t) \rho(\underline{y}, t) d\underline{y}$. This seems like a bad idea since the dimension of this partial differential equation can be very large. However, we can approximate ρ with particles where $\rho(\underline{x}, t) = \sum_{j=1}^{M} p_j(t) \delta_\epsilon \left(\underline{x} - \underline{x}_j(t) \right)$ where each of the M particles has a position $\underline{x}_j(t)$ and strength $p_j(t)$. The particles evolve according to

$$\frac{d\underline{x}}{dt} = \underline{f}(\underline{x}, t) + \underline{C}(t) - \frac{\underline{\underline{D}} \nabla \rho}{\rho}.$$

See Stinchcombe and Forger (2016) for more details.

Proper choice of particles lets M << N, making it a very efficient method. This method works best if the position of the particles is carefully monitored and redistributed to best approximate the distribution.

10.3 Frontiers: Piecewise Linear Approximations to Nonlinear Equations

Consider a nonlinear system,

$$\frac{d\underline{x}}{dt} = \underline{f}(\underline{x}),$$

that can be approximated by a linear system,

$$\frac{d\underline{x}}{dt} = \underline{\underline{A}}(\underline{x} - \underline{x}_0) + \underline{f}(\underline{x}_0)$$

with $\underline{x}(0) = \underline{x}_0$. This approximation gets better for t closer to 0. Next let's take an eigenvalue decomposition of $\underline{\underline{A}}$:

$$\underline{\underline{A}} = \underline{\underline{E}} \, \underline{\underline{D}} \, \underline{\underline{E}}^{-1}.$$

With $\underline{f}_0 \equiv \underline{f}(\underline{x}_0)$ and $\underline{x}_1 = \underline{\underline{A}}^{-1} \underline{f}_0$, our solution is

$$\underline{x}(t) = \underline{x}_0 - \underline{x}_1 + \exp(\underline{\underline{A}} t) \underline{x}_1,$$

$$\underline{x}(t) = \underline{x}_0 - \underline{\underline{A}}^{-1} \underline{f}_0 + \exp(\underline{\underline{A}} t) \underline{\underline{A}}^{-1} \underline{f}_0,$$

$$\underline{x}(t) = \underline{x}_0 + (\exp(\underline{\underline{A}} t) - \underline{\underline{I}}) \underline{\underline{A}}^{-1} \underline{f}_0,$$

$$\underline{x}(t) = \underline{x}_0 + \underline{\underline{E}} \left(exp(\underline{\underline{D}} t) - \underline{\underline{I}} \right) \underline{\underline{E}}^{-1} \underline{\underline{A}}^{-1} \underline{f}_0.$$

Now, let us look at the dynamics of this system along a single eigenvector $\underline{n}_i$. If $\underline{f}_0 = \underline{n}_i$,

$$\underline{x}(t) = \underline{x}_0 + \left(\frac{e^{\lambda_i t} - 1}{\lambda_i}\right)\underline{n}_i,$$

which gives a simple solution to solve while the linearized equations hold.

As pointed out by Aluffi-Pentini et al. (2003), another way to solve

$$\frac{d\underline{\tilde{x}}}{dt} \approx \underline{f}_0 + \underline{\underline{A}}\,\underline{\tilde{x}}$$

is to let

$$\underline{\breve{x}} = \begin{pmatrix} \underline{\tilde{x}} \\ 1 \end{pmatrix} \text{ with } \underline{\breve{x}}(0) = \begin{pmatrix} \underline{0} \\ 1 \end{pmatrix}$$

and

$$\underline{\underline{\breve{A}}} = \begin{pmatrix} \underline{\underline{A}} & \underline{f}_0 \\ \underline{0}^T & 0 \end{pmatrix}.$$

Note that

$$\underline{\underline{\breve{A}}}\,\underline{\breve{x}} = \underline{\underline{A}}\,\underline{\tilde{x}} + \underline{f}_0$$

so

$$\frac{d\underline{\breve{x}}}{dt} = \underline{\underline{\breve{A}}}\,\underline{\breve{x}}, \quad \underline{\breve{x}}(t) = \exp(\underline{\underline{\breve{A}}}t).$$

There are many ways to calculate $\exp(\underline{\underline{A}}t)$. Some do not require explicitly calculating eigenvalues and eigenvectors, and some are very efficient (Moler and Van Loan 1978).

In fact, this can be a very effective numerical method to solve models of biological clocks with multiple time scales. At each time point, we linearize the system and solve it exactly, as indicated earlier. This system is then advanced in time, and the process is repeated. This method is known as an exponential integrator. It is particularly effective (in comparison with other numerical methods for solving differential equations) in systems that have multiple time scales.

We end by briefly describing a way this can be used to reduce a large-dimensional system to a two-dimensional system. One can approximate the system by just looking at the eigenvalues with the largest real parts. In fact, the system can sometimes be further simplified by assuming the real and imaginary parts of the pair of eigenvalues are linearly related. The limit cycle occurs when the average of the real part is zero (otherwise solutions would grow or decay), and the period can then be estimated assuming the real parts are zero on average, just as in the previous section.

Just looking at the leading pair of eigenvalues, the preceding method gives an instantaneous measure of the phase and amplitude of the oscillator. The imaginary part of the eigenvalues also can be used to describe the instantaneous speed or period of the oscillator. Recent work on the Hamilton-Hopf transform provides another instantaneous measure of phase and amplitude (Huang et al. 1998). Both work and the present method provide instantaneous estimates; however, what we describe is model-based, whereas the Hamilton-Hopf method is based on data analysis.

10.3.1 Simulational Example with the Goodwin Model

To illustrate the exponential integrator method discussed in section 10.3, we simulate the Goodwin model with three methods (See code 10.1 for more details). The first method uses the Runge–Kutta 4–5 (RK45) method, which is one of the most standard methods for simulating ordinary differential equations (see figure 10.5a). The Euler method is also tried (see figure 10.5b), but this yields a significantly different and erroneous solution. The

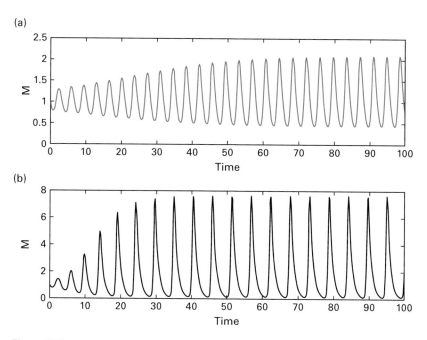

Figure 10.5
Simulations of the Goodwin model. (a) The Runge–Kutta 4-5 method (black) and the method of section 10.3 (blue) are used to simulate the Goodwin model, and give almost identical answers. The method of section 10.3 is quite simple to implement. (b) Simulating the Euler method with a similar time step yields an incorrect answer. See code 10.1 for more details.

method of section 10.3 is then also used and gives an almost identical solution to the RK45 method (see figure 10.5a). This method is particularly easy to code, much more so than RK45 (see code 10.1).

10.3.2 Analytical Example with the Van der Pol Model

It may be helpful to see an actual example of this piecewise linear approximation described in section 10.3. Here we illustrate it with the van der Pol equation:

$$\frac{dx}{dt} = ax_c + \varepsilon(x - cx^3) \quad \frac{dx_c}{dt} = -ax.$$

We linearize around the point (x_0, x_{c0}) and find

$$\underline{f_0} = \begin{pmatrix} ax_{c0} + \varepsilon(x_0 - cx_0^3) \\ -ax_0 \end{pmatrix},$$

$$\underline{\underline{A}} = \begin{pmatrix} \varepsilon(1 - 3cx_0^2) & a \\ -a & 0 \end{pmatrix},$$

$$\underline{\underline{A}}^{-1} = \begin{pmatrix} 0 & -1/a \\ 1/a & \varepsilon(1 - 3cx_0^2)/a^2 \end{pmatrix},$$

$$\underline{x_1} = \begin{pmatrix} x_0 \\ x_{c0} + 4cx_0^3 \dfrac{\varepsilon}{a} \end{pmatrix}.$$

This indicates that if $\varepsilon \ll a$, then $\underline{x_1} \approx \underline{x_0}$ and the effective origin of the system is zero. Otherwise, the origin can be shifted along the x_c direction depending on the value of x. The eigenvalues of the system are

$$\varepsilon(1 - 3cx_0^2) \pm \frac{\sqrt{\varepsilon^2 (1 - 3cx_0^2)^2 - 4a^2}}{2},$$

which with $\varepsilon \ll a$ would yield $\pm ia$. This gives the details of the piecewise linear assumption described earlier.

10.4 Frontiers: Poincaré Maps and Model Reduction

To show how one can drastically reduce the dimension of models of biological oscillators, we now consider a detailed model of the mammalian circadian clock proposed by Kim

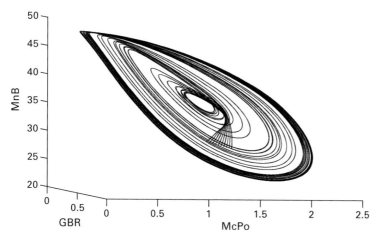

Figure 10.6
Simulations of the Kim and Forger detailed model of the mammalian circadian clock. Although the model has 180 variables, initial conditions are quickly attracted to a two-dimensional surface (manifold). This is seen by a parametric plot of three variables. See code 10.2 for more details.

and Forger (2012). This model has 180 variables, and we will reduce it down to a two-dimensional model. We first note that, as described earlier, many models quickly contract to a two-dimensional manifold where the interesting dynamics exist. We see that phenomenon in this detailed model. Code 10.2 simulates the model from 11 different initial conditions, and all quickly approach a two-dimensional manifold (see figure 10.6). We start the system in many initial conditions so that the solutions span this plane.

The method is the following. We construct a hyperplane, or a surface that the solutions of the system are approximately normal to. For our purposes, we can consider the point at which one of the variables crosses its steady-state value from below. We then look at solutions of the system, such as those shown in figure 10.6, and determine the amplitude of the model when crossings occur and the time between them. The time between these crossings is an approximate value for the period at that amplitude. How much the amplitude changes over this period is an approximate value for the rate of change of the amplitude. Code 10.2 illustrates how this can be done, and the results are shown in figure 10.7. It is surprising how similar this model looks to a van der Pol model.

10.5   Ruling out Limit Cycles

Here we describe two methods that can be used to determine if oscillations can exist. (This section closely follows section 7.2 of Strogatz 2000.) For feedback loops, also note the method based on the secant condition discussed in chapter 4.

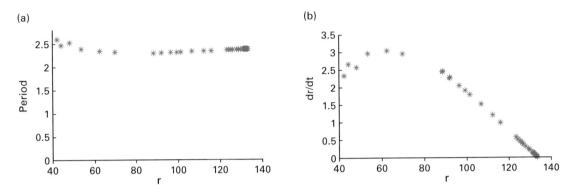

Figure 10.7
Reduction of the 180-variable detailed mammalian model proposed by Kim and Forger to an amplitude-and-phase model. (a) The plot shows the period of the oscillator at different amplitudes (*r*). Period is the indicated value times 10 in hours. (b) The plot shows the rate of change in amplitude of the clock at different amplitudes. The dynamics here look surprisingly similar to those of a van der Pol model (e.g., see figure 10.2). See code 10.2 for more details.

The most straightforward of these techniques (but one that is often difficult to implement) is the construction of a Lyapunov function. The point is to construct a function of the system's variables $V(\underline{x})$ that is >0 and that always decreases as the system evolves with time ($dV/dt < 0$) in a region of the phase space. If the system oscillates, there must be some period τ where $\underline{x}(t) = \underline{x}(t + \tau)$. If a Lyapunov function exists, $V(\underline{x}(t)) > V(\underline{x}(t + \tau))$, so there must be a contradiction. This can be illustrated in the following example taken from Strogatz (2000).

Consider the system $dx/dt = -x + 4y$, $dy/dt = -x - y^3$. Consider the function $V(x,y) = x^2 + ay^2$, where a will be determined later. Differentiating V, we have $dV/dt = 2x(-x + 4y) + 2ay(-x - y^3) = -2x^2 + (8 - 2a)xy - 2ay^4$. If $a = 4$, we have $dV/dt = -2x^2 - 8y^4$. Here $V(x,y) > 0$ and $dV/dt < 0$, so no oscillations can exist.

However, often there is no guidance as to how to construct such a function, making this less useful.

We now come to one of the most celebrated results in the theory of differential equations, the Poincaré–Bendixson theorem.

Theorem (see Hirsch and Smale 1974): A nonempty closed and bounded limit set of a continuous two-dimensional dynamical system, which contains no equilibrium point, is a cycle.

It states that any limit set of a two-dimensional system, with continuous derivatives, that contains no equilibrium point, is a cycle. This theorem is useful for several reasons.

Application 1: Existence of oscillations. If we can construct a region of the phase space where all solutions flow inward and that contains no equilibrium point, there must be a limit cycle within this region. To see this, consider any point on the boundary of the region.

Flow must be inward and can never leave the region. Therefore, by the Poincaré–Bendixson theorem, the behavior of trajectories from this point forward in time must be a cycle, and we have a limit cycle. The following is a somewhat trivial example of this:

Consider a system where we know that $d\phi/dt = 1$ and that $dr/dt > 0$ at $r = r_1$ and $dr/dt < 0$ at $r = r_2$. We know that there must be at least one attracting limit cycle between r_1 and r_2.

Application 2: No chaos in two-dimensional systems. The Poincaré–Bendixson theorem shows that if the attractor is not an equilibrium, it must be a cycle.

Application 3: There must exist a fixed point within each limit cycle in a two-dimensional system.

In the only extension of the Poincaré–Bendixson theorem to higher dimensions known to the author, Mallet-Paret and Smith (1990) showed that, for a class of systems of arbitrary dimension that includes the Goodwin oscillator, the theorem can be applied. This class is of the form

$$dx_i/dt = f_i(x_i, x_{i-1}),$$

where f is monotonic as defined by

$$\pm \frac{\partial f_i(x_i, x_{i-1})}{\partial x_{i-1}} > 0.$$

They achieve this result by considering the limit sets in the planes (x_i, x_{i-1}) and $(x_i, dx_i/dt)$. While trajectories can cross in these planes, they show that the limit sets cannot. This proof requires many pages of advanced mathematics. A sketch of a simpler proof is offered in section 10.5.1.

We note that if this property is violated, chaos can be seen. For example, consider a Goodwin model where there are two sites for transcription. The first site acts to activate and the second to repress. The transcription function and timecourse are shown in figure 10.8. The specific details of the model are given in DeWoskin et al. (2014).

10.5.1   Theory: A Sketch of a Proof of the Poincaré–Bendixson Theorem for Biochemical Feedback Loops

Here we consider biochemical feedback loops of the form

$$dp_i/dt = f_{i-1}(p_{i-1}) - g_i(p_i) \quad 1 \leq i \leq n \quad \text{with } f_0(p_0) \equiv f_n(p_n). \tag{10.5.1}$$

Each element of the system, p_i (where the elements are protein concentrations, mRNA concentrations, or levels of any other chemical species), is controlled by a production rate depending on the previous element of the system $f_{i-1}(p_{i-1})$ and a clearance rate $g_i(p_i)$. We assume that $\partial g_i/\partial p_i > 0$, or as p_i increases, the clearance rate of p_i also increases, and that

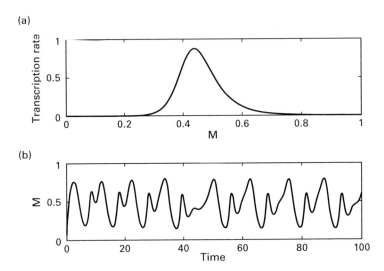

Figure 10.8
Chaos from a simple biochemical feedback loop. Simulations are for a Goodwin model with five steps and a transcription regulation function shown in (a). The model shows chaotic oscillations (b). See code 10.3 for more details.

$\partial f_{i-1}/\partial p_{i-1}$ is also of one sign. With $\partial f_{i-1}/\partial p_{i-1} < 0$ for an odd number of i, we have a negative feedback loop. We additionally assume that $f_{i-1}(p_{i-1})$ and $g_i(p_i)$ are invertible.

Claim 1: The sustained behaviors of (10.5.1) must either be oscillations or fixed points.

We consider solution on an attractor (i.e., the long-time behavior). Pigolotti, Krishna, and Jensen (2007) (PKJ) show that, on the attractor, the timing of the minimum and maximum values of p_{i-1} and p_i are interlaced. So if we denote the time of the jth minimum or maximum of $p_i(t)$ as $t_{i,j}$, we have $t_{i-1,j} < t_{i,j} < t_{i-1,j+1} < t_{i,j+1} < \cdots$. Consider the mapping of $f_{i-1}(p_{i-1}(t))$ into $p_i(t)$ piecewise with $p_{i-1}(t)$ in the region $(t_{i-1,j} < t < t_{i-1,j+1})$ and $p_i(t)$ in the region $(t_{i,j} < t < t_{i,j+1})$. First, let us consider a linear scaling of time from $t_{i-1,j} < t < t_{i-1,j+1}$ to $t_{i,j} < t < t_{i,j+1}$, $\tau_i(t) = t_{i,j} + (t - t_{i-1,j})(t_{i,j+1} - t_{i,j})/(t_{i-1,j+1} - t_{i-1,j})$. Since both $p_{i-1}(t)$ and $p_i(t)$ are monotonic with respect to time on these intervals, we can then define a function, $m_{i,j}$, so that $p_i(t) = m_{i,j}(f_{i-1}(p_{i-1}(\tau_i(t))))$.

We can proceed from p_{i-1} to p_i to p_{i+1}, ..., until we return to p_{i-1} and then go around again so that we a minimum of p_{i-1} maps to the next minimum of p_{i-1}. Since scaling time and taking a function of $s(t)$, $f(s(t))$, commute, let

$$m_{i-1,j+1}(f_{i-2}(\cdots m_{i+1,j}(f_i(m_{i,j}(f_{i-1}(p_{i-1}))))) \cdots)) \equiv m_j(p_{i-1})$$

and

$$\tau_{i-1}(\cdots \tau_{i+1}(\tau_i(t)) \cdots) \equiv \tau(t).$$

Assume that $m_{i,j} = m_{i,k}$ for all j and k, and thus $m_j(p_{i-1}) = m_k(p_{i-1}) \equiv m(p_{i-1})$, monotonic and valid over all time. Then m_i must be the identity, since iterations of the same bounded monotonic map will approach a fixed point of the map. Otherwise, $m(\ldots m(m(p_{i-1}))\ldots)$ would lead to an accumulation point, contradicting the PKJ property or the fact that we are on an attractor. With $m(p_{i-1}) = p_{i-1}$, $d\tau/dt$ must equal 1, and we must have oscillations or a fixed point. All that is left to show is that $m_{i,j} = m_{i,k}$ on the attractor.

Rather than looking at the PKJ property, we also can map p_{i-1} to itself forward in time in another way. First we take the function $f_{i-1}(p_{i-1}(t))$, which can be thought of as a monotonic scaling $p_{i-1}(t)$, and then map $f_{i-1}(p_{i-1}(t))$ into $p_i(t)$ over all time by the differential equation (10.5.1). As described in Forger (2011), if we let $g_i(p_i) = n_i r$, we have $d(g_i^{-1}(n_i r))/dt = s(t) - n_i r$. Scaling time, we can represent this as $dr/d\tau = s(\tau) - n_i r$ with n_i chosen so that, on average, $d\tau/dt = 1$ and $d\tau/dt > 0$ at all points since g_i is monotonic. The solution of $dr/d\tau = s(\tau) - n_i r$ can be expressed as

$$r(\tau) = \int_{-\infty}^{0} s(\tau + t') e^{n_i t'} dt',$$

and thus interpreted as an averaging of $s(\tau)$. Time can then be scaled back to give $p_i(t)$, and we can proceed to $p_{i+1}(t)$ and so on.

So, we have shown that we can reconstruct the solution over all time by averaging the solution, first perhaps scaling the variables and time in a monotonic way, and do this as many times as we wish. Since we can average as much as we would like, this shows that two mappings separated by k–j intervals must be the same, $m_{i,j} = m_{i,k}$, or that the map of $f_{i-1}(p_{i-1}(t))$ into $p_i(t)$ is identical whether it happens later or earlier in time.

Interesting future work would be to develop a formal proof based on this outline.

The following two remarks can help clarify our arguments:

1. The example shown in figure 10.8 does not have monotonic rates. In this case, we have an iterative map where applying f_i as in figure 10.8 looks similar to the logistic equation. The iterative logistic map is one of the classic examples of chaos (Strogatz 2000).
2. It may be helpful for the reader to consider the case where all f_i and g_i are linear. In this case, solutions can be expressed as the sum of exponential functions $e^{(a+bi)t}$. Here the mapping discussed in the proof would map $e^{(a+bi)t}$ forward by $v = 2\pi/b$ and scale its amplitude by e^{av}.

Code 10.1 Five Simulations of the Goodwin Model

(See figures 10.1, 10.3, and 10.5)

```
%This code simulates the Goodwin model in five ways. First with
%MATLAB's basic solver, then with a method described in section 10.2.1,
%then with Euler's method. We then solve the system linearized around
%the fixed point and the Goodwin model restricted to a plane determined
%by the eigenvectors around the fixed point.
```

```
clear Y YY YYY YYYY YYYYY
[T, Y] = ode45(@goodwin, 0:0.1:100, [1 1 1.1]);% Calculate the solution
%using MATLAB's standard solver and store the answer in Y
figure(1)
hold off
plot(T, Y(:, 1))
xlabel('Time')
ylabel('M')
hold on
YY(1,:) = [1 1 1.1];
for ij = 1:1000%This calculates the solution using the method of 10.3
xone = (lingood(YY(ij, 3))^(-1))*goodwin(1, YY(ij,:));
b = YY(ij,:)'-xone + expm(lingood(YY(ij, 3))*0.1)*xone;
YY(ij+1,:) = b';
end
plot(T, YY(:,1), 'k')
YYY(1,:) = [1 1 1.1];
for ij = 1:1000% This calculates the solution using Euler's method
YYY(ij+1,:) = YYY(ij,:)+ 0.1*goodwin(1, YYY(ij,:))';
end
figure(2)
plot(T, YYY(:,1))
xlabel('Time')
ylabel('M')
figure(3)
plot3(Y(:,1), Y(:,2), Y(:,3))
hold on
[V, D] = eigs(lingood(1));
[a, b] = max(imag(diag(D)));
for ij = 1:5000
bs = [1 1 1]'+0.1*real(exp(D(b,b)*ij*0.01).*V(:, b));
aa(ij) = bs(1);
ab(ij) = bs(2);
ac(ij) = bs(3);
end
plot3(aa, ab, ac, 'k')
xlabel('M')
ylabel('P')
zlabel('R')
VV(:,1) = real(V(:,b))./norm(real(V(:,b)));%Takes the real part of the
%eigenvector and scales it
VV(:,2) = imag(V(:,b)) - (VV(:,1)'*imag(V(:,b)))'*VV(:,1);%Takes the
%orthogonal part of the complex eigenvector to define the plane spanned
%by the real and complex parts of the eigenvectors
VV(:,2) = VV(:,2)./norm(VV(:,2));%scales this part too
YYYYY(1,:) = [1 1 1]+0.1*VV(:,1)';%starts on this plane, slightly away
%from the fixed point of [1 1 1]
```

```
for ij = 1:30000%This simulated the Goodwin model but restricts it to the
%plane.
YYYYY(ij+1,:) = YYYYY(ij,:)+ 0.001*((VV(:,1)'*goodwin(1,…
YYYYY(ij,:)))'*VV(:,1)+(VV(:,2)'*goodwin(1, YYYYY(ij,:)))'*VV(:,2))';
end
figure(4)
plot3(YYYYY(:,1), YYYYY(:,2), YYYYY(:,3))
hold on
plot3(aa, ab, ac, 'k')
xlabel('M')
ylabel('P')
zlabel('R')
function z = goodwin(t, x)% The Goodwin model similar to what was used in
%Chapter 2
m= 9;
z(1) = 1/(0.0000001^m + x(3)^m) - x(1);
z(2) = x(1) - x(2);
z(3) = x(2) - x(3);
z = z';
end
function B = lingood(x) %This calculates the Jacobian of the Goodwin
%model
m = 9;
B = [-1 0 -(m)*(x^(m-1))/(0.0000001^m + x^(m))^2
1-1 0
0 1-1];
end
```

Code 10.2 Poincaré Maps of a Detailed Mammalian Model

(See figures 10.6, 10.7)

```
% This code shows how a 180 variable model can be reduced to behavior
% of a 2-dimensional system
%This code requires the file DetailedModel.m that can be downloaded in
% the supplemental information for Kim and Forger 2012
incs = DetailedModel();%This gets the initial conditions
change = zeros(180, 1);%change is a vector of some of the variables
%that will be perturbed. Other choices could have been used.
change(11:24) = incs(11:24); change(43) = incs(43);
change(46) = incs(46); change(49) = incs(49); change(51) = incs(51);
% The following lines of code approximate the fixed point
[t,x]=ode15s(@DetailedModel,0:0.1:200,DetailedModel());
[t,x]=ode15s(@DetailedModel,0:0.1:200,x(225,:)+2.2*change'); %gives a
%perturbation which I've found brings the system close to the fixed
%point
xx = mean(x(101:340,:));% by taking the average of each of the
```

```
%variables over a cycle, we come closer to the fixed point
[t,x]=ode15s(@DetailedModel,0:0.1:200,xx);
xx = mean(x(1:240,:));
[t,x]=ode15s(@DetailedModel,0:0.1:200,xx);
fixpt = mean(x(1:240,:));%all the above code just approximates the
%fixed point. It is shorter than writing out the > 100 values of the
%variables.
[T, Y] = ode45(@DetailedModel, 0:0.001:200, fixpt);%Solves the model
Z = find((Y(1:200000, 7)-xx(7)).*(Y(2:200001,7)-xx(7)) < 0);% thus we
%find crossings of the 7th variable from its approximate fixed point
%value xx(7)
aa = size(Z);
aa = aa(1);
pers = Z(3:2:aa) - Z(1:2:(aa-2));%calculates the time between crossings
%to estimate the period
%note we skip every other since there are two crossings and we just
%want the upcrossing
for ij = 3:2:aa
rs((ij-1)/2) = norm(Y(Z(ij),:)-xx)*0.5+norm(Y(Z(ij-2),:)-
xx)*0.5%estimates the amplitude as the half way point between two
%crossings
rdot((ij-1)/2) = (norm(Y(Z(ij),:)-xx)-norm(Y(Z(ij-2),:)-xx))/pers((ij-…
1)/2);
end
ab = size(rs);
ab = ab(2);
figure(33)
hold on
plot(rs(2:ab), pers(2:ab), '*')
figure(34)
hold on
plot(rs(2:ab), rdot(2:ab), '*')
figure(31)
hold on
% The following code repeats what was done starting at the fixed point
%at 10 other points. These give us 10 other trajectories to estimate
%the amplitude and phase
for ij = 1:10
clear pers rs rdot
[t,x]=ode15s(@DetailedModel,0:0.1:200,DetailedModel());
[t,x]=ode15s(@DetailedModel,0:0.001:200,x(220+ij,:)+2.2*change');
[ac, ad] = min(x(170000:200000, 1));
prc(ij) = mod(ad/1000+ij, 24)
figure(31)
plot3(x(5000:200000, 7), x(5000:200000, 12), x(5000:200000,21));
```

```
Y = x;
Z = find((Y(1:200000, 7)-xx(7)).*(Y(2:200001,7)-xx(7)) < 0);
aa = size(Z);
aa = aa(1);
pers = Z(3:2:aa) - Z(1:2:(aa-2));
for ijj = 3:2:aa
rs((ijj-1)/2) = norm(Y(Z(ijj),:)-xx)*0.5+norm(Y(Z(ijj-2),:)-xx)*0.5
rdot((ijj-1)/2) = (norm(Y(Z(ijj),:)-xx)-norm(Y(Z(ijj-2),:)-…
xx))/pers((ijj-1)/2);
end
ab = size(rs);
ab = ab(2);
figure(33)
plot(rs(2:ab), pers(2:ab), '*')
figure(34)
plot(rs(2:ab), rdot(2:ab), '*')
end
```

Code 10.3 Chaotic Motions

(See figure 10.8)

```
%This gives an example of Goodwin model showing chaos.
function Y = goodchaos(t, X)
ast = 0.07; bst = 0.4; cst = 0.8; lst = 0.2; mst = 0.5;
cc = 2.8; aa = 0.44; ab = 0.423; np = 10; npp = 13;
a = X(1); b = X(2); c = X(3); l = X(4); m = X(5);
f = cc*((aa^np)/(aa^np + m^np))*((m^npp)/(ab^npp + m^npp));
Y(1) = f-a;
Y(2) = a-b;
Y(3) = b-c;
Y(4) = c-l;
Y(5) = l-m;
Y = Y';
end
%The following code should be entered at the command line.
%bottom plot
%[T, X] = ode15s(@goodchaos, 0:0.01:100, [0.07 .4 .8 .2 .5]);
%plot(T, X(:,1), 'k')
%top plot
%m = 0:0.01:1;
%plot(m, cc.*((aa^np)./(aa^np + m.^np)).*((m.^npp)./(ab^npp + m.^npp)),…
%'k');
```

Exercises

General Problems

For these problems pick a biological system of interest, preferably one you study.

1. Do the assumptions of the method of averaging hold for your system (including averaging in higher dimensions)? If so, average the equations.

2. If your answer to the previous question was no, use a piecewise linear approximation to determine the solution of your equation. Does this work better than standard ode simulation methods? Can you learn anything by looking at the changing origin?

3. Are there other simplifications (e.g., reduction to a 2-D system for phase plane analysis) that are useful in understanding the model?

4. Can you rule out oscillations for particular parameter choices, using the techniques described in the chapter?

Specific Problems

1. In a 2-D C^1 system, the Poincaré–Bendixson theorem could allow us to do which of the following? (Choose all that apply.)

   a. Rule out chaos

   b. Prove the existence of a limit cycle

   c. Provide maximal values of variables

2. Write down a system of equations that shows cycles that are not limit cycles.

3. Write down a Lyapunov function for the following system:

$dx/dt = -y - x,$

$dy/dt = x - y.$

4. Working with researchers in a lab, you model a particular biological clock by the van der Pol equation: $\dfrac{dx}{dt} = x_c + 0.1\left(x - \dfrac{4}{3}x^3\right)$, $\dfrac{dx_c}{dt} = -x$

   a. The amplitude of the limit cycle of this model is 1. Researchers discover that the amplitude of the biological process you are modeling is 2. Modify one parameter of the model to make the amplitude 2. Justify your choice through simulations or the method of averaging.

   b. We find that, when averaged, the period of this model does not have a large dependence on the amplitude (as the model settles to the limit cycle it keeps the same period). Experimentally it is found that the period depends on the amplitude of the clock. Add one term to this model to cause the period to depend on the amplitude. Justify your choice by simulations or the method of averaging.

5. Write down the averaged equations for the Pavlidis model described in section 5.11.

6. Write down the averaged equations for the following system (with ε small):

$dx/dt = x_c + \varepsilon(x^2 + x * x_c + x_c^2),$

$dx_c/dt = -x + \varepsilon(x^2 + x * x_c + x_c^2).$

What does this tell you about these terms?

Glossary

This glossary serves to provide intuition to the non-mathematical reader about mathematical terms and intuition to the mathematical reader about biological terms.

I point the reader to sections of the text or to other texts where more in-depth and/or rigorous definitions can be found. My goal here is to explain specifically how each term is used in this text, which may be oversimplified from its more general use.

action potential   a brief rise in the voltage of a neuron that is used to signal other neurons.

activation energy   the energy required for a reaction to take place.

activators   proteins that increase the rate of transcription of a gene, for example, by binding directly to the gene and recruiting the cellular machinery that generates mRNA.

actogram   a plot showing the activity of an organism over time.

alkalinize   to increase the pH of a solution.

all-to-all (mean field) coupling   coupling where each oscillator receives signals from all other oscillators.

amplitude recovery terms   terms in the equations of a model that affect how quickly the system converges to a limit cycle, as typically measured by the method of averaging.

angular variance   a measure of the variance defined for circular variables. See 1.12.

Arrhenius relationship   an equation, developed by Svante Arrhenius, for calculating how reaction rates depend on temperature.

asymptotic phase   a way to measure the phase of a system by comparing its phase as time approaches infinity with the state of another reference system.

attractor   a set of states to which a model tends toward as time increases. See Strogatz (2000), section 9.3.

axiom of parenthood   the requirement that clearance terms in many biophysical models must be proportional to the species they are clearing.

bang-bang   piecewise constant solutions of an optimization problem where the value of the control is found at its minimum or maximum values. See Bryson and Ho (1975), section 3.9.

basin of attraction   a region of the phase space of a model that tends toward an attractor. See Strogatz (2000), section 6.4.

bifurcation   a qualitative change in the behavior of a model, for example increasing or decreasing the number of attractors. See Strogatz (2000), chapter 3.

bifurcation diagrams   diagrams that plot the attractors of a system as a parameter is varied. See Strogatz (2000), chapter 3.

bistability   the ability of a differential equation to show different steady states for different initial conditions. See Strogatz (2000), section 3.7.

cascade   a type of pulse-coupling where a pulse from one element triggers pulses from other oscillators.

chi-squared   a measure for how different measurements are calculated from a known distribution. See 1.12.

circadian rhythms   rhythms that have a period that lasts around a day.

circular data   data that are defined modulo a period (e.g., measured phases of an oscillator).

clock   a mechanism that can be used to time events.

closed   a region of the phase space where every sequence of points within the region has its limit within the region. See Hirsch and Smale (1974), section 5.1.

clustering   a steady state of a coupled system where elements form multiple groups whose oscillators all have the same phase.

coding sites   regions of DNA that encode mRNA.

coefficient of variation   a measurement of the noise in a system as determined by the standard deviation of measurements divided by their mean.

conformational   related to the structure of a protein.

continuous coupling   a type of coupling where oscillators can send signals at any time. This is contrasted with pulse-coupling.

coordinate search   a method of searching a parameter space where each parameter is individually varied to determine an optimal set of parameters. See 8.1.5 for a sample algorithm.

critical resetting   the case where a stimulus brings a system to a critical state, for example the origin, where phase is undefined.

cumulative distribution function   a function that determines the probability that a random number is less than a fixed value.

current clamp   a technique used in the study of the electrical activity of cells where the current applied to a cell is fixed.

cytoplasm   the material in a cell besides what is found in the nucleus.

dimerization   the process by which two proteins bind together.

Dirac delta function   a generalized function that is zero everywhere except the origin and whose integral over the real line is one.

dose response curve   a plot of the strength of a signal and its effect.

Drosophila   fruit fly.

dual problem   a secondary problem in optimization whose solution also gives the solution of a primary problem. See Winston (1994), chapter 6.

eclosion   the act of an insect emerging from its pupal case.

eigenbasis   a basis of eigenvectors.

eigenspace   a space spanned by eigenvectors.

entrained   when an oscillator changes its period to match that of an external signal.

equilibrium potential   the voltage difference across the membrane of an electrically active cell where no net current flows.

eukaryote   a cell with a nucleus.

excitatory coupling   a type of coupling where the signal from another element increases the state of other oscillators.

exponential integrator   a numerical method for solving differential equations that uses the exact solution of linearized versions of the differential equations. See 10.3 for an algorithm.

FASPS   familial advanced sleep phase syndrome; patients with this syndrome typically wake early in the morning.

first-order term   a term in a differential equation whose magnitude decreases proportionally to a small parameter or timestep.

Fourier transform   a method to decompose a signal into sinusoidal components. See Körner (1988), chapter 46.

Gaussian random variables a random variable drawn from a normal distribution.

Gillespie's direct method a stochastic simulation method used for biochemical reactions. See 3.4 for details.

Gillespie's first-reaction method a stochastic simulation method used for biochemical reactions. See 3.4 for details.

global optimal the best possible choice of parameters over a region.

Goldman–Hodgkin–Katz equation an equation predicting the current that flows through a membrane channel. See 2.11 for details.

gradient methods an optimization method where the gradient of a function is used to determine an improved set of parameters.

hard excited a system that has an attracting fixed point and other stable states.

higher-order synchrony index a measure to determine if coupled oscillators are synchronized.

homodimerization the process by which two proteins of the same species bind together.

Hopf bifurcation a bifurcation where a steady state loses stability and produces oscillations. See Strogatz (2000), section 8.2.

hyperpolarized decreased voltage.

identifiable the ability to determine the value of a parameter from a measurement. See Eisenberg and Hayashi (2014), Eisenberg et al. (2013).

independent random variables Two random events are independent if the probability of both occurring equals the product of each event individually occurring.

influence functions functions used in optimization to determine how changes in a system's state at a particular time translate to changes in the final cost of the system. They are a generalization of Lagrange multipliers. See Bryson and Ho (1975), section 2.3.

inhibitory coupling a type of coupling where the signal from another element decreases the state of other oscillators.

isochron a set of states of a model which all have the same phase. See Winfree (1980), chapter 6.

kinase an enzyme that adds a phosphate group to other proteins.

Kuiper's test a statistical test to determine if samples come from a known distribution. See 1.12 for details.

Lagrange multiplier a function added to optimization problems to ensure that a constraint is met. See Winston (1994), section 12.7.

Langevin equation a stochastic differential equation where a noise term is added to an ordinary differential equation. See van Kampen (2011), chapter 9.

law of mass action the statement that reactions in well mixed media have a rate proportional to the product of the reactant concentrations.

limit cycle a cycle which nearby trajectories spiral toward or away from. See Hirsch and Smale (1974), section 11.5.

limit set the set of states of a system that occur after all transients have passed. See Hirsch and Smale (1974), section 11.1.

local optimum a parameter choice where any small change to the parameters yields a less good outcome.

Lyapunov function a function that can be used to prove that a system tends toward a steady state. See 10.5 for details.

manifold a generalization of Euclidean space. See Hirsch and Smale (1974), section 10.5.

mean average.

membrane potential the difference in voltage between the outside and inside of the cell.

method of averaging a method used to approximate the behavior of a rhythmic system on a slower timescale than the rhythms. See 10.1.

method of variation of parameters a method that uses the solution of a linearized system to determine slower dynamics. See 10.1.1.

minimal time problems   optimization problems that seek to bring a system from one state to another in minimal time.

mixed-mode oscillations   systems that show two concurrent rhythms typically on very different timescales.

multiplicative noise   a stochastic differential equation where the noise term also depends on the state of the system. Cases where this is not true are called additive noise. See Gardiner (2004), section 10.5.

nearest-neighbor coupling   coupling of an ordered set of elements where oscillators only receive signals from their neighbors.

neutral stability   stability that is neither attracting nor repelling. See Strogatz (2000), section 5.1.

nondimensionalize   to scale the variables of a model to remove parameters.

nucleus   a separated part of the cell of higher organisms that contains the DNA.

observable   the ability to determine the value of a parameter from a measurement.

oscillator   a system that shows oscillations.

phase-amplitude resetting map   a two-dimensional plot showing the initial phase and amplitude as well as the final phase and amplitude of a system after a stimulus has been applied.

phase-desynchronized   oscillators that have the same period but different phases.

phase response curves   a plot of the effect of a stimulus with respect to when the stimulus is applied.

phase-synchronized oscillators   oscillators that have the same period and phase.

phosphatases   enzymes that reverse the action of kinases.

Picard iteration   a method for determining the solution of a differential equation. Here an initial solution of the system is guessed, and this solution is used in the differential equation iteratively to develop better approximations of the solution.

Poincaré map   a map from an n to an n-1 dimensional space transverse to the flow that follows the trajectory of a system. See Strogatz (2000), section 8.5.

Poincaré–Bendixson theorem   a theorem that is often used to show that typically two-dimensional continuous systems show only cycles or fixed points as attractors. See Hirsch and Smale (1974), chapter 11.

Poisson random variable   a random variable that counts the number of events given by a Poisson distribution.

postsynaptic current (PSC)   the current received by a neuron when a synaptic signal is received from other neurons. This current can be excitatory (EPSC) or inhibitory (IPSC).

posttranslational modification   a change to a protein after it is been created.

prediction   the use of previous measurements to estimate the state of a system in the future.

probability distribution function   a function that assigns the probability of any event occurring.

prokaryotes   lower organisms whose cells do not contain a nucleus.

pseudoinverse   a matrix that acts like the inverse of a matrix when no actual inverse can be found. See Trefethen and Bau (1997), chapter 11.

pulse-coupling   a type of coupling where oscillators can send signals only at a particular phase. This is contrasted with continuous-coupling.

pursuit problem   an optimization problem where the final state of the system depends on when that state is reached.

Q_{10} a measure of how a reaction rate changes with increasing temperature. It is the ratio of the reaction rate when the temperature is increased by 10°C to the original rate.

quasilinear system   a set of differential equations where the nonlinear terms are small.

random coupling   coupling where which oscillators are coupled is determined randomly.

Rao's spacing test   a test to determine if there is a preferred phase in a set of measured phases. See 1.12 for details.

Rayleigh test   a test to determine if measured phases come from a uniform distribution. See 1.12 for details.

reentrainment   when an oscillator changes its period to match that of an external signal after a perturbation.

relaxation oscillation   an oscillation that is generated by a system switching (or relaxing) between two distinct states.

repressors   proteins that decrease the rate of transcription of a gene, for example, by binding directly to the gene and stopping the cellular machinery that generates mRNA.

retinohypothalamic tract (RHT)   a collection of cells in the retina that project to the hypothalamus, and, in particular, the SCN.

rhythms   a recurring set of states of a system.

ribosomes   the part of a cell that creates proteins.

run test   a test to determine if two sets of measured phases are different. See 1.12.

Runge–Kutta 4–5   a popular numerical method for solving differential equations on a computer.

saddle-node on an invariant circle (SNIC) bifurcation   a bifurcation where two steady states (nodes) that lie on a circle coalesce. See 5.3 for details.

scale-free networks coupling   a coupling where which oscillators are coupled is determined randomly.

search methods   a generalization of coordinate search methods where a parameter space is searched.

secant condition   a mathematical test to determine if oscillations are possible in a biological feedback loop. See Forger (2011) for more details.

second-order term   a term in a differential equation whose magnitude decreases proportionally to the square of a small parameter or timestep.

sensitivity   the derivative of a function scaled by the ratio of the function to its input.

serial correlation test   a test to determine if a measurement that is shorter than the mean is typically followed by one that is longer than the mean or vice versa. See 1.12.

shooting method   a numerical method for solving a differential equation where conditions are placed on both the initial state and final state. The solution is guessed and refinements are made based on the accuracy of these guesses. See Bradie (2006).

small-world network coupling   a variation of nearest-neighbor coupling where some connections between oscillators are removed and replaced by a connection chosen randomly.

smoothing   a technique used to estimate the state of a system at a previous time based on measurements of the system at multiple timepoints.

stable limit cycles/unstable limit cycle   limit cycle to which nearby states attract or repel.

standard form   the form for solutions used in the method of averaging. See 10.1.1.

stoichiometry   the ratio of reactants and/or products in a chemical reaction.

suprachiasmatic nucleus (SCN)   the site of the central circadian clock in the mammalian brain.

temperature compensation   the ability of some biological clocks to keep the same period regardless of the external temperature after enough time has passed.

transcription   the act of making mRNA.

transcription factors   activators or repressors.

translation   the act of making a protein.

type 1/type 2 oscillators   two classifications of biological oscillators depending on the mathematical mechanism by which oscillations arise. See 5.3–5.6.

V test   a test to determine if phases cluster around a particular value. See 1.12.

variable end condition optimization   an optimization problem where the end state is not specified.

voltage clamp   a technique used in the study of the electrical activity of cells where the membrane voltage of a cell is fixed.

von Mises distribution   a probability distribution similar to a Gaussian distribution used for circular data. See 1.12.

Wiener process   a stochastic process; Brownian motion. See van Kampen (2011), section 4.2.

zeitgeber   an external signal that confers time.

Bibliography

Akaike, H. 1974. A new look at the statistical model identification. *IEEE Transactions on Automatic Control* 19:716–723.

Aluffi-Pentini, F., and V. De Fonzo. 2003. V, P.: A novel algorithm for the numerical integration of systems of ordinary differential equations in chemical problems. *Journal of Mathematical Chemistry* 33:1–15.

Anderson, D. F. 2007. A modified next reaction method for simulating chemical systems with time dependent propensities and delays. *Journal of Chemical Physics* 127:214107.

Andronov, A. A., S. E. Khaikin, and S. Lefschetz. 1949. *Theory of oscillations.* Princeton, NJ: Princeton University Press.

Aschoff, J. 1960. Exogenous and endogenous components in circadian rhythms. *Cold Spring Harbor Symposia on Quantitative Biology* 25:11–28.

Atkinson, M. R., M. A. Savageau, J. T. Myers, and A. J. Ninfa. 2003. Development of genetic circuitry exhibiting toggle switch or oscillatory behavior in Escherichia coli. *Cell* 113 (5): 597–607.

Ay, A., J. Holland, A. Sperlea, G. S. Devakanmalai, S. Knierer, S. Sangervasi, et al. 2014. Spatial gradients of protein-level time delays set the pace of the traveling segmentation clock waves. *Development* 141 (21): 4158–4167. doi:10.1242/dev.111930.

Ay, A., S. Knierer, A. Sperlea, J. Holland, and E. M. Ozbudak. 2013. Short-lived Her proteins drive robust synchronized oscillations in the zebrafish segmentation clock. *Development* 140 (15): 3244–3253. doi:10.1242/dev.093278.

Barik, D., W. T. Baumann, M. R. Paul, B. Novak, and J. J. Tyson. 2010. A model of yeast cell-cycle regulation based on multisite phosphorylation. *Molecular Systems Biology* 6:405. doi:10.1038/msb.2010.55.

Batschelet, E. 1981. *Circular statistics in biology.* New York: Academic Press.

Belle, M. D., C. O. Diekman, D. B. Forger, and H. D. Piggins. 2009. Daily electrical silencing in the mammalian circadian clock. *Science* 326 (5950): 281–284. doi:10.1126/science.1169657.

Berry, H. 2002. Monte Carlo simulations of enzyme reactions in two dimensions: Fractal kinetics and spatial segregation. *Biophysical Journal* 83 (4): 1891–1901. doi:10.1016/S0006-3495(02)73953-2.

Best, E. N. 1979. Null space in the Hodgkin-Huxley equations: A critical test. *Biophysical Journal* 27 (1): 87–104. doi:10.1016/S0006-3495(79)85204-2.

Bieler, J., R. Cannavo, K. Gustafson, C. Gobet, D. Gatfield, and F. Naef. 2014. Robust synchronization of coupled circadian and cell cycle oscillators in single mammalian cells. *Molecular Systems Biology* 10:739. doi:10.15252/msb.20145218.

Bodova, K., D. Paydarfar, and D. B. Forger. 2015. Characterizing spiking in noisy type II neurons. *Journal of Theoretical Biology* 365C:40–54. doi:10.1016/j.jtbi.2014.09.041.

Bonev, B., P. Stanley, and N. Papalopulu. 2012. MicroRNA-9 Modulates Hes1 ultradian oscillations by forming a double-negative feedback loop. *Cell Reports* 2:10–18. doi:10.1016/j.celrep.2012.05.017.

Bradie, B. 2006. *A friendly introduction to numerical analysis.* Upper Saddle River, NJ: Pearson Prentice Hall.

Briggs, G. E., and J. B. Haldane. 1925. A note on the kinetics of enzyme action. *Biochemical Journal* 19 (2): 338–339.

Brown, E. N., Y. Choe, H. Luithardt, and C. A. Czeisler. 2000. A statistical model of the human core-temperature circadian rhythm. *American Journal of Physiology. Endocrinology and Metabolism* 279 (3): E669–E683.

Brown, E. N., and H. Luithardt. 1999. Statistical model building and model criticism for human circadian data. *Journal of Biological Rhythms* 14 (6): 609–616.

Bryson, A. E., and Y.-C. Ho. 1975. *Applied optimal control: optimization, estimation, and control.* Rev. ed. New York: Hemisphere Pub. Corp.

Bünning, E. 1964. *The physiological clock: Endogenous diurnal rhythms and biological chronometry*. 2nd ed. New York: Academic Press.

Burnham, K. P., and D. R. Anderson. 1998. *Model selection and inference: A practical information-theoretic approach*. New York: Springer.

Chance, B., E. K. Pye, A. K. Ghosh, and B. Hess. 1973. *Biological and biochemical oscillators*. New York: Academic Press.

Chang, D. E., S. Leung, M. R. Atkinson, A. Reifler, D. Forger, and A. J. Ninfa. 2010. Building biological memory by linking positive feedback loops. *Proceedings of the National Academy of Sciences of the United States of America* 107 (1): 175–180. doi:10.1073/pnas.0908314107.

Chang, J., and D. Paydarfar. 2014. Switching neuronal state: Optimal stimuli revealed using a stochastically-seeded gradient algorithm. *Journal of Computational Neuroscience* 37 (3): 569–582. doi:10.1007/s10827-014-0525-5.

Chouvet, G., J. Mouret, J. Coindet, M. Siffre, and M. Jouvet. 1974. Periodicité bicircadienne du cycle veille-sommeil dans des conditions hors du temps. *Electroencephalography and Clinical Neurophysiology* 37:367.

Chung, K. L. 2001. *A course in probability theory*. 3rd ed. San Diego: Academic Press.

Clay, J. R. 2009. Determining k channel activation curves from k channel currents often requires the Goldman-Hodgkin-Katz equation. *Frontiers in Cellular Neuroscience* 3:20. doi:10.3389/neuro.03.020.2009.

Clay, J. R., and L. J. DeFelice. 1983. Relationship between membrane excitability and single channel open-close kinetics. *Biophysical Journal* 42 (2): 151–157. doi:10.1016/S0006-3495(83)84381-1.

Clay, J. R., D. B. Forger, and D. Paydarfar. 2012. Ionic mechanism underlying optimal stimuli for neuronal excitation: Role of Na+ channel inactivation. *PLoS One* 7 (9): e45983. doi:10.1371/journal.pone.0045983.

Clay, J. R., D. Paydarfar, and D. B. Forger. 2008. A simple modification of the Hodgkin and Huxley equations explains type 3 excitability in squid giant axons. *Journal of the Royal Society, Interface* 5 (29): 1421–1428. doi:10.1098/rsif.2008.0166.

Cold Spring Harbor. 1960. Cold Spring Harbor symposia on quantitative biology: Biological clocks, 25.

Conrad, E., A. E. Mayo, A. J. Ninfa, and D. B. Forger. 2008. Rate constants rather than biochemical mechanism determine behaviour of genetic clocks. *Journal of the Royal Society, Interface* 5 (Suppl 1): S9–S15. doi:10.1098/rsif.2008.0046.focus.

Cosnard, M., J. Demongeot, and A. Le Breton. 1983. *Rhythms in biology and other fields of application: Deterministic and stochastic approaches*. New York: Springer.

Cyran, S. A., A. M. Buchsbaum, K. L. Reddy, M. C. Lin, N. R. Glossop, P. E. Hardin, et al. 2003. vrille, Pdp1, and dClock form a second feedback loop in the Drosophila circadian clock. *Cell* 112 (3): 329–341.

Czeisler, C. A., J. F. Duffy, T. L. Shanahan, E. N. Brown, J. F. Mitchell, D. W. Rimmer, et al. 1999. Stability, precision, and near-24-hour period of the human circadian pacemaker. *Science* 284 (5423): 2177–2181.

Dean, D. A., II, D. B. Forger, and E. B. Klerman. 2009. Taking the lag out of jet lag through model-based schedule design. *PLoS Computational Biology* 5 (6): e1000418. doi:10.1371/journal.pcbi.1000418.

Del Vecchio, D., A. J. Ninfa, and E. D. Sontag. 2008. Modular cell biology: Retroactivity and insulation. *Molecular Systems Biology* 4:161. doi:10.1038/msb4100204.

DeWoskin, D., W. Geng, A. R. Stinchcombe, and D. B. Forger. 2014. It is not the parts, but how they interact that determines the behaviour of circadian rhythms across scales and organisms. *Interface Focus* 4 (3): 20130076. doi:10.1098/rsfs.2013.0076.

DeWoskin, D., J. Myung, M. D. Belle, H. D. Piggins, T. Takumi, and D. B. Forger. 2015. Distinct roles for GABA across multiple timescales in mammalian circadian timekeeping. *Proceedings of the National Academy of Sciences of the United States of America* 112:E3991–E3999. doi:10.1073/pnas.1420753112

Diekman, C. O., M. D. Belle, R. P. Irwin, C. N. Allen, H. D. Piggins, and D. B. Forger. 2013. Causes and consequences of hyperexcitation in central clock neurons. *PLoS Computational Biology* 9 (8): e1003196. doi:10.1371/journal.pcbi.1003196.

Diekman, C. O., and D. B. Forger. 2009. Clustering predicted by an electrophysiological model of the suprachiasmatic nucleus. *Journal of Biological Rhythms* 24 (4): 322–333. doi:10.1177/0748730409337601.

Duffy, J. F., R. E. Kronauer, and C. A. Czeisler. 1996. Phase-shifting human circadian rhythms: Influence of sleep timing, social contact and light exposure. *Journal of Physiology* 495 (Pt 1): 289–297.

Dunlap, J. C., J. J. Loros, and P. J. DeCoursey. 2004. *Chronobiology: Biological timekeeping*. Sunderland, MA: Sinauer Associates.

Dyson, F. 2004. A meeting with Enrico Fermi. *Nature* 427 (6972): 297. doi:10.1038/427297a.

Edelstein-Keshet, L. 1988. *Mathematical models in biology*. Philadelphia: SIAM Press.

Eisenberg, M. C., and M. A. Hayashi. 2014. Determining identifiable parameter combinations using subset profiling. *Mathematical Biosciences* 256:116–126. doi:10.1016/j.mbs.2014.08.008.

Eisenberg, M. C., S. L. Robertson, and J. H. Tien. 2013. Identifiability and estimation of multiple transmission pathways in cholera and waterborne disease. *Journal of Theoretical Biology* 324:84–102. doi:10.1016/j.jtbi.2012.12.021.

Elowitz, M. B., and S. Leibler. 2000. A synthetic oscillatory network of transcriptional regulators. *Nature* 403 (6767): 335–338. doi:10.1038/35002125.

Enright, J. T. 1980a. Temporal precision in circadian systems: A reliable neuronal clock from unreliable components? *Science* 209 (4464): 1542–1545.

Enright, J. T. 1980b. *The timing of sleep and wakefulness: On the substructure and dynamics of the circadian pacemakers underlying the wake-sleep cycle*. New York: Springer.

Ermentrout, G. B., L. Glass, and B. E. Oldeman. 2012. The shape of phase-resetting curves in oscillators with a saddle node on an invariant circle bifurcation. *Neural Computation* 24:3111–3125.

Ermentrout, B., and N. Kopell. 1986. Parabolic bursting in an excitable system coupled with a slow oscillation. *SIAM Journal on Applied Mathematics* 46:233–253.

Ermentrout, B., and D. H. Terman. 2010. *Mathematical foundations of neuroscience*. New York: Springer.

Ermentrout, G. B., and N. Kopell. 1984. Frequency plateaus in a chain of weakly coupled oscillators, I. *SIAM Journal on Mathematical Analysis* 15:215–237.

Feillet, C., P. Krusche, F. Tamanini, R. C. Janssens, M. J. Downey, P. Martin, et al. 2014. Phase locking and multiple oscillating attractors for the coupled mammalian clock and cell cycle. *Proceedings of the National Academy of Sciences of the United States of America* 111 (27): 9828–9833. doi:10.1073/pnas.1320474111.

Fisher, N. I. 1993. *Statistical analysis of circular data*. Cambridge: Cambridge University Press.

Fitzhugh, R. 1961. Impulses and physiological states in theoretical models of nerve membrane. *Biophysical Journal* 1 (6): 445–466.

Forger, D. B. online.kitp.ucsb.edu/online/bioclocks07/forger/.

Forger, D. B. 2011. Signal processing in cellular clocks. *Proceedings of the National Academy of Sciences of the United States of America* 108 (11): 4281–4285. doi:10.1073/pnas.1004720108.

Forger, D. B., M. E. Jewett, and R. E. Kronauer. 1999. A simpler model of the human circadian pacemaker. *Journal of Biological Rhythms* 14 (6): 532–537.

Forger, D. B., and R. E. Kronauer. 2002. Reconciling mathematical models of biological clocks by averaging on approximate manifolds. *SIAM Journal on Applied Mathematics* 62:1281–1296.

Forger, D. B., and D. Paydarfar. 2004. Starting, stopping, and resetting biological oscillators: In search of optimum perturbations. *Journal of Theoretical Biology* 230 (4): 521–532. doi:10.1016/j.jtbi.2004.04.043.

Forger, D. B., D. Paydarfar, and J. R. Clay. 2011. Optimal stimulus shapes for neuronal excitation. *PLoS Computational Biology* 7 (7): e1002089. doi:10.1371/journal.pcbi.1002089.

Forger, D. B., and C. S. Peskin. 2003. A detailed predictive model of the mammalian circadian clock. *Proceedings of the National Academy of Sciences of the United States of America* 100 (25): 14806–14811. doi:10.1073/pnas.2036281100.

Forger, D. B., and C. S. Peskin. 2005. Stochastic simulation of the mammalian circadian clock. *Proceedings of the National Academy of Sciences of the United States of America* 102 (2): 321–324. doi:10.1073/pnas.0408465102.

Fox, R. F., and Y. Lu. 1994. Emergent collective behavior in large numbers of globally coupled independently stochastic ion channels. *Physical Review E: Statistical Physics, Plasmas, Fluids, and Related Interdisciplinary Topics* 49:3421.

Gallego, M., E. J. Eide, M. F. Woolf, D. M. Virshup, and D. B. Forger. 2006. An opposite role for tau in circadian rhythms revealed by mathematical modeling. *Proceedings of the National Academy of Sciences of the United States of America* 103 (28). 10618–10623. doi:10.1073/pnas.0604511103.

Gardiner, C. W. 2004. *Handbook of stochastic methods*. 3rd ed. New York: Springer.

Garmendia-Torres, C., A. Goldbeter, and M. Jacquet. 2007. Nucleocytoplasmic oscillations of the yeast transcription factor Msn2: Evidence for periodic PKA activation. *Current Biology* 17 (12): 1044–1049. doi:10.1016/j.cub.2007.05.032.

Gedeon, T. 1998. *Cyclic feedback systems*. Vol. 134, no. 637, Memoirs of the American Mathematical Society. Providence, RI: AMS.

Gelfand, I. M., and S. V. Fomin. 1963. *Calculus of variations*. Rev. English ed. Englewood Cliffs, NJ: Prentice-Hall.

Geva-Zatorsky, N., N. Rosenfeld, S. Itzkovitz, R. Milo, A. Sigal, E. Dekel, T. Yarnitzky, Y. Liron, P. Polak, G. Lahav, and U. Alon. 2006. Oscillations and variability in the p53 system. *Molecular Systems Biology* 2, 2006 0033. doi:10.1038/msb4100068.

Gibson, M. A., and J. Bruck. 2000. Efficient exact stochastic simulation of chemical systems with many species and many channels. *Journal of Physical Chemistry A* 104:1876–1889.

Gillespie, D. T. 1977. Exact stochastic simulation of coupled chemical reactions. *Journal of Chemical Physics* 81:2340.

Glass, L., and M. C. Mackey. 1988. *From clocks to chaos: The rhythms of life*. Princeton, NJ: Princeton University Press.

Glass, L., and A. T. Winfree. 1984. Discontinuities in phase-resetting experiments. *American Journal of Physiology* 246 (2 Pt 2): R251–R258.

Gleit, R. D., C. G. Diniz Behn, and V. Booth. 2013. Modeling interindividual differences in spontaneous internal desynchrony patterns. *Journal of Biological Rhythms* 28 (5): 339–355. doi:10.1177/0748730413504277.

Goldbeter, A. 1995. A model for circadian oscillations in the *Drosophila* period protein (PER). *Proceedings of the Royal Society. Biological Sciences* 261 (1362): 319–324. doi:10.1098/rspb.1995.0153.

Goldbeter, A. 1996. *Biochemical oscillations and cellular rhythms: The molecular bases of periodic and chaotic behaviour*. Cambridge: Cambridge University Press.

Goldbeter, A. 2002. Computational approaches to cellular rhythms. *Nature* 420 (6912): 238–245. doi:10.1038/nature01259.

Goldman, D. E. 1943. Potential, impedance, and rectification in membranes. *Journal of General Physiology* 27 (1): 37–60.

Goldobin, D., J. Teramae, H. Nakao, and G. Ermentrout. 2010. Dynamics of limit-cycle oscillators subject to general noise. *Physical Review Letters* 105:154101.

Golomb, D., D. Hansel, B. Shraiman, and H. Sompolinsky. 1992. Clustering in globally coupled phase oscillators. *Physical Review A.* 45:3516–3530.

Golubitsky, M., I. Stewart, P. L. Buono, and J. J. Collins. 1999. Symmetry in locomotor central pattern generators and animal gaits. *Nature* 401 (6754): 693–695. doi:10.1038/44416.

Gonzalez, O. R., C. Kuper, K. Jung, P. C. Naval, Jr., and E. Mendoza. 2007. Parameter estimation using simulated annealing for S-system models of biochemical networks. *Bioinformatics (Oxford, England)* 23 (4): 480–486. doi:10.1093/bioinformatics/btl522.

Goodwin, B. C. 1966. An entrainment model for timed enzyme syntheses in bacteria. *Nature* 209 (5022): 479–481.

Goodwin, F. K., K. R. Jamison, and S. N. Ghaemi. 2007. *Manic-depressive illness: Bipolar disorders and recurrent depression*. 2nd ed. New York: Oxford University Press.

Goriki, A., F. Hatanaka, J. Myung, J. K. Kim, T. Yoritaka, S. Tanoue, et al. 2014. A novel protein, CHRONO, functions as a core component of the mammalian circadian clock. *PLoS Biology* 12 (4): e1001839. doi:10.1371/journal.pbio.1001839.

Grima, R. 2010. Intrinsic biochemical noise in crowded intracellular conditions. *Journal of Chemical Physics* 132:185102.

Grodins, F. S. 1963. *Control theory and biological systems.* New York: Columbia University Press.

Guantes, R., and J. F. Poyatos. 2006. Dynamical principles of two-component genetic oscillators. *PLoS Computational Biology* 2 (3): e30. doi:10.1371/journal.pcbi.0020030.

Guevara, M. R., and L. Glass. 1982. Phase locking, period doubling bifurcations and chaos in a mathematical model of a periodically driven oscillator: A theory for the entrainment of biological oscillators and the generation of cardiac dysrhythmias. *Journal of Mathematical Biology* 14:1–23.

Guckenheimer, J. 1975. Isochrons and phaseless sets. *Journal of Mathematical Biology* 1:259–273.

Guckenheimer, J., and P. Holmes. 1983. *Nonlinear oscillations, dynamical systems and bifurcations of vector fields.* New York: Springer.

Hale, J. K. 1963. *Oscillations in nonlinear systems.* New York: McGraw-Hill.

Hammar, P., P. Leroy, A. Mahmutovic, E. G. Marklund, O. G. Berg, and J. Elf. 2012. The lac repressor displays facilitated diffusion in living cells. *Science* 336 (6088): 1595–1598. doi:10.1126/science.1221648.

Harris, A. L., and V. J. Vitzthum. 2013. Darwin's legacy: An evolutionary view of women's reproductive and sexual functioning. *Journal of Sex Research* 50 (3–4): 207–246. doi:10.1080/00224499.2012.763085.

Hastings, J. W., and B. M. Sweeney. 1957. On the mechanism of temperature independence in a biological clock. *Proceedings of the National Academy of Sciences of the United States of America* 43 (9): 804–811.

Hayashi, C. 1964. *Nonlinear oscillations in physical systems.* Rev. ed. New York: McGraw-Hill.

Hill, S., and G. Tononi. 2005. Modeling sleep and wakefulness in the thalamocortical system. *Journal of Neurophysiology* 93 (3): 1671–1698. doi:10.1152/jn.00915.2004.

Hirata, H., S. Yoshiura, T. Ohtsuka, Y. Bessho, T. Harada, K. Yoshikawa, et al. 2002. Oscillatory expression of the bHLH factor Hes1 regulated by a negative feedback loop. *Science* 298 (5594): 840–843. doi:10.1126/science.1074560.

Hirsch, M. W., and S. Smale. 1974. *Differential equations, dynamical systems, and linear algebra.* New York: Academic Press.

Hodgkin, A. L. 1948. The local electric changes associated with repetitive action in a non-medullated axon. *Journal of Physiology* 107:165–181.

Hodgkin, A. L., and A. F. Huxley. 1952. A quantitative description of membrane current and its application to conduction and excitation in nerve. *Journal of Physiology* 117 (4): 500–544.

Hodgkin, A. L., and B. Katz. 1949. The effect of sodium ions on the electrical activity of giant axon of the squid. *Journal of Physiology* 108 (1): 37–77.

Hoppensteadt, F. C., C. S. Peskin, and F. C. Hoppensteadt. 2002. *Modeling and simulation in medicine and the life sciences.* 2nd ed. New York: Springer.

Horsthemke, W., and R. Lefever. 1984. *Noise-induced transitions: Theory and applications in physics, chemistry, and biology.* New York: Springer.

Huang, N. E., Z. Shen, S. R. Long, M. C. Wu, H. H. Shih, Q. Zheng, et al. 1998. The empirical model decomposition and Hilbert spectrum for nonlinear and non-stationary time series analysis. *Proceedings of the Royal Society of London. Series A* 454:903–995.

Indic, P., D. B. Forger, M. A. St Hilaire, D. A. Dean, II, E. N. Brown, R. E. Kronauer, et al. 2005. Comparison of amplitude recovery dynamics of two limit cycle oscillator models of the human circadian pacemaker. *Chronobiology International* 22 (4): 613–629. doi:10.1080/07420520500180371.

Isaacson, S. A., and C. S. Peskin. 2006. Incorporating diffusion in complex geometries into stochastic chemical kinetics simulators. *SIAM Journal on Scientific Computing* 28:47.

Izhikevich, E.M. 2001. Resonate-and-fire neurons. *Neural Networks* 14 (6–7), 883–894.

Jahnke, W., W. E. Skaggs, and A. T. Winfree. 1989. Chemical vortex dynamics in the Belousov-Zhabotinskii reaction and in the two-variable oregonator model. *Journal of Physical Chemistry* 93:740–749.

Jewett, M. E., D. B. Forger, and R. E. Kronauer. 1999. Revised limit cycle oscillator model of human circadian pacemaker. *Journal of Biological Rhythms* 14 (6): 493–499.

Jewett, M. E., and R. E. Kronauer. 1998. Refinement of a limit cycle oscillator model of the effects of light on the human circadian pacemaker. *Journal of Theoretical Biology* 192 (4): 455–465.

Jewett, M. E., and R. E. Kronauer. 1999. Interactive mathematical models of subjective alertness and cognitive throughput in humans. *Journal of Biological Rhythms* 14 (6): 588–597.

Jewett, M. E., R. E. Kronauer, and C. A. Czeisler. 1994. Phase-amplitude resetting of the human circadian pacemaker via bright light: A further analysis. *Journal of Biological Rhythms* 9 (3–4): 295–314.

Johnson, C. H. 1999. Forty years of PRCs: What have we learned? *Chronobiology International* 16 (6): 711–743.

Jones, R. H. 1980. Maximum likelihood fitting of ARMA models to time series with missing observations. *Technometrics* 22:389–395.

Jordan, D. W., and P. Smith. 1987. *Nonlinear ordinary differential equations*. 2nd ed. New York: Oxford University Press.

Kalman, R. E. 1960. A new approach to linear filtering and prediction problems. *Journal of Basic Engineering* 82:35–45.

Kaplan, D., and L. Glass. 1995. *Understanding nonlinear dynamics*. New York: Springer.

Keener, J. P., and L. Glass. 1984. Global bifurcations of a periodically forced nonlinear oscillator. *Journal of Mathematical Biology* 21:175–190.

Keener, J. P., and J. Sneyd. 2009. *Mathematical physiology*. 2nd ed. New York: Springer.

Kim, J. K., and D. B. Forger. 2012a. A mechanism for robust circadian timekeeping via stoichiometric balance. *Molecular Systems Biology* 8:630. doi:10.1038/msb.2012.62.

Kim, J. K., and D. B. Forger. 2012b. On the existence and uniqueness of biological clock models matching experimental data. *SIAM Journal on Applied Mathematics* 72:1842–1855.

Kim, J. K., D. B. Forger, M. Marconi, D. Wood, A. Doran, T. Wager, et al. 2013. Modeling and validating chronic pharmacological manipulation of circadian rhythms. *CPT: Pharmacometrics & Systems Pharmacology* 2:e57. doi:10.1038/psp.2013.34.

Kim, J. K., and T. L. Jackson. 2013. Mechanisms that enhance sustainability of p53 pulses. *PLoS One* 8 (6): e65242. doi:10.1371/journal.pone.0065242.

Kirkpatrick, S., C. D. Gelatt Jr., and M. P. Vecchi. 1983. Optimization by simulated annealing. *Science* 220 (4598): 671–680. doi:10.1126/science.220.4598.671.

Kloeden, P. E., and E. Platen. 1992. *Numerical solution of stochastic differential equations*. New York: Springer.

Knoerchen, R., and G. Hildebrandt. 1976. Tagesrhythmische Schwankungen der visuellen Lichtempfindlichkiet beim Menschen. *Journal of Interdisciplinary Cycle Research* 7, 51.

Ko, C. H., Y. R. Yamada, D. K. Welsh, E. D. Buhr, A. C. Liu, E. E. Zhang, et al. 2010. Emergence of noise-induced oscillations in the central circadian pacemaker. *PLoS Biology* 8 (10): e1000513. doi:10.1371/journal.pbio.1000513.

Kobayashi, T., H. Mizuno, I. Imayoshi, C. Furusawa, K. Shirahige, and R. Kageyama. 2009. The cyclic gene Hes1 contributes to diverse differentiation responses of embryonic stem cells. *Genes & Development* 23:1870–1875. doi:10.1101/gad.1823109.

Kodadek, T., D. Sikder, and K. Nalley. 2006. Keeping transcriptional activators under control. *Cell* 127 (2): 261–264. doi:10.1016/j.cell.2006.10.002.

Körner, T. W. 1988. *Fourier analysis*. Cambridge: Cambridge University Press.

Koukkari, W. L., and R. B. Sothern. 2006. *Introducing biological rhythms: A primer on the temporal organization of life, with implications for health, society, reproduction and the natural environment*. New York: Springer.

Krebs, C. J., R. Boonstra, S. Boutin, and A. R. E. Sinclair. 2001. What drives the 10-year cycle of snowshoe hares? *Bioscience* 51:25–35.

Kronauer, R. E. 1987. Temporal subdivision of the circadian cycle. In *Some mathematical questions in biology: Circadian rhythms*. vol. 19. Ed. G. A. Carpenter. New York: American Mathematical Society.

Kronauer, R. E., C. A. Czeisler, S. F. Pilato, M. C. Moore-Ede, and E. D. Weitzman. 1982. Mathematical model of the human circadian system with two interacting oscillators. *American Journal of Physiology* 242 (1): R3–R17.

Kronauer, R. E., D. B. Forger, and M. E. Jewett. 1999. Quantifying human circadian pacemaker response to brief, extended, and repeated light stimuli over the phototopic range. *Journal of Biological Rhythms* 14 (6): 500–515.

Kruse, K., M. Howard, and W. Margolin. 2007. An experimentalist's guide to computational modelling of the Min system. *Molecular Microbiology* 63 (5): 1279–1284. doi:10.1111/j.1365-2958.2007.05607.x.

Kuramoto, Y. 1984. *Chemical oscillations, waves, and turbulence.* New York: Springer.

Kurtz, T. G. 1972. The relationship between stochastic and deterministic models for chemical reactions. *Journal of Chemical Physics* 57:2976.

Kuznetsov, I. U. A. 2004. *Elements of applied bifurcation theory.* 3rd ed. New York: Springer.

Lagarias, J. C., J. A. Reeds, M. H. Wright, and P. E. Wright. 1998. Convergence properties of the Nelder-Mead simplex method in low dimensions. *SIAM Journal on Optimization* 9:112–147.

Lakin-Thomas, P. L., S. Brody, and G. G. Cote. 1991. Amplitude model for the effects of mutations and temperature on period and phase resetting of the Neurospora circadian oscillator. *Journal of Biological Rhythms* 6 (4): 281–297.

Lander, A. D. 2010. The edges of understanding. *BMC Biology* 8:40. doi:10.1186/1741-7007-8-40.

Langfield, P., B. Krauskopf, and H. M. Osinga. 2014. Solving Winfree's puzzle: The isochrons in the FitzHugh-Nagumo model. *Chaos* 24 (1): 013131. doi:10.1063/1.4867877.

Latham, P. E., B. J. Richmond, P. G. Nelson, and S. Nirenberg. 2000. Intrinsic dynamics in neuronal networks. I. Theory. *Journal of Neurophysiology* 83 (2): 808–827.

Leloup, J. C., and A. Goldbeter. 1998. A model for circadian rhythms in Drosophila incorporating the formation of a complex between the PER and TIM proteins. *Journal of Biological Rhythms* 13 (1): 70–87.

LeMasson, G., E. Marder, and L. F. Abbott. 1993. Activity-dependent regulation of conductances in model neurons. *Science* 259 (5103): 1915–1917.

Lev Bar-Or, R., R. Maya, L. A. Segel, U. Alon, A. J. Levine, and M. Oren. 2000. Generation of oscillations by the p53-Mdm2 feedback loop: A theoretical and experimental study. *Proceedings of the National Academy of Sciences of the United States of America* 97 (21): 11250–11255. doi:10.1073/pnas.210171597.

Levins, R. 1966. The strategy of model building in population biology. *American Scientist* 54:4.

Lindner, B., A. Longtin, and A. Bulsara. 2003. Analytic expressions for rate and CV of a type I neuron driven by white Gaussian noise. *Neural Computation* 15 (8): 1760–1787. doi:10.1162/08997660360675035.

Linkens, D. A. 1976. Stability of entrainment conditions for a particular form of mutually coupled van der Pol oscillators. *IEEE Transactions on Circuits and Systems* CAS-23:113–121.

Liu, C., D. R. Weaver, S. H. Strogatz, and S. M. Reppert. 1997. Cellular construction of a circadian clock: Period determination in the suprachiasmatic nuclei. *Cell* 91 (6): 855–860.

Lloyd, D., K. M. Lemar, L. E. Salgado, T. M. Gould, and D. B. Murray. 2003. Respiratory oscillations in yeast: Mitochondrial reactive oxygen species, apoptosis and time—a hypothesis. *FEMS Yeast Research* 3:333–339.

Locker, A. 1973. *Biogenesis, evolution, homeostasis: A symposium by correspondence.* New York: Springer.

Luce, G. G. 1971. *Biological rhythms in human and animal physiology.* New York: Dover Publications.

MacLulich, D.A. 1937. *Fluctuations in the numbers of the varying hare (Lepus americanus).* Toronto: University of Toronto Press.

Mallet-Paret, J., and H. L. Smith. 1990. The Poincaré-Bendixson theorem for monotone cyclic feedback systems. *Journal of Dynamics and Differential Equations* 2:367–421.

Marder, E., and J. M. Goaillard. 2006. Variability, compensation and homeostasis in neuron and network function. *Nature Reviews. Neuroscience* 7 (7): 563–574. doi:10.1038/nrn1949.

Marmarelis, V. Z. 2004. *Nonlinear dynamic modeling of physiological systems.* Hoboken, NJ: Wiley-Interscience.

Maroto, M., and N. A. M. Monk. 2008. *Cellular oscillatory mechanisms.* New York: Springer.

Mees, A. I., and P. E. Rapp. 1978. Periodic metabolic systems: Oscillations in multiple-loop negative feedback biochemical control networks. *Journal of Mathematical Biology* 5:99–114.

Meier, E. B., and A. E. Bryson. 1990. Efficient algorithm for time-optimal control of a two-link manipulator. *Journal of Guidance, Control, and Dynamics* 13:859–866.

Michaelis, L., and M. I. Menten. 1913. Die Kinetik der Invertinwirkund. *Biochemische Zeitschrift* 49, 333

Mirollo, R. E., and S. H. Strogatz. 1990a. Amplitude death in an array of limit-cycle oscillators. *Journal of Statistical Physics* 60:245–262.

Mirollo, R. E., and S. H. Strogatz. 1990b. Synchronization of pulse-coupled biological oscillators. *SIAM Journal on Applied Mathematics* 50:1645–1662.

Moler, C., and C. Van Loan. 1978. Nineteen dubious ways to compute the exponential of a matrix. *SIAM Review* 20:801–836.

Monk, N. A. 2003. Oscillatory expression of Hes1, p53, and NF-kappaB driven by transcriptional time delays. *Current Biology* 13 (16): 1409–1413.

Morre, D. J., P. J. Chueh, J. Pletcher, X. Tang, L. Y. Wu, and D. M. Morre. 2002. Biochemical basis for the biological clock. *Biochemistry* 41 (40): 11941–11945.

Morris, C., and H. Lecar. 1981. Voltage oscillations in the barnacle giant muscle fiber. *Biophysical Journal* 35 (1): 193–213. doi:10.1016/S0006-3495(81)84782-0.

Murray, J. D. 2002. *Mathematical biology*. 3rd ed. New York: Springer.

Myung, J., S. Hong, D. DeWoskin, E. De Schutter, D. B. Forger, and T. Takumi. 2015. GABA-mediated repulsive coupling between circadian clock neurons in the SCN encodes seasonal time. *Proceedings of the National Academy of Sciences of the United States of America* 223:E3920–E3929. doi:10.1073/pnas.1421200112

Nabi, A., and J. Moehlis. 2012. Time optimal control of spiking neurons. *Journal of Mathematical Biology* 64:981–1004. doi:10.1007/s00285-011-0441-5.

Nelder, J. A., and R. Mead. 1965. A simplex method for function minimization. *Computer Journal* 7:308–313.

Novak, B., and J. J. Tyson. 2008. Design principles of biochemical oscillators. *Nature Reviews. Molecular Cell Biology* 9 (12): 981–991. doi:10.1038/nrm2530.

Panda, S., M. P. Antoch, B. H. Miller, A. I. Su, A. B. Schook, M. Straume, et al. 2002. Coordinated transcription of key pathways in the mouse by the circadian clock. *Cell* 109 (3): 307–320.

Pavlidis, T. 1969. Populations of interacting oscillators and circadian rhythms. *Journal of Theoretical Biology* 22 (3): 418–436.

Pavlidis, T. 1973. *Biological oscillators: Their mathematical analysis*. New York: Academic Press.

Pavlidis, T. 1978. Qualitative similarities between the behavior of coupled oscillators and circadian rhythms. *Bulletin of Mathematical Biology* 40 (6): 675–692.

Paydarfar, D., D. B. Forger, and J. R. Clay. 2006. Noisy inputs and the induction of on-off switching behavior in a neuronal pacemaker. *Journal of Neurophysiology* 96 (6): 3338–3348. doi:10.1152/jn.00486.2006.

Paydarfar, D., and W. J. Schwartz. 2001. An algorithm for discovery. *Science* 292 (5514): 13. doi:10.1126/science.1058915.

Peskin, C.S. 1975. *Mathematical aspects of heart physiology*. New York: Courant Institute of Mathematical Sciences.

Peskin, C. S. 2000. Mathematical aspects of neurophysiology. Lecture notes posted on author's website: http://www.math.nyu.edu/faculty/peskin/

Petzold, L. R., and D. T. Gillespie. 2003. Improved leap-size selection for accelerated stochastic simulation. *Journal of Chemical Physics* 119:8229.

Pigolotti, S., S. Krishna, and M. H. Jensen. 2007. Oscillation patterns in negative feedback loops. *Proceedings of the National Academy of Sciences of the United States of America* 104 (16): 6533–6537. doi:10.1073/pnas.0610759104.

Pikovsky, A., M. Rosenblum, and J. Kurths. 2003. *Synchronization: A universal concept in nonlinear sciences*. Cambridge: Cambridge University Press.

Pittendrigh, C. S. 1954. On temperature independence in the clock system controlling emergence time in *Drosophila*. *Proceedings of the National Academy of Sciences of the United States of America* 40 (10): 1018–1029.

Pittendrigh, C. S., and S. Daan. 1976. A functional analysis of circadian pacemakers in nocturnal rodents, I. The stability and lability of spontaneous frequency. *Journal of Comparative Physiology* 106:223–252.

Pittendrigh, C. S., W. T. Kyner, and T. Takamura. 1991. The amplitude of circadian oscillations: Temperature dependence, latitudinal clines, and the photoperiodic time measurement. *Journal of Biological Rhythms* 6 (4): 299–313.

Pontryagin, L. 1957. Asymptotic behavior of solutions of systems of differential equations with a small parameter in the derivatives of highest order. *Izvestiya rossiiskoi adademii nauk.* Seriya mathematicheskaya 21:605.

Pravitha, R., P. Indic, V. P. Nampoori, and R. Pratap. 2001. Effect of time scales on the unfolding of neural attractors. *International Journal of Neuroscience* 111 (3–4): 175–186.

Prigogine, I., and R. Lefever. 1968. Symmetry breaking instabilities in dissipative systems. II. *Journal of Chemical Physics* 48:1695.

Rajapakse, I., M. Groudine, and M. Mesbahi. 2011. Dynamics and control of state-dependent networks for probing genomic organization. *Proceedings of the National Academy of Sciences of the United States of America* 108 (42): 17257–17262. doi:10.1073/pnas.1113249108.

Rand, R. H., and P. J. Holmes. 1980. Bifurcation of periodic motions in two weakly coupled van der Pol oscillators. *International Journal of Non-linear Mechanics* 15:387–399.

Rapp, P. E. 1979. Bifurcation theory, control theory and metabolic regulation. In *Biological Systems, Modeling and Control*, ed. D. A. Linkens, 1–83. Stevenage: Peter Peregrinus.

Raskin, D. M., and P. A. de Boer. 1999. Rapid pole-to-pole oscillation of a protein required for directing division to the middle of Escherichia coli. *Proceedings of the National Academy of Sciences of the United States of America* 96:4971–4976.

Rensing, L., S. Mohsenzadeh, P. Ruoff, and U. Meyer. 1997. Temperature compensation of the circadian period length—a special case among general homeostatic mechanisms of gene expression? *Chronobiology International* 14 (5): 481–498.

Richter, C. P. 1965. *Biological clocks in medicine and psychiatry.* Springfield, IL: Thomas.

Rimmer, D. W., D. B. Boivin, T. L. Shanahan, R. E. Kronauer, J. F. Duffy, and C. A. Czeisler. 2000. Dynamic resetting of the human circadian pacemaker by intermittent bright light. *American Journal of Physiology. Regulatory, Integrative and Comparative Physiology* 279 (5): R1574–R1579.

Rinzel, J., and B. Ermentrout. 1998. Analysis of neuronal excitability and oscillations. In *Methods in neuronal modeling: From ions to networks*, ed. C. Koch and I. Segev. Cambridge, MA: MIT Press.

Ruoff, P., J. J. Loros, and J. C. Dunlap. 2005. The relationship between FRQ-protein stability and temperature compensation in the Neurospora circadian clock. *Proceedings of the National Academy of Sciences of the United States of America* 102 (49): 17681–17686. doi:10.1073/pnas.0505137102.

Ruoff, P., S. Mohsenzadeh, and L. Rensing. 1996. Circadian rhythms and protein turnover: The effect of temperature on the period lengths of clock mutants simulated by the Goodwin oscillator. *Naturwissenschaften* 83 (11): 514–517.

Ruoff, P., L. Rensing, R. Kommedal, and S. Mohsenzadeh. 1997. Modeling temperature compensation in chemical and biological oscillators. *Chronobiology International* 14 (5): 499–510.

Sanders, J. A., F. Verhulst, and J. A. Murdock. 2007. *Averaging methods in nonlinear dynamical systems.* 2nd ed. New York: Springer.

Savageau, M. A. 1995. Michaelis-Menten mechanism reconsidered: Implications of fractal kinetics. *Journal of Theoretical Biology* 176 (1): 115–124. doi:10.1006/jtbi.1995.0181.

Schättler, H. M., and U. Ledzewicz. 2012. *Geometric optimal control theory: Methods and examples.* New York: Springer.

Schnell, S., and T. E. Turner. 2004. Reaction kinetics in intracellular environments with macromolecular crowding: Simulations and rate laws. *Progress in Biophysics and Molecular Biology* 85 (2–3): 235–260. doi:10.1016/j.pbiomolbio.2004.01.012.

Serkh, K., and D. B. Forger. 2014. Optimal schedules of light exposure for rapidly correcting circadian misalignment. *PLoS Computational Biology* 10 (4): e1003523. doi:10.1371/journal.pcbi.1003523.

Siegel, M., E. Marder, and L. F. Abbott. 1994. Activity-dependent current distributions in model neurons. *Proceedings of the National Academy of Sciences of the United States of America* 91 (24): 11308–11312.

Siepka, S. M., S. H. Yoo, J. Park, W. Song, V. Kumar, Y. Hu, et al. 2007. Circadian mutant Overtime reveals F-box protein FBXL3 regulation of cryptochrome and period gene expression. *Cell* 129 (5): 1011–1023. doi:10.1016/j.cell.2007.04.030.

Sim, C. K., and D. B. Forger. 2007. Modeling the electrophysiology of suprachiasmatic nucleus neurons. *Journal of Biological Rhythms* 22 (5): 445–453. doi:10.1177/0748730407306041.

Simon, W. 1977. *Mathematical techniques for biology and medicine*. Cambridge, MA: MIT Press.

Smolen, P., J. Rinzel, and A. Sherman. 1993. Why pancreatic islets burst but single beta cells do not: The heterogeneity hypothesis. *Biophysical Journal* 64 (6): 1668–1680. doi:10.1016/S0006-3495(93)81539-X.

Smoluchowski, M. 1917. Versuch einer mathematischen Theorie der koagulationshkinetik kolloider Lösungen. *Zeitschrift für physikaliche Chemie* 9:129.

Sneyd, J., K. Tsaneva-Atanasova, V. Reznikov, Y. Bai, M. J. Sanderson, and D. I. Yule. 2006. A method for determining the dependence of calcium oscillations on inositol trisphosphate oscillations. *Proceedings of the National Academy of Sciences of the United States of America* 103:1675–1680. doi:10.1073/pnas.0506135103.

Sontag, E. D. 2006. Passivity gains and the "secant condition" for stability. *Systems & Control Letters* 55:171–183.

Still, H. 1972. *Of time, tides, and inner clocks: Taking advantage of the natural rhythms of life*. Harrisburg, PA: Stackpole Books.

Stinchcombe, A., and D. B. Forger. 2016. An efficient method for simulation of noisy coupled multi-dimensional oscillators. *Journal of Computational Physics* 321:932–946

Stoker, J. J. 1950. *Nonlinear vibrations in mechanical and electrical systems*. New York: Interscience.

Stricker, J., S. Cookson, M. R. Bennett, W. H. Mather, L. S. Tsimring, and J. Hasty. 2008. A fast, robust and tunable synthetic gene oscillator. *Nature* 456 (7221): 516–519. doi:10.1038/nature07389.

Strogatz, S. H. 1986. *The mathematical structure of the human sleep-wake cycle*. New York: Springer.

Strogatz, S. H. 2000. *Nonlinear dynamics and chaos, with applications to physics, biology, chemistry, and engineering*. 1st ed. Cambridge, MA: Westview Press.

Strogatz, S. H. 2003. *Sync: The emerging science of spontaneous order*. 1st ed. New York: Theia.

Suter, D. M., N. Molina, D. Gatfield, K. Schneider, U. Schibler, and F. Naef. 2011. Mammalian genes are transcribed with widely different bursting kinetics. *Science* 332 (6028): 472–474. doi:10.1126/science.1198817.

Sweeney, B. M., and J. W. Hastings. 1960. Effects of temperature upon diurnal rhythms. *Cold Spring Harbor Symposia on Quantitative Biology* 25:87–104.

Thoke, H. S., A. Tobiesen, J. Brewer, P. L. Hansen, R. P. Stock, L. F. Olsen, et al. 2015. Tight coupling of metabolic oscillations and intracellular water dynamics in Saccharomyces cerevisiae. *PLoS One* 10:e0117308. doi:10.1371/journal.pone.0117308.

Thomas, R., and R. D'Ari. 1990. *Biological feedback*. Boca Raton: CRC Press.

Thron, C. D. 1991. The secant condition for instability in biochemical feedback control –1: The role of cooperativity and saturability. *Bulletin of Mathematical Biology* 53:383–401.

Tiana, G., S. Krishna, S. Pigolotti, M. H. Jensen, and K. Sneppen. 2007. Oscillations and temporal signalling in cells. *Physical Biology* 4 (2): R1–R17. doi:10.1088/1478-3975/4/2/R01.

Toh, K. L., C. R. Jones, Y. He, E. J. Eide, W. A. Hinz, D. M. Virshup, et al. 2001. An hPer2 phosphorylation site mutation in familial advanced sleep phase syndrome. *Science* 291 (5506): 1040–1043.

Trefethen, L. N., and D. Bau. 1997. *Numerical linear algebra*. Philadelphia: SIAM.

Turing, A. M. 1990. The chemical basis of morphogenesis. 1953. *Bulletin of Mathematical Biology* 52 (1–2): 153–197, discussion 119–152.

Tyson, J. J. 2004. A precarious balance. *Current Biology* 14 (7): R262–R263. doi:10.1016/j.cub.2004.03.017.

Tyson, J. J., and B. Novak. 2001. Regulation of the eukaryotic cell cycle: Molecular antagonism, hysteresis, and irreversible transitions. *Journal of Theoretical Biology* 210 (2): 249–263. doi:10.1006/jtbi.2001.2293.

Tyson, J. J., and H. G. Othmer. 1978. The dynamics of feedback control circuits in biochemical pathways. *Progress in Theoretical Biology* 5:1–62.

van der Pol, B., and J. van der Mark. 1928. The heartbeat considered as a relaxation oscillation, and an electrical model of the heart. *Philosophical Magazine* 6:763–775.

Vandewalle, G., B. Middleton, S. M. Rajaratnam, B. M. Stone, B. Thorleifsdottir, J. Arendt, et al. 2007. Robust circadian rhythm in heart rate and its variability: Influence of exogenous melatonin and photoperiod. *Journal of Sleep Research* 16:148–155. doi:10.1111/j.1365-2869.2007.00581.x.

Ventura, A. C., P. Jiang, L. Van Wassenhove, D. Del Vecchio, S. D. Merajver, and A. J. Ninfa. 2010. Signaling properties of a covalent modification cycle are altered by a downstream target. *Proceedings of the National Academy of Sciences of the United States of America* 107 (22): 10032–10037. doi:10.1073/pnas.0913815107.

Virshup, D. M., E. J. Eide, D. B. Forger, M. Gallego, and E. V. Harnish. 2007. Reversible protein phosphorylation regulates circadian rhythms. *Cold Spring Harbor Symposia on Quantitative Biology* 72:413–420. doi:10.1101/sqb.2007.72.048.

Vitaterna, M. H., C. H. Ko, A. M. Chang, E. D. Buhr, E. M. Fruechte, A. Schook, et al. 2006. The mouse Clock mutation reduces circadian pacemaker amplitude and enhances efficacy of resetting stimuli and phase-response curve amplitude. *Proceedings of the National Academy of Sciences of the United States of America* 103:9327–9332. doi:10.1073/pnas.0603601103.

Wang, W., and J. J. Slotine. 2005. On partial contraction analysis for coupled nonlinear oscillators. *Biological Cybernetics* 92 (1): 38–53. doi:10.1007/s00422-004-0527-x.

Ward, R. R. 1972. *The living clocks*. London: Collins.

Wasserman, L. 2015. Asymptotic theory. Available at http://www.stat.cmu.edu/~larry/=stat705/Lecture9.pdf.

Watts, D. J., and S. H. Strogatz. 1998. Collective dynamics of "small-world" networks. *Nature* 393 (6684): 440–442. doi:10.1038/30918.

Wiener, N. 1976. *Collected works*. Cambridge, MA: MIT Press.

Winfree, A. T. 1967. Biological rhythms and the behavior of populations of coupled oscillators. *Journal of Theoretical Biology* 16 (1): 15–42.

Winfree, A. T. 1980. *The geometry of biological time*. New York: Springer.

Winfree, A. T. 1984. The prehistory of the Belousov-Zhabotinsky oscillator. *Journal of Chemical Education* 61:661.

Winfree, A. T. 1987a. *The timing of biological clocks*. New York: W.H. Freeman.

Winfree, A. T. 1987b. *When time breaks down: The three-dimensional dynamics of electrochemical waves and cardiac arrhythmias*. Princeton, NJ: Princeton University Press.

Winston, W. L. 1994. *Operations research: Applications and algorithms*. 3rd ed. Belmont, CA: Duxbury Press.

Wu, C. F., N. S. Savage, and D. J. Lew. 2013. Interaction between bud-site selection and polarity-establishment machineries in budding yeast. *Philosophical Transactions of the Royal Society of London. Series B, Biological Sciences* 368:20130006. doi:10.1098/rstb.2013.0006.

Wu, F., B. G. van Schie, J. E. Keymer, and C. Dekker. 2015. Symmetry and scale orient Min protein patterns in shaped bacterial sculptures. *Nature Nanotechnology* 10:719–726. doi:10.1038/nnano.2015.126.

Xue, M., H. Momiji, N. Rabbani, G. Barker, T. Bretschneider, A. Shmygol, D. A. Rand, and P. J. Thornalley. 2014. Frequency modulated translocational oscillations of Nrf2 mediate the antioxidant response element cytoprotective transcriptional response. *Antioxidants & Redox Signaling*. doi:10.1089/ars.2014.5962

Index